From Semiconductors to Proteins:

Beyond the Average Structure

FUNDAMENTAL MATERIALS RESEARCH

Series Editor: M. F. Thorpe, *Michigan State University*
East Lansing, Michigan

ACCESS IN NANOPOROUS MATERIALS
Edited by Thomas J. Pinnavaia and M. F. Thorpe

DYNAMICS OF CRYSTAL SURFACES AND INTERFACES
Edited by P. M. Duxbury and T. J. Pence

ELECTRONIC PROPERTIES OF SOLIDS USING CLUSTER METHODS
Edited by T. A. Kaplan and S. D. Mahanti

FROM SEMICONDUCTORS TO PROTEINS: BEYOND THE AVERAGE STRUCTURE
Edited by S. J. L. Billinge and M. F. Thorpe

LOCAL STRUCTURE FROM DIFFRACTION
Edited by S. J. L. Billinge and M. F. Thorpe

PHASE TRANSITIONS AND SELF-ORGANIZATION IN ELECTRONIC AND MOLECULAR NETWORKS
Edited by J. C. Phillips and M. F. Thorpe

PHYSICS OF MANGANITES
Edited by T. A. Kaplan and S. D. Mahanti

RIGIDITY THEORY AND APPLICATIONS
Edited by M. F. Thorpe and P. M. Duxbury

SCIENCE AND APPLICATION OF NANOTUBES
Edited by D. Tománek and R. J. Enbody

A Continuation Order Plan is available for this series. A continuation order will bring delivery of each new volume immediately upon publication. Volumes are billed only upon actual shipment. For further information please contact the publisher.

From Semiconductors to Proteins:

Beyond the Average Structure

Edited by

S. J. L. Billinge and M. F. Thorpe

Michigan State University
East Lansing, Michigan

Kluwer Academic / Plenum Publishers
New York, Boston, Dordrecht, London, Moscow

Proceedings from the workshop, From Semiconductors to Proteins: Beyond the Average Structure, held July 28–August 1, 2001, in Traverse City, Michigan

ISBN 0-306-47239-2

233 Spring Street, New York, New York 10013

http://www.wkap.nl/

10 9 8 7 6 5 4 3 2 1

A C.I.P. record for this book is available from the Library of Congress

Printed in the United States of America

SERIES PREFACE

This series of books, which is published at the rate of about one per year, addresses fundamental problems in materials science. The contents cover a broad range of topics from small clusters of atoms to engineering materials and involve *chemistry*, *physics*, *materials science,* and *engineering*, with length scales ranging from Ångstroms up to millimeters. The emphasis is on basic science rather than on applications. Each book focuses on a single area of current interest and brings together leading experts to give an up-to-date discussion of their work and the work of others. Each article contains enough references that the interested reader can access the relevant literature. Thanks are given to the Center for Fundamental Materials Research at Michigan State University for supporting this series.

M.F. Thorpe, Series Editor
E-mail: thorpe@pa.msu.edu
East Lansing, Michigan, November 2001

PREFACE

The study of the atomic structure of crystalline materials began at the beginning of the twentieth century with the discovery by Max von Laue and by W.H. and W.L. Bragg that crystals diffract x-rays. At that time, even the existence of atoms was controversial. Since then our understanding of solid and liquid matter, and our ability to manipulate and control matter, has led to technological developments that would have been far beyond the imaginations of Laue and the Braggs. These advances are completely underpinned by our ability to study atomic structure. As techniques and sources have improved, it is becoming possible to study the structure of more and more complex systems. Now it is relatively straightforward to solve the structure of proteins with many thousands of atoms in the unit cell, and even the structures of viruses are being revealed at the atomic level. Advances in theory and computing are also allowing calculations to be made on systems with comparable numbers of atoms and degrees of complexity. At the same time, it is becoming possible to manipulate materials and make artificial structures, for example in semiconductor devices, that have nanometer scale dimensions similar to natural proteins and viruses. This confluence of theory, characterization and manipulation that has happened on the nanometer length-scale is the basis of the current excitement around the world for *nanotechnology* and how this might affect our lives in the upcoming decades.

The nanometer length-scale is also one that presents very special challenges to be studied experimentally. Periodic crystals have order that extends over hundreds of Ångstroms. However, the nano-revolution suggests that features need to be studied that are on a shorter length-scale. For example, functioning proteins exist in the folded state in solution, not crystallized and frozen into an extensive periodic lattice. This implies being able to go *beyond the average structure* when studying structure – precisely the topic of this workshop. Recent developments in sources, techniques and computing are allowing us to extract structural information from nano-scale fluctuations and inhomogeneities in unprecedented detail from what Krivoglaz, a pioneer of diffuse scattering, called *real crystals*. This workshop was organized to bring together leading researchers studying local structure using various techniques. The purpose was not only to share information about state of the art results, but also to educate ourselves and each other about progress in areas that are not necessarily so close to our own research. This was the reason for the broad scope of the meeting: *from semiconductors to proteins*. This intimate and intensive workshop was held at the historic and picturesque Park Place Hotel in Traverse City, Michigan, USA from July 28 to August 1, 2001. It was the second workshop of a similar theme following the successful workshop *Local Structure from Diffraction* that was held at the same location in the summer of 1997. All the attendees were by invitation only, 24 of whom gave oral presentations at the meeting. There were also many animated and detailed private discussions.

The invited speakers were asked to produce a manuscript containing a pedagogical account of their work which would be of lasting value as a text for newcomers to the field and as a reference for established researchers. As such, this book brings up to date and extends the book from the previous conference, *Local Structure from Diffraction*, published by Plenum Press in 1998, by bringing into one place descriptions of the various approaches which are used to collect, analyze and interpret data giving information beyond the average structure. This present volume contains 15 articles by many of the leading researchers in the field of local structure.

We would like to thank Michigan State University for financing the meeting and the *Center for Fundamental Materials Research* at MSU for contributing to the cost of producing the proceedings. Also, the efforts of Lorie Neuman and Janet King, who organized the workshop and proceedings, are deeply appreciated as was the advice and efforts of the *Advisory Committee members*: Dr. Duncan McRee, Professor Simon C. Moss, Dr. David. L. Price and Dr. Alan. P. Soper.

Simon J.L. Billinge
Michael F. Thorpe
East Lansing, Michigan, November 2001

CONTENTS

1. TECHNIQUES TAKING US BEYOND THE AVERAGE STRUCTURE

2. HOW DO SEMICONDUCTORS AND INORGANIC MATERIALS TAKE THE STRAIN?

3. TAKING CHARGE OF DISORDERLINESS IN PHYSICS AND CHEMISTRY

ANALYSIS OF SINGLE-CRYSTAL DIFFUSE X-RAY SCATTERING VIA AUTOMATIC REFINEMENT OF A MONTE CARLO MODEL

T.R.Welberry*

1. INTRODUCTION

Although diffuse scattering from disordered crystals has been known since the earliest days of X-ray diffraction its interpretation and analysis is still far from being a routine process and quantitative studies are relatively rare. The most extensive and quantitative methods have been developed in the field of alloys and simple oxides where the number of parameters needed to define a disordered structure is small and the diffracted patterns can be attributed unambiguously to basic interatomic interactions[1-4]. For structures containing more than a few atoms, however, this type of detailed analysis is simply not feasible, and alternative approaches must be adopted (see e.g. the reviews by Welberry & Butler[5] and Frey[6,7]). For large (macromolecular) structures interpreting diffuse scattering, when the structure contains so many parameters, would appear to be a quite intractible problem; but even here useful information can be extracted[8-11]. The present work forms part of a long-term study to develop methods that will allow routine analysis of diffuse scattering for a diverse range of materials.

In numerous papers[12-16] we have reported investigations of disordered crystals in which we have used Monte Carlo (MC) simulation of a computer model to aid in the interpretation of observed diffuse X-ray scattering patterns. First a model is set up in the computer in terms of sets of random variables representing the atomic occupancies and positions or molecular orientations. A relatively small number of energy parameters describes the way in which these atoms or molecules interact. Monte Carlo simulation is then carried out for a time sufficient to allow the system to approach equilibrium. The final atomic coordinates of this model crystal realisation are then used to calculate diffraction patterns, which may be compared to the observed X-ray patterns. After assessing the points of agreement and disagreement from this comparison the model parameters are adjusted and the whole process is repeated iteratively until a satisfactory agreement between observed and calculated patterns is obtained.

Although convincing results have been obtained by this method for a variety of quite different systems, the crucial step of comparing the observed and calculated patterns has, until recently, been performed visually and adjustment of the system parameters has re-

* T.R.Welberry, Research School of Chemistry, Australian National University, Canberra, ACT 0200, Australia.

From Semiconductors to Protein, edited by S.J.L. Billinge and M.F. Thorpe
Kluwer Academic/Plenum Publishers, 2002

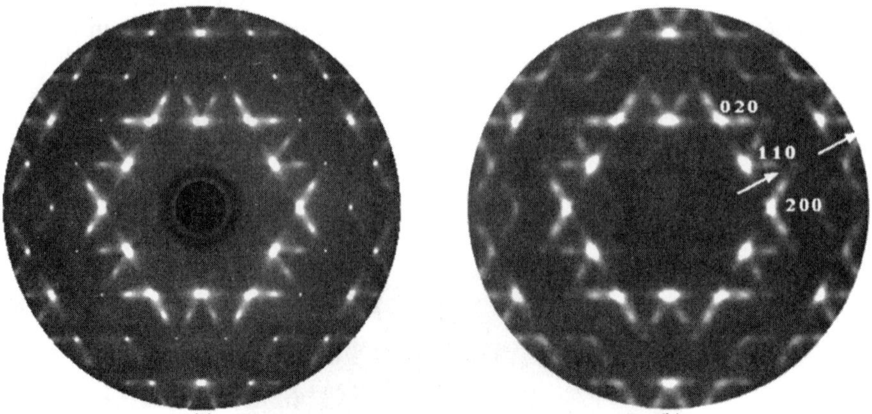

(a) (b)

Figure 1. Comparison of (a) observed and (b) calculated diffraction patterns for dibromodecane/urea inclusion compound. The fit was obtained by manual adjustment of the parameters of a Monte Carlo model using trial and error methods (see Welberry & Mayo[15]).

lied heavily on an accumulation of experience, gained over a number of years. More recently we have moved towards a more objective and quantitative approach in which we attempt to perform this iterative MC methodology solely by computer, using quantitative rather than visual comparison of observed and calculated diffraction patterns and with automatic updating of model parameters, using a least-squares algorithm. It should be stressed at the outset that this represents a formidable computational task. At each stage of iteration complete MC simulations are carried out for the current set of model parameters, diffraction patterns calculated and the goodness-of-fit parameter, χ^2, obtained as a quantitative measure of the agreement with the observed data. In addition, complete MC simulations, together with accompanying calculations of their diffraction patterns, are carried out for sets of parameters in which each of the parameters has, in turn, been changed by a small amount. From these, numerical estimates of the differentials of χ^2 with respect to each of the system variables are obtained, and are used to form the least-squares matrix, **A** (see later). Although computer performance per dollar has increased rapidly over the last few years, it is still not feasible to perform such a calculation using extensive three dimensional diffuse scattering data nor using models which represent the real-space structure in as much detail as would ideally be desired. In order to make any progress a number of approximations have had to be made, and these are discussed in the sections that follow.

A brief summary of experimental data collection is given in section 2. Details of the Monte Carlo method are presented in section 3 and section 4 describes how calculated diffraction patterns are obtained from these. Section 5 describes the least-squares process by which the comparison between observed and calculated patterns can be used to update the system parameters. Finally, results for three different examples of different degrees of complexity are presented in sections 6,7 and 8. The first example is that of the iron carbonyl complex, $Fe_3(CO)_{12}$[17]. Crystals of this compound display very strong diffuse scattering originating from the fact that the triangle of Fe atoms occupies one of two different orientations in each molecular site. The carbonyls, though also disordered, do not appear to contribute significantly to the observed scattering. The example thus represents a fairly simple system capable of being described by relatively few parameters. The second example is that of in *p*-methyl-*N*-(*p*-chlorobenzylidene) aniline, $C_{14}H_{12}ClN$,

(ClMe)[18]. In this compound each molecular site is occupied by a molecule in one of *four* basically different orientations and so represents a considerable increase in complexity from the first example. For this compound the diffraction pattern is very rich in detail showing numerous distinctive features (see Fig. 2). For the third example, benzil $C_{14}H_{10}O_2$[19], each molecular site contains the same basic molecular orientation but the molecule has three degrees of internal flexibility and this results in strong and distinctive diffuse scattering patterns.

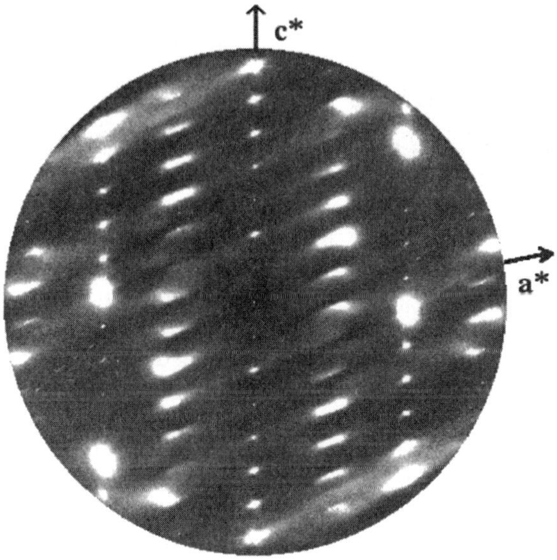

Figure 2. The (*h0l*) section of data of ClMe as recorded on the initial scale of 400×400 pixels. Note that the resolution is such that Bragg peaks are only ~1 pixel in width.

2. EXPERIMENTAL DATA COLLECTION

Complete sections of diffuse X-ray scattering data normal to particular zone axes of the example materials were recorded using the Position-Sensitive Detector (PSD) system described by Osborn and Welberry[20]. This system uses a flat-cone Weissenberg geometry. The raw recorded data are obtained as a two-dimensional array of points in the Weissenberg space (2θ vs. ω). Before use these are re-binned into a square array of points (400 × 400 pixels) in the undistorted reciprocal lattice plane. The experimental resolution due to beam divergence, crystal size and mosaic spread then corresponds approximately to a single pixel in this plane. Each data point is renormalised by dividing by the number of raw data points contributing to it, so that no correction due to the Lorenz effect is required. In the examples described in the present paper corrections due to X-ray polarisation and specimen absorption were neglected. Graphite monochromated CoKα (λ=1.790Å) (for the Fe compound) or CuKα (λ=1.542Å) radiation was used. Though the measured data are obtained in digital form it is convenient for comparison purposes to display the data as false-colour or grey-scale images. For the analysis of the $Fe_3(CO)_{12}$ compound described in section 6 it was decided to use only small subsections of the data where the strongest diffuse scattering was seen to occur. In the subsequent studies of MeCl and benzil whole diffuse sections were utilised. In this case, to minimize the total number of data points to be handled, the raw data were rebinned onto a grid of lower resolution comprising 200 × 200 pixels.

3. MONTE CARLO SIMULATION

Of crucial importance in carrying out a Monte Carlo simulation for use in the analysis is the need to have a model system of sufficient size that statistical variations are small and the resulting calculated patterns are relatively noise-free and of a comparable quality to the observed data. Our experience is that a minimum system size of ~32×32×32 or 32768 unit cells is required (see e.g. Welberry & Proffen[21]). The actual choice depends on the relative lengths of the primary lattice vectors. In cases where one cell vector is particularly long the number of cells in that direction is reduced and similarly for particularly short cell vectors the number of cells is increased, so that the model crystal sample has approximately the same overall size in all three dimensions.

It is also important for the Monte Carlo simulation to be carried out for a sufficient number of cycles that the system approaches equilibrium reasonably closely. In order to keep computational requirements within reasonable bounds it has been found necessary to restrict the number of cycles to ~200-400. Though this usually means the system falls somewhat short of true equilibrium it has been found to be sufficiently close that any further changes that occur in the diffraction pattern are long-range in nature and only affect regions very close to the Bragg positions.

The Monte Carlo model is completely defined by an energy, E_{tot}. This energy is formulated in terms of variable parameters which describe the way in which the constituent atoms and molecules interact with each other. At one extreme the model may be formulated to describe the system in as detailed and realistic a way as possible (using for example a complete set of atom-atom potentials). At the other extreme it may be a much simplified model which attempts to encapsulate the basic physics and chemistry using a few well-chosen parameters which define 'effective interactions'. At this point in time the need to keep the computational effort within reasonable bounds dictates that a strategy close to the latter scenario be adopted.

The Monte Carlo models used in the examples described in sections 6-8 are all based on the following simple picture of the disordered crystal.

3.1 Ordering of molecular orientations

First it is assumed that in each molecular site a molecule exists in one of a number of different basic orientations. These basic molecular orientations are represented in the computer by occupational variables, $s_{i,j,k,l}$. Here i,j,k specify the unit cell and l one of the molecular sites within the unit cell. In the case of $Fe_3(CO)_{12}$ these occupational variables are binary with $s_{i,j,k,l}$=+1 representing the Fe_3 triangle in one orientation and with $s_{i,j,k,l}$=-1 representing the Fe_3 triangle rotated through 180°. In the case of *p*-methyl-*N*-(*p*-chlorobenzylidene) aniline, $C_{14}H_{12}ClN$, (ClMe), two different occupancy variables are required, one to described the end-to-end flipping of the molecule and one to describe the side-to-side flipping. In the third example of benzil no occupational disorder was assumed.

The Monte Carlo energy used to specify local ordering of these molecular orientations is in the form of an Ising model:-

$$E_1 = \sum_n a_n s_{i,j,k,l} s_{i_n,j_n,k_n,l_n} \tag{1}$$

Here the summation is over all n-types of neighbour of a particular site. The a_n are interaction parameters which govern whether particular neighbouring molecular orientations are correlated. One such parameter is required for each symmetry-independent intermolecular vector along which significant correlation might be expected to occur.

Similarly, in the case of ClMe where two different occupational variables are used, an energy term allowing the possibility of cross-correlation between the two types of molecular flip can be invoked. E.g.,

$$E_2 = \sum_n b_n s_{i,j,k,l} r_{i_n,j_n,k_n,l_n} \tag{2}$$

3.2 Centre-of-mass displacements and orientational relaxation

The molecules occupying the molecular sites are assumed to be held in place by a set of harmonic springs which represent 'effective interactions' between neighbouring molecules.

Any local ordering of site occupancies such as that described above is invariably accompanied by local relaxation displacements which depend on the particular type of intermolecular contact, i.e. if neighbouring sites are occupied by molecules in orientations (+ +), (+ -), (- +) or (- -). In the case of a simple binary alloy the phenomenon is generally known as the 'atomic size effect'[22-24]. Though such 'size-effect' displacements are generally small they can contribute substantially to the diffuse scattering signal. In the present case we are dealing not with single atoms but with complex molecular shapes. Even if the molecule is assumed to be rigid, local relaxation may involve not only a shift of the centre of mass, but also a rotation away from its average orientation. Two different approaches have been used, and these are described below.

3.2.1 Springs acting on centres of mass and torsional force constants.

The centres of mass of different molecules are linked directly by a set of springs with force constants $k_{n,m}$. In addition, to take account of orientational relaxations, further sets of torsional springs are placed between neighbours.

For the centre of mass shifts an energy of the form,

$$E_3 = \sum_{n,m} k_{n,m} \left[d_{n,m} - d_{ave}\left(1 + \varepsilon_{n,m}\right)\right]^2 \tag{3}$$

is used. Here, as before, the summation is over all intermolecular vectors from a molecule at site n to its neighbours at sites m. $d_{n,m}$ is the instantaneous value of the distance between the centres of mass and d_{ave} is the corresponding equilibrium distance in the average structure. $\varepsilon_{n,m}$ are 'size-effect' parameters which take different values according to whether the two sites joined by the vector are (+ +), (- +), (+ -) or (- -). Though four different combinations of neighbour exist only three variables per inter-molecular vector are needed in general since the average centre of mass separation must be maintained. In some cases further relationships between the four values of $\varepsilon_{n,m}$ may be dictated by symmetry considerations.

For the orientational relaxations we consider small rotations away from the equilibrium orientations about the principal molecular axes. If the random variable $\phi_{i,j,k,l}$ is used to represent the angular variation about a particular molecular axis we can express the way in which the rotations of neighbouring molecules are linked by a size-effect like expression analogous to that for the centre of mass shifts given by Eq. (3) above.

$$E_4 = \sum_{n,m} L_{n,m}\left(\Delta\phi_{n,m} - \xi_{n,m}\right)^2 \tag{4}$$

Here $\Delta\phi_{n,m}$ is the difference between the value of $\phi_{i,j,k,l}$ in neighbouring molecules at sites n and m. Just as for the centre-of-mass 'size-effect' parameters $\varepsilon_{n,m}$, $\xi_{n,m}$ may take different values according to whether the sites n and m are (+ +), (+ -), (- +) or (- -), and again in order to keep the average constant this implies three independent values.

3.2.2 Springs acting between several points on a molecular motif

In situations where the molecule has not got a compact shape it is considered more appropriate that the mutual orientation of neighbouring molecules be modelled by using springs along a multiplicity of distinct inter-molecular vectors that more closely resemble the real intermolecular contacts. A schematic view of such an arrangement is shown in Fig. 3. Note that to maintain the average symmetry, springs which are related by symmetry must have identical lengths and force constants. The arrangement shown allows for the possibility that the mutual orientation of neighbouring molecules might change, while maintaining the average structure. In this case the molecular motif is just a simple rod coincident with the long axis of the molecule and springs are joined at the centre and ends of this rod. In a real 3D example such as the benzil example described in section 8 this general concept is developed further and the motif in this case is required to have internal degrees of flexibility. In general it has been found to be a reasonable approximation to represent each inter-molecular contact by ~3 springs which reasonably define both the separation and mutual orientation of neighbours. With inter-molecular contacts defined in this way, size-effect-like relaxations as given by Eq. (3) are assumed to operate on all such vectors.

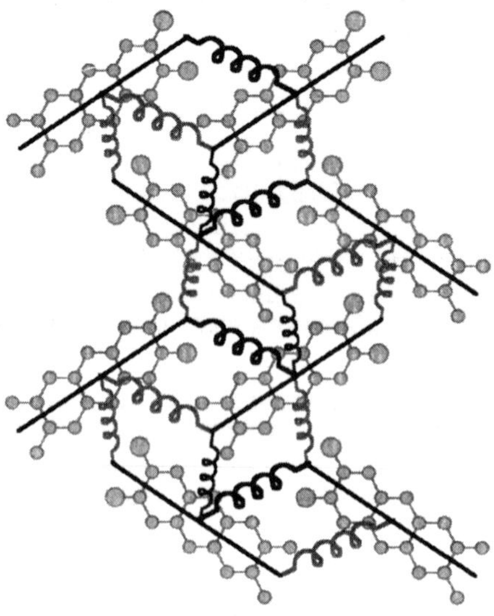

Figure 3. Schematic diagram (2D) showing how a multiplicity of intermolecular springs are used to define the mutual spacing and orientation of neighbouring molecules.

In Fig. 3 four different types of spring are depicted. If it assumed that each of the molecules may occupy the molecular site in an alternative orientation related to that shown by a 180° rotation in the plane of the paper, then there will be relationships between the spring constants. For example the two types of spring printed in thick lines are seen to connect the middle of one molecule to one end of a neighbouring molecule in the horizontal direction. The black spring joins the centre of one molecule with the end of the neighbouring molecule which has two large atom substituents (Cl say) while the grey spring joins the centre of one molecule with the end of the molecule which has two small atom substituents (methyl say). These two springs then are just two versions of the same spring with different values of $\varepsilon_{n,m}$.

4. CALCULATION OF DIFFRACTION PATTERNS

For calculating the diffraction pattern from the output of the MC runs we used the program DIFFUSE [25]. This algorithm obtains the diffraction pattern by taking the average of a large number diffraction patterns calculated from small regions (or lots), picked at random, from the simulated real-space array. This has the effect of removing the high-frequency noise, which necessarily occurs if a single calculation from the whole simulated array is performed. It is only necessary that the 'lot size' is large enough to include all significant non-zero correlations. For a crystal of 32×32×32 unit cells and a 'lot size' of 7×7×7 the whole crystal is sampled once on average if about 95 'lots' are used. It has been found that optimum results are obtained for a given simulation if an oversampling by a factor of about 2 is used (in this case say 200 lots). No improvement in the pattern is obtained by further oversampling and if the patterns are still unsatisfactorily noisy the only remedy is to increase the size of the crystal.

Before being used to compare to the observed patterns the quality of the calculated patterns can be further improved by averaging over symmetry related regions. For example in the case of benzil the patterns have 3m symmetry so each pixel in the diffraction pattern can be obtained as an average of six independently calculated pixels, resulting in a much smoother and less noisy pattern.

5. LEAST SQUARES

The basic least squares method seeks to minimise the sum of squares of the differences between a set of observed and calculated quantities. In our present case the set of observed quantities consists of diffuse scattering intensities measured at individual pixels in the diffuse sections such as that shown in Fig. 2. The corresponding calculated quantities are the suitably scaled values of the intensity obtained at corresponding points in reciprocal space from a MC simulation of a model system having system parameters p_i. These p_i correspond to the interaction parameters a_n & b_n in Eqs. (1)-(2), force constants $k_{n,m}$ & $L_{n,m}$ and size-effect parameters $\varepsilon_{n,m}$ & $\xi_{n,m}$ in Eqs. (3)-(4). That is we minimise the goodness-of-fit χ^2:-

$$\chi^2 = \sum_{h,k,l,m} \omega_{hklm} (\Delta I)^2 \tag{5}$$

where,

$$\Delta I = I_{obs} - (b_m + f_m I_{calc})$$

Here the summation is over all are non-integral reciprocal points h,k,l corresponding to individual pixels in the m measured sections of data. f_m is a scale and b_m a backgound correction applied to section m. [Note, f_m and b_m are determined as described in Proffen & Welberry[26] and are not included as variables in the least squares matrix]. ω_{hklm} is the weight for the corresponding data point h,k,l of data plane m. The weights used in the work described here were taken as $\omega_{hklm} = 1/I_{obs}$ or unity.

Increments Δp_i to be applied to the model parameters are calculated using the following expression:

$$\Delta p_i = \sum_l \mathbf{A}_{il}^{-1}\mathbf{B}_l \tag{6}$$

Here the summation is over the number of parameters, l. The matrix **A** and the vector **B** involve the differentials of ΔI with respect to each of the variables p_i. **A** is a symmetric matrix and **B** is a vector,

$$A_{ij} = \sum_{hklm} \omega_{hklm} \frac{\partial \Delta I}{\partial p_i} \frac{\partial \Delta I}{\partial p_j} \tag{7}$$

$$B_i = -\sum_{hklm} \omega_{hklm} \Delta I_{\text{trial}} \frac{\partial \Delta I}{\partial p_i} \tag{8}$$

It is also convenient to define a correlation matrix as a measure of how dependent the parameters used in the least-squares process are with each other. The correlation matrix **C** is defined as: $C_{ij} = A_{ij}^{-1} \Big/ \sqrt{A_{ii}^{-1} A_{jj}^{-1}}$. Note that these correlations are not to be confused with the occupational correlations present in the disordered structure.

5.1 Estimation of the differentials, $\partial \Delta I / \partial p_i$

The differentials $\partial \Delta I / \partial p_i$ may be computed as follows. If $p = (p_0, p_1, p_2, p_3 .. p_i ... p_n)$ is the current set of system parameters, the differential can be estimated by performing two complete MC simulation and diffraction pattern calculations using parameter sets, $p^+ = (p_0, p_1, p_2, p_3 .. p_i + \delta_i, ... p_n)$ and $p^- = (p_0, p_1, p_2, p_3 .. p_i - \delta_i, ... p_n)$, where δ_i is a suitably chosen small increment. The differential is then taken as:

$$\frac{\partial \Delta I}{\partial p_i} = \sum_{hklm} \frac{\left(\Delta I_{p+} - \Delta I_{p-}\right)}{2\delta_i} \tag{9}$$

If the calculated diffraction patterns were infinitely accurate then it would be best to choose δ_i as small as possible. However, each calculation of ΔI is only an approximation to the true intensity corresponding to a given model parameter set. Both the MC simulation itself and the ensuing diffraction pattern calculation result in inaccuracies. The size of the model crystal is one limiting factor. For a chosen system size of 32×32×32×2 molecular sites the normal statistical variations in the MC simulation lead to lattice averages such as an occupancy correlation coefficient c_n, say, having an accuracy of $\sim 1/\sqrt{32 \times 32 \times 32 \times 2} \approx 0.004$. Clearly the increment in the parameter p_i corresponding to such a correlation needs to be large enough to induce a change substantially larger than 0.004 otherwise the difference between the two patterns will only contain noise. Though using a larger system size would improve this accuracy, a corresponding increase in computer time (and memory) would be incurred. Fig. 4 shows an example of the difference between the diffraction patterns calculated from two MC runs for $Fe_3(CO)_{12}$ in which one of the size-effect parameters $\varepsilon_{n,m}$ was changed by an amount $\delta_i = 0.03$.

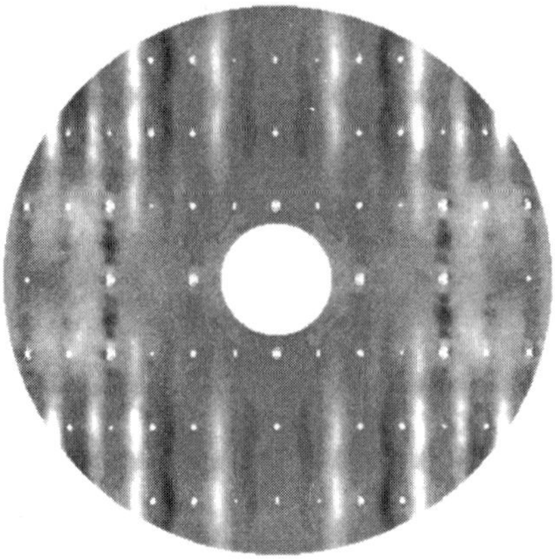

Figure 4. The difference between the diffraction patterns calculated from two MC runs of the [101] section of $Fe_3(CO)_{12}$ in which one of the size-effect parameters $\varepsilon_{n,m}$ was changed by an amount $\delta_i = 0.03$. The Bragg peaks were artificially superimposed on the figure as a guide to position in reciprocal space.

Since Eq. (9) involves two complete MC runs for each system variable an attractive alternative is to compute the differential using the MC run corresponding to the current set of system variables which is constant for all variables, and only one of the two incremented sets. I.e.

$$\frac{\partial \Delta I}{\partial p_i} = \sum_{hklm} \frac{\left(\Delta I_{p+} - \Delta I_{p0}\right)}{\delta_i} \tag{10}$$

$$\frac{\partial \Delta I}{\partial p_i} = \sum_{hklm} \frac{\left(\Delta I_{p-} - \Delta I_{p0}\right)}{-\delta_i} \tag{11}$$

Although Eq. (10) has been used extensively it is clearly biased towards positive parameter shifts. Consequently we have recently adopted the strategy of using Eqs. (10) and (11) in alternate cycles.

6. APPLICATION TO DODECACARBONYLTRIIRON, $Fe_3(CO)_{12}$

The compound $Fe_3(CO)_{12}$ was chosen for the first study to test the efficacy of the least-squares fitting method because it exhibits strong and fairly simple diffuse scattering which arises predominantly from the orientational disorder of the triangle of iron atoms (see Fig. 5). The system could thus be expected to be simply represented by the centre-of-mass positions and orientations of the iron triangles in each of the two sites in the unit cell. The average crystal structure determination indicated that the Fe_3 triangles appeared to have large displacement amplitudes corresponding to librations about an axis passing through one Fe atom and bisecting the bond joining the other two. The diffraction patterns for the zero-level sections normal to [101] and [100] are shown in Fig. 6.

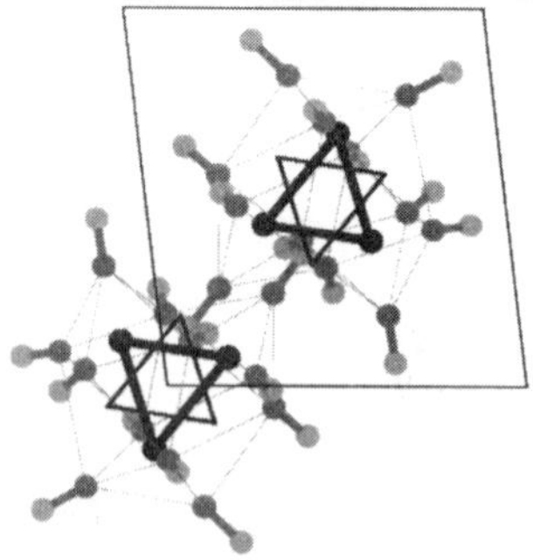

Figure 5. The projection of the $Fe_3(CO)_{12}$ structure down **b**, showing the alternative orientations of the Fe_3 triangles and the surrounding cage of carbonyls.

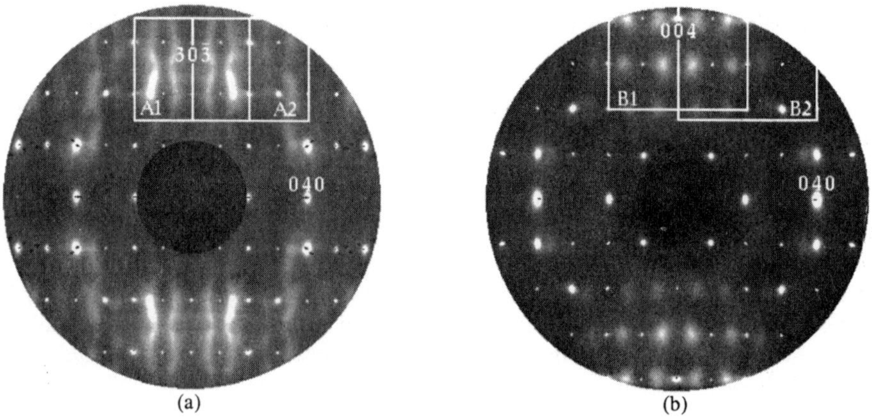

Figure 6. The zero-level sections of X-ray diffuse scattering data for $Fe_3(CO)_{12}$. (a) normal to [101]; (b) normal to [100]. The rectangular regions labeled A1, A2, B1 and B2 were used in the analysis.

In Fig. 7 we show the eight different types of near-neighbour intermolecular contacts that were used in the analysis, as viewed down the [101] and [100] axes. These projections correspond to the zero-level diffraction patterns shown in Fig. 6. The MC model assumed that interactions of the form given by Eq. (1) took place only along these eight types of vector. Initially only occupational disorder was considered and the fit was restricted to the data contained in regions A1 and B1 seen in Fig. 6. In these regions the diffuse data does not appear to have substantial asymmetry about the lines containing Bragg peaks. Subsequently additional displacement/orientation parameters as given by Eqs. (3) & (4) were included and then the fit was made to the regions of data labeled A2 and B2 in Fig. 6. In order to keep the number of parameters to a minimum these displacement/orientation parameters were restricted to the vectors 1 and 2 which appeared by inspection to be the most important. In addition, along each of these only two different values of $\varepsilon_{n,m}$ and two of $\xi_{n,m}$ were used, making eight relaxational parameters in all.

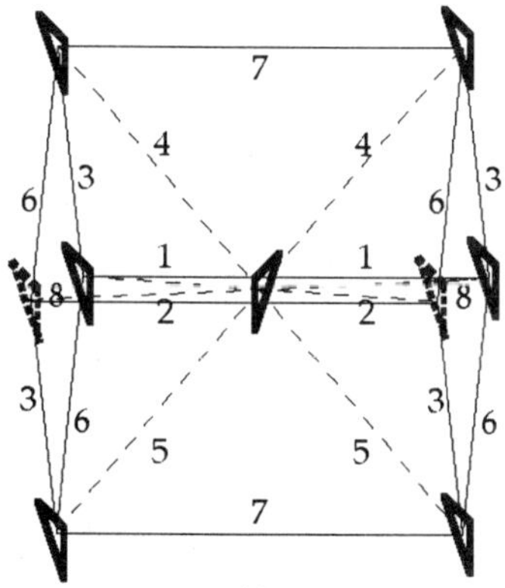

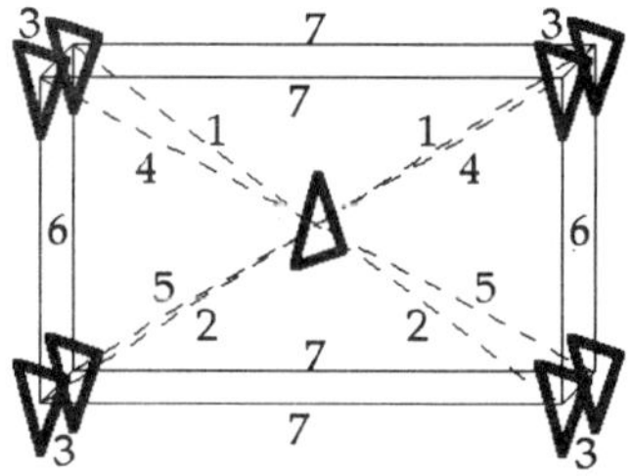

Figure 7. The definition of the 8 types of vectors, linking near-neighbour Fe_3 sites, that were used in the analysis: (a) View down [101]; (b) View down [100].

6.1 Results for $Fe_3(CO)_{12}$.

The final MC model contained 16 variables: 8 occupancy correlation variables a_n along the vectors 1 to 8; 4 centre-of-mass 'size-effect' variables $\varepsilon_{n,m}$ along vectors 1 and 2; and 4 orientational variables $\xi_{n,m}$ also along vectors 1 and 2. The fitting was carried out in two stages, with the relaxation variables being adjusted only after the occupancy correlations had been fixed. Values for the determined parameters are given in Table 1.

The resulting fit of the calculated patterns to the observed data appears quite good, given that the model is such a simplified one (see Figs. 6 & 8). The overall R-factor for the fitted data, $R = \left(\chi^2 / \sum \omega I_{obs}^2\right)^{1/2}$, was 21% for ~40,000 data points from selected regions of three reciprocal sections. Of particular significance is the reduction in intensity of the vertical band of intensity labeled '3' in Fig. 8a. This results directly from the 'size-effect' relaxation parameters. See also Fig. 4.

Table 1. Values of the MC parameters determined in the analysis. The c_n are correlation coefficients resulting from the interactions a_n in Eq. (1). Note that though 3 values per vector are given for $\varepsilon_{n,m}$ and $\xi_{n,m}$ only two of the values are independent.

Parameter	Vector 1	Vector 2	Vector 3	Vector 4	Vector 5	Vector 6	Vector 7	Vector 8
c_n	–0.120	+0.243	–0.097	+0.114	–0.155	+0.202	–0.156	+0.099
$\varepsilon_{n,m}$ (+ +)	–0.042	–0.027	-	-	-	-	-	-
$\varepsilon_{n,m}$ (+ –)/(– +)	+0.014	+0.023	-	-	-	-	-	-
$\varepsilon_{n,m}$ (– –)	+0.014	–0.019	-	-	-	-	-	-
$\xi_{n,m}$ (+ +)	–0.085	+0.236	-	-	-	-	-	-
$\xi_{n,m}$ (+ –)/(– +)	+0.124	–0.128	-	-	-	-	-	-
$\xi_{n,m}$ (– –)	–0.163	+0.020	-	-	-	-	-	-

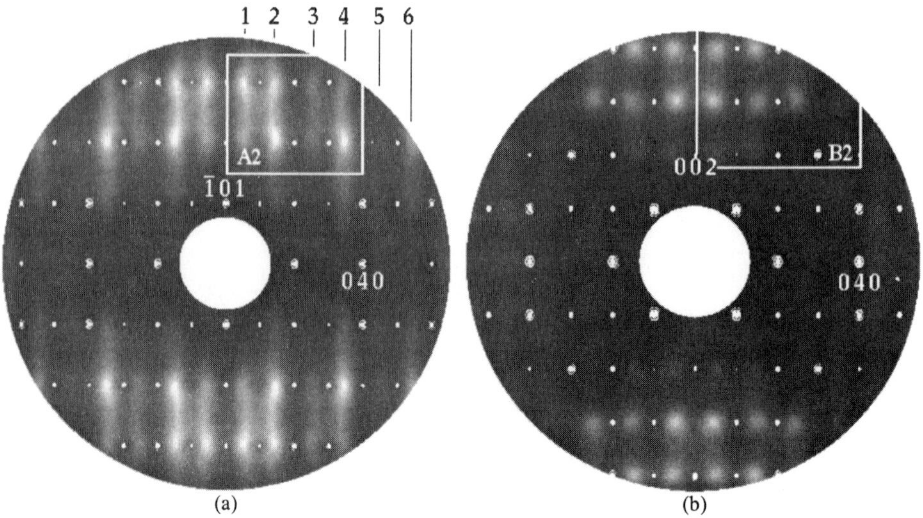

Figure 8. The diffraction patterns calculated from the final MC model for the zero-level sections: (a) normal to [101]; (b) normal to [100]. Only the data enclosed in the rectangles labeled A2 and B2 was used in the final fitting together with similar regions in a third section of data (*h 0.5 l*) not shown here.

Once the MC model has been established the results of the simulation can be explored to obtain insight into what is happening in the disordered structure. In Fig. 9 we show plots obtained from the coordinates of the final MC simulation. Fig. 9a shows a superposition of all Fe atoms from every unit cell in the structure and clearly shows the large anisotropic mean-square displacement amplitudes of two of the three Fe atoms in each triangle with the third much smaller and isotropic. This information is essentially the same as obtained from the Bragg average structure determination. On the other hand Figs. 9b, 9c & 9d show plots of only those unit cells which contain particular combinations of the basic Fe_3 triangle orientations. Here it is clearly seen that there is substantial cooperative displacement of neighbouring molecules. For example in Fig. 9b when the origin triangle is in the (+) orientation the triangle at site 2 rotates clockwise while in Fig. 9c with the origin molecule in the (–) orientation the triangle at site 2 rotates anticlockwise.

7. APPLICATION TO ClMe, $C_{14}H_{12}ClN$

In the study of $Fe_3(CO)_{12}$, which was the first attempt to use the automatic refinement method, a number of choices and approximations had to be made to keep the computational problem within reasonable bounds. For this second example, *p*-methyl-*N*-(*p*-chlorobenzylidene) aniline, $C_{14}H_{12}ClN$, (ClMe), some rather different choices were made. First of all the problem was a much more complex one with each molecular site being occupied by a molecule in one of *four* basically different orientations. In addition the length of the molecule was such that treating the molecular entity in terms of a centre-of-mass plus an orientation was not a sensible option. Consequently the second option described in section 3.2.2 was adopted. A simple molecular *motif* comprising only four coplanar pseudoatoms as shown in Fig. 10a was used in the MC simulation. The four different molecular orientations were then defined relative to this *motif* as shown in

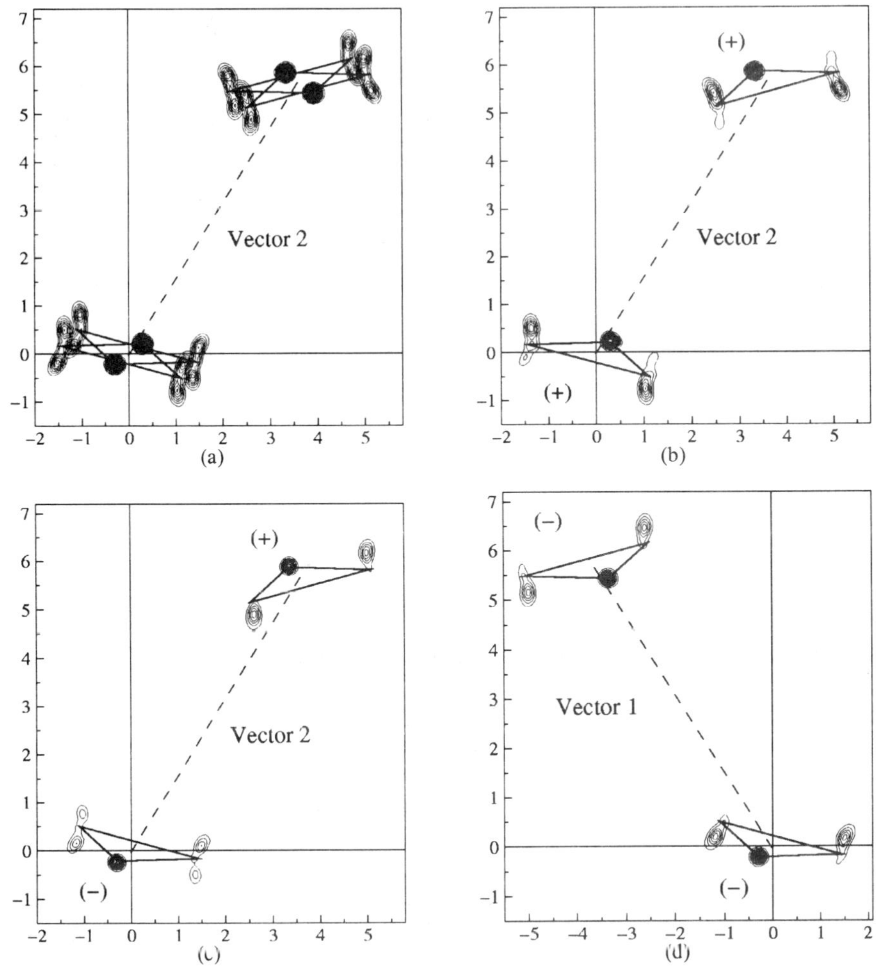

Figure 9. (a) Superposition plot of the distribution of all Fe atom positions from all unit cells in the MC simulation. (b) Superposition plot of only those unit cells in which both sites 1 and 2 are occupied by the (+) orientation of the Fe triangle. (c) Superposition plot of only those unit cells in which site 1 is occupied by a (–) orientation and site 2 a (+) orientation. (d) Similar plot for pairs of sites separated by vector 1 with Fe atoms in the (–) orientation on both sites. The black triangles mark the average position of the atoms determined from the Bragg structure solution. Axes are marked in Å.

Fig. 10b. The whole MC calculation was carried out using only the *motif* and the atomic positions defined relative to this were only used at the point of computing the diffraction pattern. An additional approximation that was made was that the C-C=N-C bridge at the centre of the molecule was assumed to be rigid.

Fig. 2 shows the (*h0l*) section of the diffraction pattern of ClMe and this clearly indicates the rich detail that is present in this section. Moreover the projection of the structure (down **b**), to which this diffraction pattern corresponds, clearly reveals all the molecules without overlap. Thus in this study, although a full 3D MC model was constructed, this single section of data was used and consequently refinement was restricted to vari-

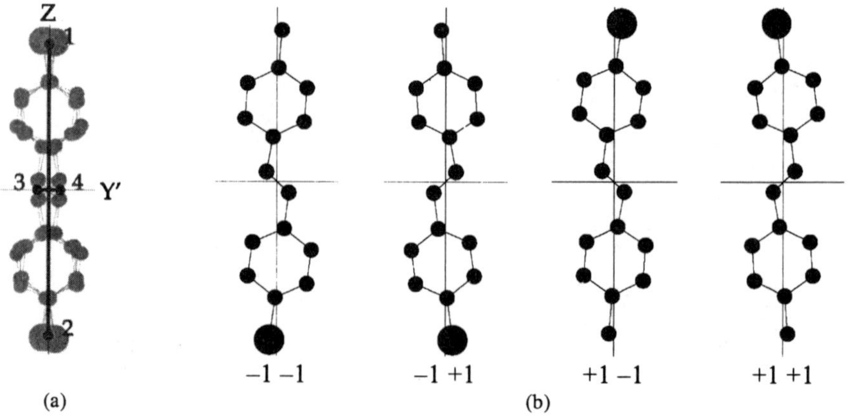

Figure 10. (a) the average molecular site showing the superposition of four different molecular orientations and the cross-shaped *motif* defined by the four pseudoatoms labelled 1-4. (b) The four different orientations of the molecule together with the values of the random variables σ and ρ that define them.

ables clearly resolved in the plane normal to **b**. As mentioned earlier the data used in the refinement was rebinned from that shown in Fig. 2 to fit in an array of 200×200 pixels.

In Fig. 11 we show the projection of the structure together with a corresponding plot of the representation using the molecular *motifs*. In this latter figure are shown all the intermolecular vectors that were used in the MC simulation. There were 26 such vectors in all. These were made up of 8 symmetry-inequivalent types of vector which occurred either in pairs (e.g. 1 & 2 or 3 & 4) or sets of 4 (e.g. 5, 6, 7, 8 or 13, 14, 15, 16).

As in the case of $Fe_3(CO)_{12}$ it was found impossible to refine occupancy variables at the same time as displacement variables and again the refinement was carried out in two stages. For the occupancy variable refinement a small region of the pattern was used which contained a strong short-range-order peak which was judged to be not significantly affected by size-effect displacements. Moreover while this occupancy fitting was able to determine the correlation structure for the end-to-end flipping of the molecule (σ variables) no similar region dependent solely on the side-to-side flipping (ρ variables) was available and in the final model it was assumed that the correlation structure of the latter was the same as the former. A cross-correlation parameter as given by Eq. (2) was refined however.

Of the eight independent types of intermolecular force-constants it was found that one continually refined to a very small value. This was for the vectors 13, 14, 15 and 16 seen in Fig. 11b and so this was omitted from the refinement altogether. It is possible that this occurred because these vectors are superimposed on the similar set comprising vectors 17, 18 ,19 and 20 and the single section of data contained no information to distinguish them. For the other seven types of vector each was allowed a variable force constant $k_{n,m}$ and 3 independent size-effect variables, $\varepsilon_{n,m}$, as defined by Eq. (3). $\varepsilon_{n,m}$ was assumed to depend on the σ occupancy variables for vectors involving the ends of the molecular *motif* and on ρ for vectors involving the sides of the molecular *motif*.

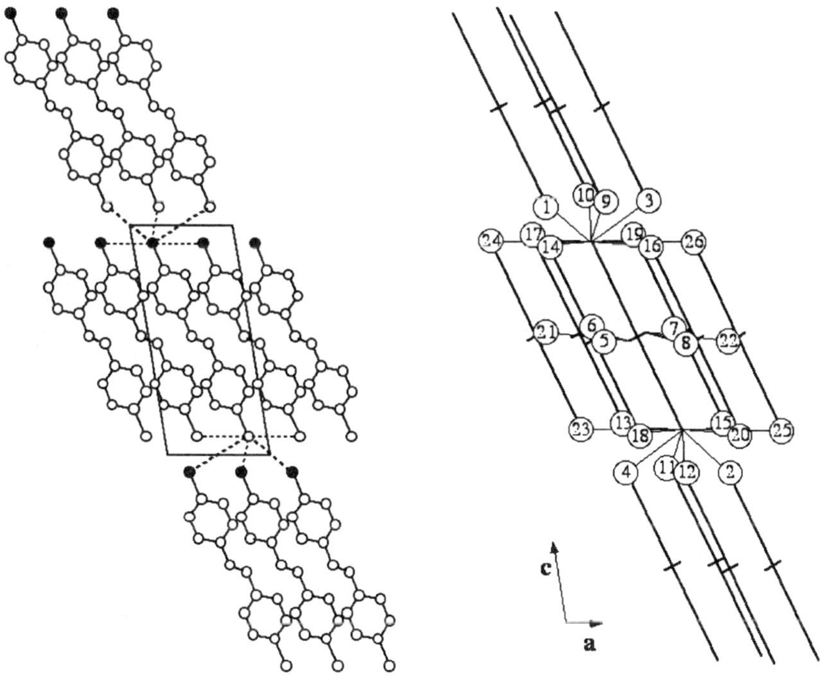

(a) (b)

Figure 11. (a) Plot of the structure of ClMe viewed down **b**. In this figure all molecules are in the orientation corresponding to $\sigma = -1$, $\rho = -1$. (b) Schematic diagram of the environment of a central molecule of ClMe showing the 26 different vectors along which 'springs' linking the central molecule to its 14 neighbours are placed to represent the effective molecular interactions. In (b) the view is slightly rotated to reveal the neighbouring molecular sites which occur above and below the height of the central molecule.

7.1 Results for ClMe

The final model for ClMe consisted of 4 variables defining the short-range occupancy correlations, 7 intermolecular force constants and 21 'size-effect' parameters. A comparison of observed and calculated diffraction patterns is shown in Fig. 12. The fit appears to be very good considering the numerous approximations and simplifications that were made. The *R*- factor (intensities) for the agreement was 32.8% for 28,500 data points. This is rather higher than the value given for the $Fe_3(CO)_{12}$ study but it should be remembered that in that case only the very strong regions of scattering were included whereas in this case the figure is for a complete reciprocal lattice section.

As for the $Fe_3(CO)_{12}$ example the final MC results can be explored to obtain insight into what is happening in the disordered structure. In this case we show in Fig. 13 a plot of the distribution of the lengths of the vectors 3 and 19 which seem to be most significant. In both cases the size-effect variables acting on these vectors was dependent on the end-to-end disorder variables σ. In each plot the distributions corresponding to the four different combinations (+ +), (+ –), (– +), (– –) are drawn separately.

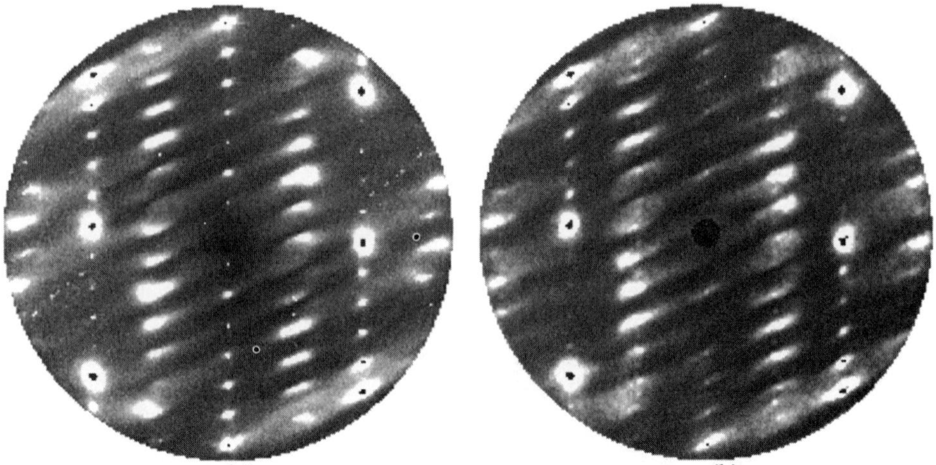

(a) (b)

Figure 12. (a) observed and (b) calculated diffraction patterns for the (*h0l*) section of ClMe.

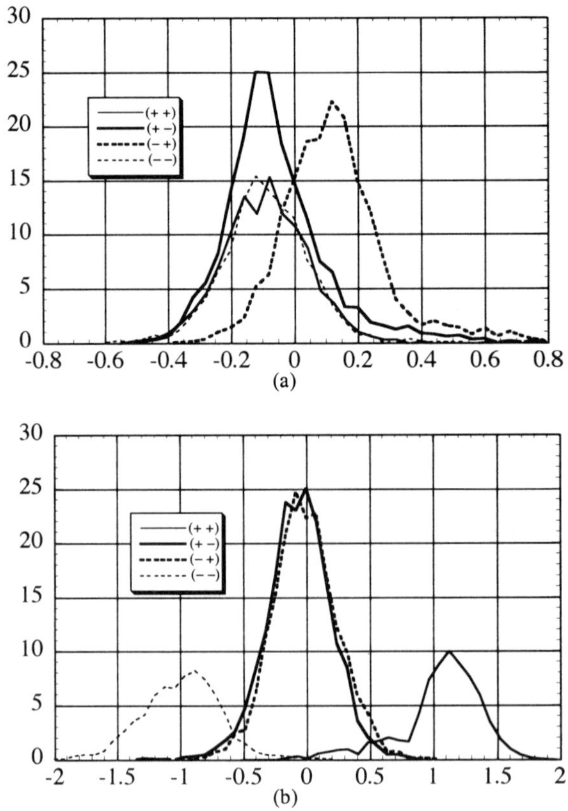

Figure 13. Plots of the distribution of lengths for (a) vector 3 and (b) vector 19. The horizontal axis is the deviation from the mean length in Å. Note that in (a) the curve corresponding to (– +) which has the greatest mean separation corresponds to when the vector is essentially between two Cl atoms. Similarly in (b) the curve with the greatest separation corresponds to a Cl-Cl contact and the smallest to Me-Me.

8. APPLICATION TO BENZIL, $C_{14}H_{10}O_2$

In study of ClMe we attempted to apply the automatic refinement method to a complex disordered system in which each molecular site was occupied by one of four basically different molecular orientations. In view of this degree of occupational complexity it was assumed for simplicity that for the accompanying relaxational displacements the molecules behaved as rigid bodies, even though they possessed some internal degrees of flexibility. Although the model obtained provided a reasonable description of the disordered system it was clear that further information was present in the experimental data, and this provided the impetus for us to attempt to incorporate the possibility of internal degrees of freedom into future models. In this section we describe, as a first step, results for a simpler system in which the molecule has internal degrees of freedom and still gives strong diffuse X-ray scattering patterns, but has only one basic molecular orientation for each site. This is the organic molecular crystal benzil, (**I**, diphenylethanedione, C_6H_5-CO-CO-C_6H_5).

I

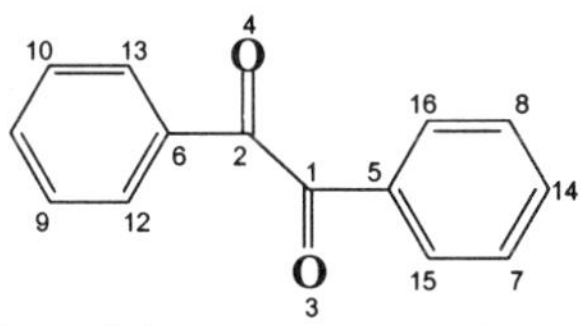

Here the unconventional numbering of the atoms was chosen for convenience and is referred to later in the text.

In order to allow for the internal flexibility of the benzil molecule a z-matrix representation was used. This method, commonly used for example in *ab initio* molecular orbital calculations (see e.g. Hehre et al[27]), allows the geometry of the molecule to be specified in terms of bond-lengths, bond-angles and dihedral angles (see also David[28]). In the present work we used a model in which all bond-lengths, bond angles and dihedral angles were fixed with the exception of the dihedral angles corresponding to rotation about the central C1-C2 bond (ϕ_1) and rotations of the phenyl groups about the C1-C5 and C2-C6 bonds respectively (ϕ_2 & ϕ_3).

To represent the intermolecular interactions we again use the idea of 'effective' interactions. Each molecule is in contact with 15 neighbouring molecules and if all atom-atom contacts were to be included the number of these would be prohibitively large to be contemplated for the MC simulations. With each molecule in its equilibrium position all intermolecular atom-atom contact distances less than 4.5Å were computed. A smaller number of contacts was obtained by pruning this list. Each intermolecular contact was represented by a small number of atom-atom contacts (2 to 3) which were sufficient to define the equilibrium separation and mutual orientation of neighbouring molecular fragments. Since each molecule was made up of rigid fragments with only 3 internal degrees of freedom, the atom-atom contacts involving only the atoms O3, O4 and C7, C8, C9 and C10 were considered sufficient for this purpose. These contact vectors, numbering 34 in total around each molecule, are shown in Fig. 14. The 34 vectors comprise 10 symmetry-inequivalent vector types. At a later stage, 4 additional contact vectors (1 additional vector type) involving atoms C11 and C14 were included.

Simulation was carried out using an energy of the form Eq. (3) in which all size-effect parameters were zero. The system variables, therefore, were the spring constants $k_{n,m}$ for the 11 symmetry inequivalent types of vector. In addition an overall isotropic Debye-Waller factor B was refined. Four sections of data were used: (*hk0*), (*hk0.5*), (*hk1*), (*hk2*).

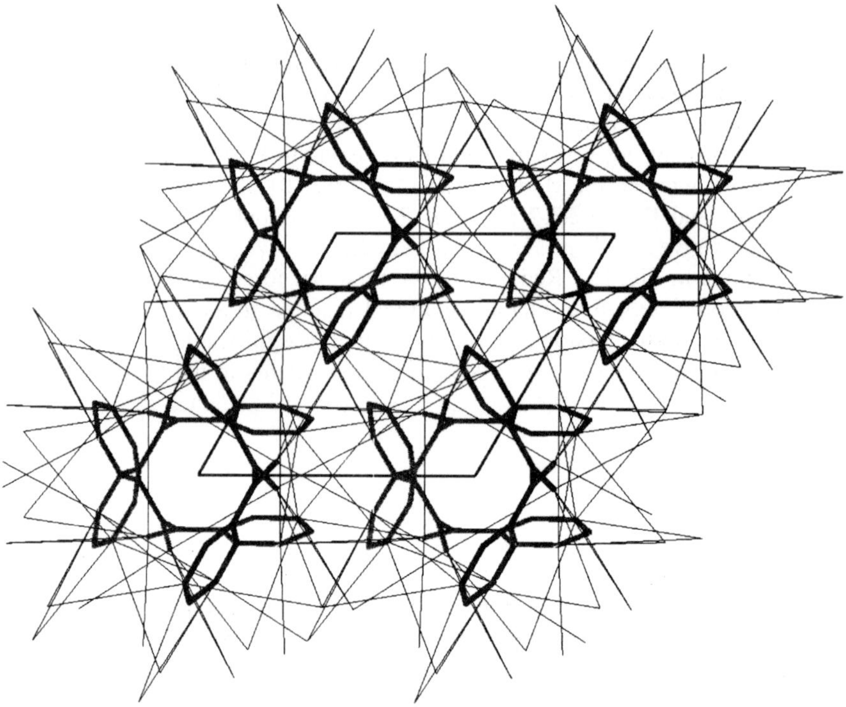

Figure 14. The structure of benzil showing the 34 intermolecular vectors surrounding each molecule that were used in the analysis. The contacts involve only the atoms O3, O4 and C7, C8, C9 and C10.

The internal flexibility of the molecule was initially assumed to involve no energy penalty but in later stages of the analysis it became necessary to include such intra-molecular forces. Two types of term were considered. First, simple torsional forces were included which depended on the deviations, $\Delta\phi_i$, of the angles ϕ_1, ϕ_2, ϕ_3 from their equilibrium values. Secondly, terms to account for the possibility of correlated motion of the different intra-molecular fragments were added. These intra-molecular contributions to the energy were,

$$E_{\text{intra}} = \sum_{\substack{\text{all}\\ \text{molecules}}} F_1(\Delta\phi_1)^2 + F_2\left[(\Delta\phi_2)^2 + (\Delta\phi_3)^2\right] + F_3\Delta\phi_2\Delta\phi_3 + F_4\left[\Delta\phi_1\Delta\phi_2 + \Delta\phi_1\Delta\phi_3\right] \tag{12}$$

The total energy of the crystal is then expressed in the form,

$$E_{\text{tot}} = E_{\text{inter}} + E_{\text{intra}} \tag{13}$$

and this was used in the application of the normal Monte Carlo algorithm[29]. All MC simulations were carried out at a temperature corresponding to the Boltzmann temperature kT=1.0 so that the values derived for the force constants $k_{n,m}$ and torsional constants F_1-F_4 were in units measured relative to this.

8.1 Results for benzil

Final cycles of refinement were carried out with two different models. Model 1 had 16 refineable parameters: 11 inter-molecular constants, one overall Debye-Waller factor and 4 intra-molecular torsional constants as defined in Eq. (12). In Model 2 only a single torsional force constant, F_1, was used, with F_2 set equal to F_1. F_3 and F_4 were set to zero. The final R-value obtained for Model-1 was 14.2% for 101,324 data points from the four sections of data, and for Model-2 the final R-value was only a little higher at 14.5%. Both refinements give excellent agreement between observed and calculated diffraction patterns, as shown in Fig. 15. Removal of the torsional force constants from the model (i.e. assuming completely free internal rotations) resulted in an R-value of about 18% indicating that the internal torsional constants are highly significant.

At first sight the values of the torsional force constants in the two refinements appeared quite different but the four values F_1- F_4 in model 1 in fact make up the elements of a force-constant tensor,

$$\mathbf{F} = \begin{pmatrix} F_1 & F_4 & F_4 \\ F_4 & F_2 & F_3 \\ F_4 & F_3 & F_2 \end{pmatrix} \tag{14}$$

A detailed investigation revealed that this tensor has two eigenvalues which are quite small and a third which is very large. Moreover, the eigenvector corresponding to the smallest eigenvalue corresponds approximately to a rotation of the angle ϕ_1 with smaller counter-rotations of ϕ_2 and ϕ_3, and the second smallest corresponds approximately to a mode in which ϕ_2 and ϕ_3 rotate in opposite directions with ϕ_1 not involved at all. The Model 1 solution is therefore not very different from Model 2 in its essentials. The third eigenvalue (the highest) in Model 1 represents a direction in the ϕ_1, ϕ_2, ϕ_3 phase space (all three angles change in phase by approximately equal amounts) in which the molecule has become very stiff. Since this does not appear to affect the level of agreement very much it is clear that the diffuse scattering data must be relatively insensitive to this particular kind of stiffness. It was thus concluded that the general form, Eq. 14, for the torsional force constants contains parameters not derivable from diffuse scattering data, and so should not be used.

The results of the refinement indicated that two of the force constants $k_{n,m}$ are much larger than any of the others. The vectors along which these act are shown in Fig. 16. The analysis has shown quite clearly that the diffuse lines that feature so prominently in the observed diffraction patterns are due to strong longitudinal displacement correlations transmitted from molecule to molecule via the network of vectors possessing these strong values of $k_{n,m}$. Note here that by *longitudinal correlation* we mean that the direction of displacement and the direction of correlation are the same. It seems likely that the molecular basis for this is hydrogen bonding between the O_3 or O_4 atom on one molecule and the *para* H atom (i.e. attached to C_{11} or C_{14}) of the phenyl ring of a neighbouring molecule.

In Fig. 17 we show a plot of the distribution of the torsional angles in the final Model-2 simulation for which the off-diagonal terms in the force-constant tensor, Eq. 14, were zero. The plot of ϕ_1 versus ϕ_2 indicates that, despite having no direct interaction linking them, ϕ_1 and ϕ_2 are correlated (*albeit* weakly). This correlation must result from the constraints imposed by the surrounding *inter*-molecular interactions. As the central bridge angle rotates the phenyl rings tend also to rotate to compensate. However, despite

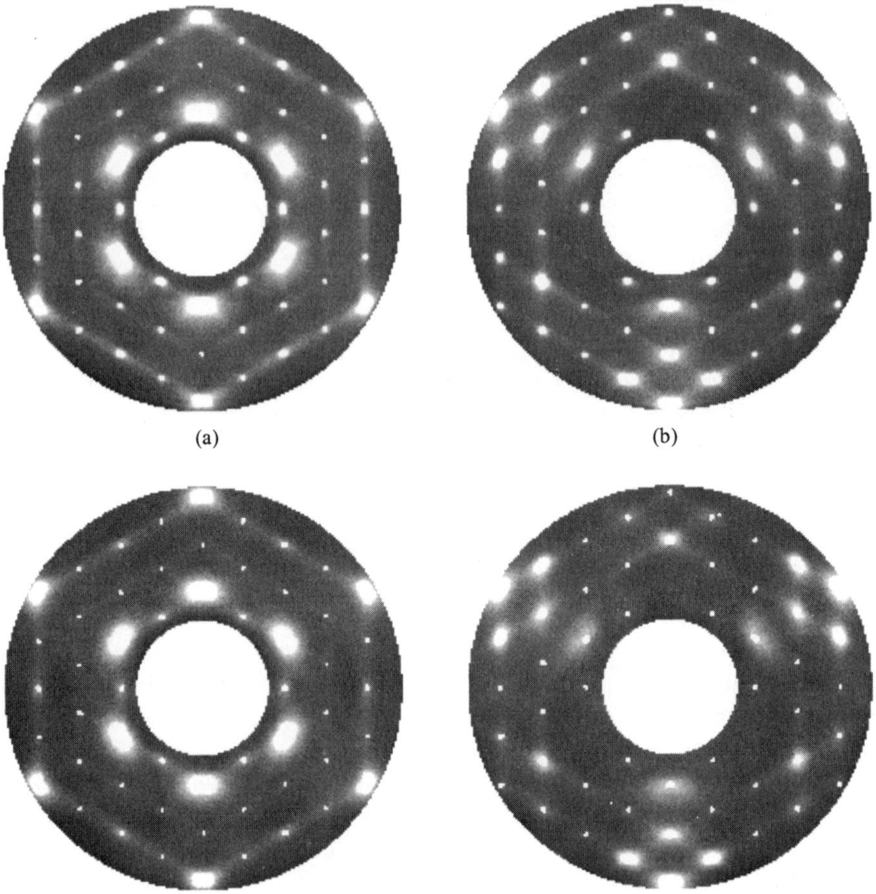

(a) (b) (c) (d)

Figure 15. Comparison of the observed and calculated diffraction patterns for two sections of benzil. (a) Observed (*hk0*) data. (b) Observed (*hk2*) data. (c) Calculated (*hk0*) section. (d) Calculated (*hk2*) section. The MC model fit was made to these two sections and two others (not shown).

this correlation between ϕ_1 and ϕ_2 (and similarly between ϕ_1 and ϕ_3), the plot of ϕ_2 versus ϕ_3 shows virtually no correlation, indicating that the rotations of the two phenyl rings are virtually independent.

9. CONCLUSION

In this paper we have described studies of the X-ray diffuse scattering in three quite different disordered systems[17-19], using a recently developed method in which a Monte Carlo model of the structure is automatically fitted to the observed data via a least-squares technique. Despite the compromises and simplifications that have had to be made these studies have successfully shown that the refineable MC methodology can be made

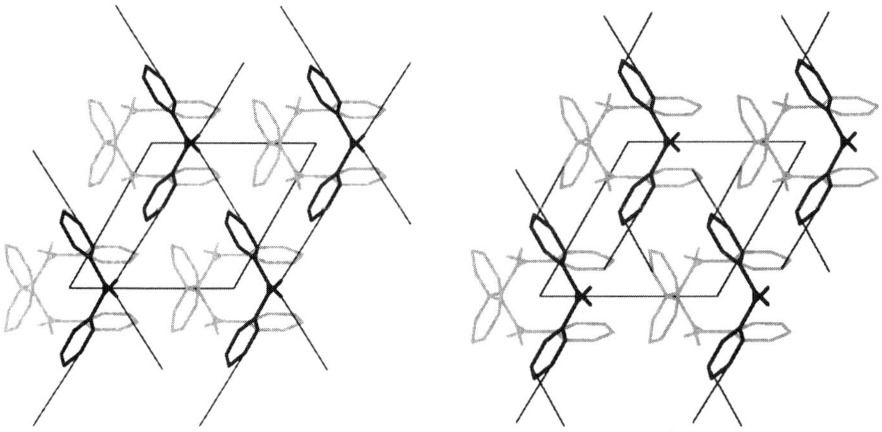

Figure 16. Diagram showing the two strongest force constants, $k_{n,m}$. For clarity, vectors to only one of the three molecules per cell are shown. The first force constant (a) produces an extended network of vectors connecting molecules in one layer and is responsible for the strong longitudinal correlation giving rise to the narrow diffuse lines seen in all of the diffraction patterns. The second force constant, (b), connects molecules in one layer with those in the layers above and below.

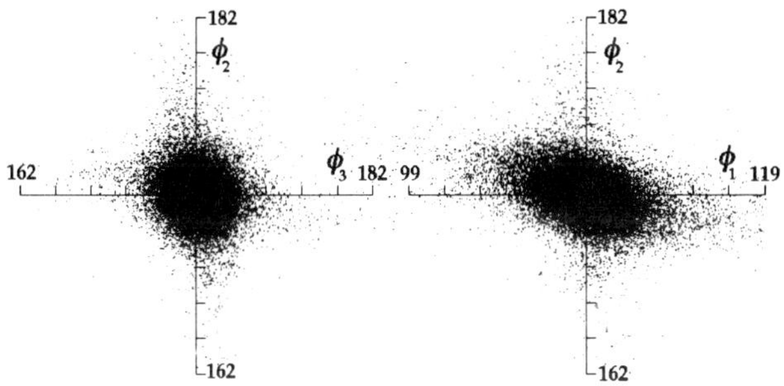

Figure 17. Plot of the distribution of torsion angles in the final MC simulation. Note that ϕ_2, ϕ_3 are virtually independent but each is weakly correlated with the central torsion angle, ϕ_1.

to work for even quite complex systems and has yielded detailed local ordering information not otherwise obtainable. In each case it has been found possible to develop a model which encapsulates the basic physics and chemistry of the problem using relatively few parameters and which gives a good agreement between observed and calculated diffraction patterns. Each study represented a considerable computational task which stretched the available computational resources to the limit. By use of even greater computational resources it is clear that even better solutions could be found in each case. Even in the short space of time since the first of these studies[17] the computational resources available have more than doubled. With the ever increasing availability of fast and inexpensive computers it is expected that the methodology will increasingly become viable as a means of analysis for a wide range of disorder problems.

REFERENCES

1. J.E. Epperson, J.P. Anderson and H. Chen. *Metall. Mater. Trans.* **A25,** 17-35. (1994).
2. B. Borie, and C.J. Sparks. *Acta Cryst.* **A27,** 198-201. (1971).
3. J.E. Tibballs. *J. Appl. Cryst.* **8,** 111-114. (1975).
4. M. Hayakawa and J.B. Cohen. *Acta Cryst.* **A31,** 635-645. (1975).
5. T.R. Welberry and B.D. Butler. *Chem. Rev.* **95,** 2369-2403. (1995).
6. F. Frey. *Acta Cryst.* **B51,** 592-603. (1995).
7. F. Frey. *Zeits. f. Kristall.* **212,** 257-282. (1997).
8. J.B. Clarage, M.S. Clarage, W.C. Phillips, R.M. Sweet and D.L.D. Caspar. *Proteins: Struct., Funct. Genet.* **12,** 145-157. (1992).
9. J. Doucet, J.-P. Benoit, P. Faure and D. Durand. *J. de Physique I.* **2,** 981-993. (1992).
10. A.R. Kolatkar, J.B. Clarage and G.N. Phillips. Acta Cryst. **D50,** 210-218. (1994).
11. J.B. Clarage and G.N. Phillips. Methods in Enzymology. **277,** 407-432. (1997).
12. T.R. Welberry, B.D. Butler, J.G. Thompson and R.L. Withers. *J. Solid State Chem.* **106,** 461-475. (1993).
13. T.R. Welberry, B.D. Butler and A.P. Heerdegen. *Acta Chimica Hungarica - Models in Chemistry*, **130,** 327-345. (1993).
14. T.R. Welberry and A.M. Glazer. *J. Appl. Crystallogr.* , **27,** 733-741. (1994).
15. T.R. Welberry and S.C. Mayo. *J. Appl. Crystallogr.* , **29,** 353-364. (1996).
16. T.R. Welberry and A.G. Christy. *Phys. Chem. Minerals.* **24,** 24-38. (1997).
17. T.R. Welberry, Th. Proffen and M. Bown. *Acta Crystallogr., Sect. A: Found. Crystallogr.* , **A54,** 661-674. (1998).
18. T.R. Welberry. *Acta Crystallogr., Sect. A: Found. Crystallogr.* , **A56,** 348-358. (2000).
19. T.R. Welberry, D.J. Goossens, A.J. Edwards and W.I.F. David. *Acta Crystallogr., Sect. A: Found. Crystallogr.* , **A57,** 101-109. (2001).
20. J.C. Osborn and T.R. Welberry. *J. Appl. Cryst.* **23,** 476-484. (1990).
21. T.R. Welberry and Th. Proffen. *J. Appl. Cryst.* **31,** 309-317. (1998).
22. B.E. Warren, B.L. Averbach and B.W. Roberts. *J. Appl. Phys.* **22,** 1493-1496. (1951).
23. T.R. Welberry. *J. Appl. Cryst.* **19,** 382-389. (1986).
24. T.R. Welberry and B.D. Butler. *J. Appl. Cryst.* **27,** 205-231. (1994).
25. B.D. Butler and T.R. Welberry. *J. Appl. Cryst.* **25,** 391-399. (1992).
26. Th. Proffen and T.R. Welberry. *Acta Cryst.* **A53,** 202-216. (1997).
27. W.J. Hehre, L. Radom, P.v.R. Schleyer and J.A.Pople. *Ab Initio Molecular Orbital Theory.* John Wiley & Sons, New York. (1986).
28. W.I.F. David, this volume.
29. N. Metropolis, A.W. Rosenbluth, M.N. Rosenbluth, A.H. Teller and E. Teller. *J. Chem. Phys.* **22,** 1493-1496. (1953).

MEDIUM-RANGE ATOMIC CORRELATION FROM THE PDF ANALYSIS

T. Egami*

1. INTRODUCTION

While well-established methods exist for determining the long-range or short-range atomic order, very few techniques are available to assess the nanometer scale medium-range (4 – 20 Å) order (MRO). Yet, properties of materials often are significantly affected by the MRO. In this paper we discuss how the MRO can be determined by the method of atomic pair-density function (PDF) analysis, and how the information relevant to understanding the properties can be deduced from the results, using the results on colossal magnetoresistive (CMR) manganites as an example.

Characterizing the atomic structure of a solid usually means carrying out an x-ray diffraction study and the crystallographic analysis. However, the crystal structure represents only the long-range order and the average unit cell structure while, except for extremely perfect crystals, the real structure is deviated from the average structure due to lattice defects, mixed site-occupation in alloys, and other self-organization instigated by electrons. Often these deviations are important in determining the properties of the solid. These deviations, the difference between the local structure and the average structure, are the main theme of the workshop.

Usually by local structure one implies the nearest neighbor environment of the atom that can be determined, for instance, by the x-ray absorption fine structure (XAFS) analysis. But the immediate local environment is largely determined by the chemistry, and for given local chemistry it is remarkably invariant. Such invariance is the basis for concepts in crystal chemistry such as the bond length and ionic radius. This is because the energy of a chemical bond is of the order of 1 eV, greater by an order of magnitude than the typical energy supplied for reorganization of the structure, for instance thermal energy of melting and crystallization. Thus, except for the case of chemically similar elements resulting in similar bond energies, such as the systems that show order-disorder transition, manipulation of the immediate local environment by processing is difficult.

In contrast the energy necessary to reorganize the medium-range order is of the order of kT at melting or annealing temperatures. Therefore MRO offers much wider and greater opportunities for engineering by processing. For this reason materials processing

* T. Egami, Laboratory for Research on the Structure of Matter and Department of Materials Science and Engineering, University of Pennsylvania, Philadelphia, PA 19104.

From Semiconductors to Protein, edited by S.J.L. Billinge and M.F. Thorpe
Kluwer Academic/Plenum Publishers, 2002

to alter the MRO is expected to become more popular, and this trend is expected to increase the value of the technique to determine MRO. The knowledge of MRO is of critical importance also in understanding materials with strong interactions of the lattice with charge or spin that leads to various local lattice distortions. As an example manganites that exhibit colossal magnetoresistivity (CMR) will be discussed in this paper.

In spite of the increased attention to MRO and nano-scale structure in recent years there is a dearth of techniques that can determine MRO with high accuracy. The XAFS method quickly loses reliability as one tries to go beyond the nearest neighbor. Surface techniques such as STM and AFM can resolve short- and medium-range structure at the surface, but cannot assess the bulk structure. In transmission electron microscopy (TEM) the structure is averaged over the thickness of the sample, which is of the order of 10 nanometers. The method we focus here is the method of atomic pair-density function (PDF) analysis [1]. This technique takes advantage of two quickly developing technologies, large accelerator based facilities such as pulsed neutron sources and synchrotron radiation sources, and fast computers.

2. THE PDF METHOD

2.1 Issue of Q-Resolution

The PDF, $\rho_0 g(r)$, describes the distribution of interatomic distances, and is obtained from the normalized structure function $S(Q)$, $Q = 4\pi\sin\theta/\lambda$, where Q is the diffraction vector, θ is the diffraction angle and λ is the wavelength of the scattering probe, by

$$G(r) = 4\pi r\rho_0\left[g(r)-1\right] = \frac{2}{\pi}\int_0^\infty \left[S(Q)-1\right]\sin QrQdQ \qquad (1)$$

where ρ_0 is the average number density of atoms [2]. This technique has been widely used for the study of liquids and glasses, but owing to the availability of modern synchrotron based radiation facilities (pulsed neutron sources and synchrotron x-radiation sources) that extended the Q range over which $S(Q)$ can be determined, it is now possible to apply it to crystalline materials as well [1].

In discussing describing the MRO with the PDF the first issue we have to touch upon is the issue of resolution in the Q space. If the Q-resolution is constant of Q, the observed structure function, $S_{obs}(\boldsymbol{Q})$, is a convolution of the true structure function, $S_{true}(\boldsymbol{Q})$, by the resolution function, $W(\boldsymbol{Q})$,

$$S_{obs}(\boldsymbol{Q}) = \int S_{true}(\boldsymbol{Q}')W(\boldsymbol{Q}-\boldsymbol{Q}')d\boldsymbol{Q}' \qquad (2)$$

Its Fourier transform is, because of the convolution theorem, a product of the Fourier transforms of each,

$$G_{obs}(r) = G_{true}(r)\widetilde{W}(r) \qquad (3)$$

Thus, if the resolution function is a Gaussian function with the standard deviation of σ_Q, the envelope of the PDF will also be a Gaussian function,

$$\widetilde{W}(r) = e^{-\frac{r^2}{2\sigma_r^2}} \qquad (4)$$

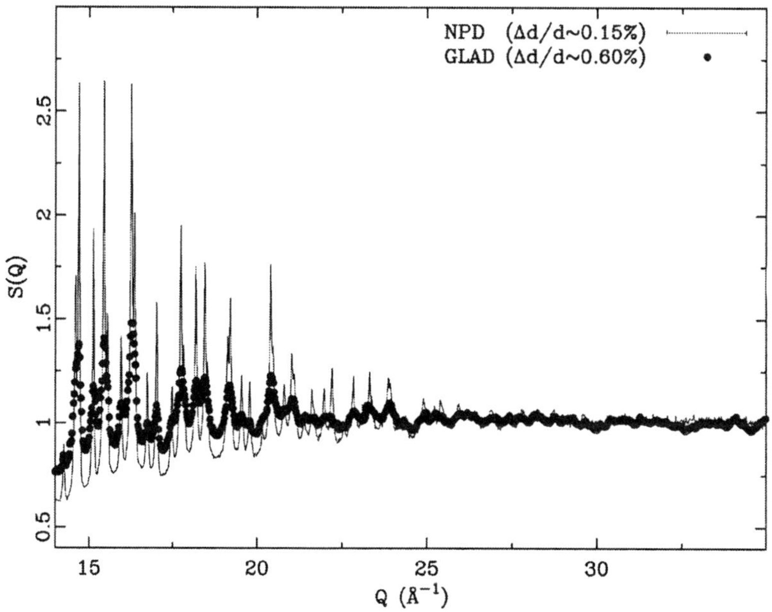

Fig. 1. Structure function, $S(Q)$, of Ni powder at room temperature in the high Q range, for NPD (Los Alamos) and GLAD (Argonne) [3].

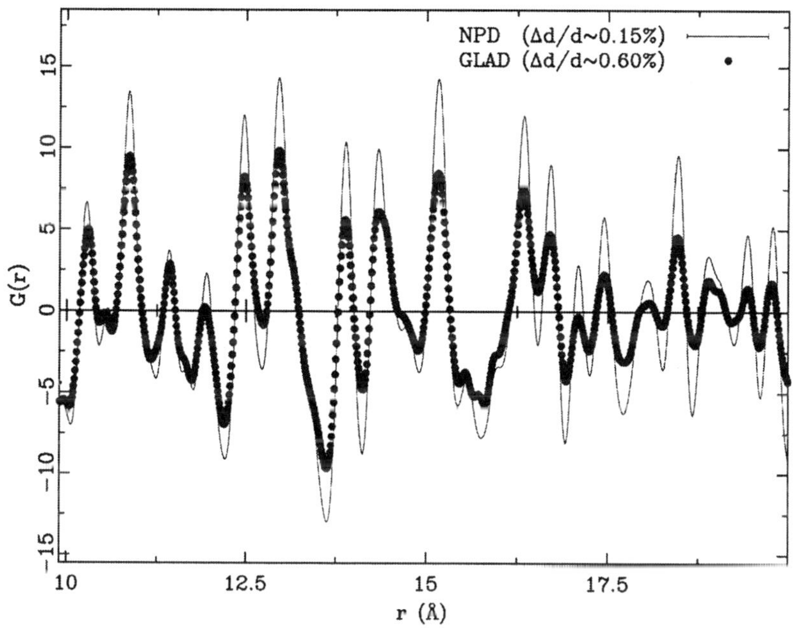

Fig. 2. Atomic pair-density function, $G(r)$, of Ni obtained by NPD and GLAD from the structure function in Fig. 1. $Q_{max} = 35$ Å^{-1} for both cases [3].

where $\sigma_r = 1/\sigma_Q$ For instance, if the Q-resolution (FWHM) is 0.04 $Å^{-1}$, which is only a medium-high resolution, the FWHM value of the envelope function will be 139 Å, long enough for most purposes. Thus, unless the resolution is particularly poor the Q-resolution does not present a problem for the PDF.

However, care has to be taken for the case of pulsed neutron sources for which ΔQ is proportional to Q. In this case the resolution quickly worsens at high Q, causing peaks to overlap, effectively damping $S(Q) - 1$. This resembles increasing the Debye-Waller factor, and consequently the peaks in the PDF become broader. Figs. 1 and 2 show the comparison of $S(Q)$ and PDF for two pulsed neutron powder diffractometers, the NPD of the Lujan Center of LANSCE, Los Alamos National Lab. ($\Delta Q/Q$ = 0.0015) and the GLAD of the IPNS, Argonne National Lab. ($\Delta Q/Q$ = 0.007) [3]. It is evident that the poorer resolution of the GLAD results in a compromised PDF. While the PDF peak position and the coordination number are correctly given, the peak width is overestimated by GLAD due to poorer resolution. The inaccuracy increases with r, since the PDF peak width approximately increases as $\sigma_r^0 + br^2$ [4].

2.2 Issue of accuracy

It is a common perception that the accuracy of the PDF degrades with increasing distance. However, this perception is mistaken, except for the resolution problem we just discussed. The statistical error in $S(Q)$, $\Delta S(Q)$, affects the PDF through [4],

$$\rho_0^{\,2} \Delta g(r)^2 = \frac{1}{4\pi^4 r^2} \sum_{\nu} \left[\Delta S(Q_\nu) Q_\nu dQ_\nu \sin(Q_\nu r)\right]^2 \tag{5}$$

where ν enumerates data points, and dQ_ν is the Q-spacing between the data points. If the noise is purely due to statistics, this can be approximated as

$$\rho_0 \Delta g(r) = \frac{1}{2\sqrt{2}\pi^2 r} \left[\int \frac{Q^2 dQ}{I(Q)} \right]^{1/2} \tag{6}$$

where $I(Q) = N(Q)/dQ$, $N(Q)$ being the detector count [5]. Thus the statistical error actually decreases as $1/r$. Other errors such as the normalization errors decrease even more quickly than $1/r$, so that the accuracy of the PDF in the medium-range is no worse than at a short-range. At a long-range the accuracy of the PDF method is equivalent to the accuracy of the crystallographic methods of analysis, such as the Rietveld analysis.

2.3 Comparison with the XAFS method

The same Fourier transformation as in (1) is used in deriving the radial distribution function (RDF) of the XAFS method,

$$R(r) = \int \chi(k) e^{ikr} k^n dk\,, \tag{7}$$

where $\chi(k)$ is the XAFS oscillation. Partly for this reason the XAFS method and the PDF method tend to be seen in the same light with respect to their meaning and accuracy. This, however, does a gross injustice to the PDF method. The PDF obtained by neutron diffraction does represent the distribution of atoms, since neutrons are scattered only by nuclei. Even in the case of x-ray scattering the total intensity, corrected for absorption, background, multiple-scattering and incoherent scattering and normalized to the electron unit is,

$$I(Q)=\sum_{i,j} f_i(Q) f_j(Q) e^{iQ\cdot r_{ij}} \tag{8}$$

where $f_i(\boldsymbol{Q})$ is the atomic scattering factor of the i-th atom. Since the Q-dependence of $f_i(\boldsymbol{Q})$ for various elements is similar, $I(\boldsymbol{Q})$ is divided by the square of the compositionally averaged scattering factor, $\langle f(\boldsymbol{Q})\rangle^2$, to give $S(\boldsymbol{Q})$ (Warren-Krutter-Morningstar approximation [6]). Thus to a very good approximation the PDF describes the *atomic* distribution. Note that if we Fourier-transform $I(\boldsymbol{Q})$, it will give the *electronic* pair-distribution function.

In the case of XAFS, however, it is difficult to calculate $f_i(\boldsymbol{Q})$ reliably, and the RDF is calculated without the form factor correction. Consequently the XAFS RDF represents the distribution of the electron back-scattering centers, which are distributed in the vicinity of the atom. That is why the XAFS RDF has wider peaks than the neutron PDF, and the peak position does not necessarily represent the atomic separation. For instance in the case of CMR manganites discussed below the pulsed neutron PDF clearly shows the split of the Mn-O peak due to the Jahn-Teller (JT) distortion, but the JT distortion is not resolved in the XAFS RDF. For these reasons usually the result of the XAFS is analyzed by comparing $\chi(k)$ with the one calculated for the model, which makes the analysis less transparent. Also because the XAFS back-scattering amplitude decreases rapidly with distance, it is difficult to observe correlations beyond the first nearest neighbors.

2.4 Debye-Waller factor determined by crystallographic analysis

Another myth about the local structure that is again incorrect is that the magnitude of local structural deviations can be accurately assessed by the enlarged Debye-Waller (DW) factor. The reason why this is wrong is that the DW approximation assumes the Gaussian distribution of the local displacements, while in complex materials the distribution is often spatially inhomogeneous and strongly non-Gaussian. When a small portion of the sample has a large DW factor, the average DW factor does not correctly represent this minority and the displacements are underestimated [1,7]. If the sample is spatially inhomogeneous, the magnitude of the DW factor depends upon the Q range to determine the DW factor. When this occurs the estimate of the magnitude of displacements by the DW factor is unreliable.

3. EXAMPLES

3.1 Medium-Range order in CMR manganites

Doped carriers in silicon become delocalized at even a very low level of doping, and provide high conductivity. In contrast in manganites that show colossal magnetoresistivity, such as $La_{1-x}Sr_xMnO_3$, doped carriers are immobile up to a high level of doping, typically $x \approx 0.16$. This is because at a low level of doping the doped carriers are localized as spin-lattice polarons. When polarons are present in the sample the lattice becomes locally distorted, and the PDF deviates from the PDF of a perfectly periodic lattice. Observation of such local distortion related to the JT distortion by the pulsed neutron PDF gave a strong evidence of lattice polaron [8-10]. An important question is whether polarons are randomly distributed, or they have some spatial correlation. A direct examination of the pulsed neutron PDF at a medium-range gave an answer to this question. As shown in Fig. 3 for $x = 0.4$ polarons are not expected, and the PDF is close to the one calculated for the periodic crystal. However, when polarons are present, as in the case of $x = 0.2$, the observed PDF deviates from the PDF of the average structure up to about 15 Å, as shown in Fig. 4 [11]. This suggests that polarons are not totally random, but are correlated up to 15 Å. This range is similar to the stripe periodicity [12], although the presence of stripes in this system is still controversial.

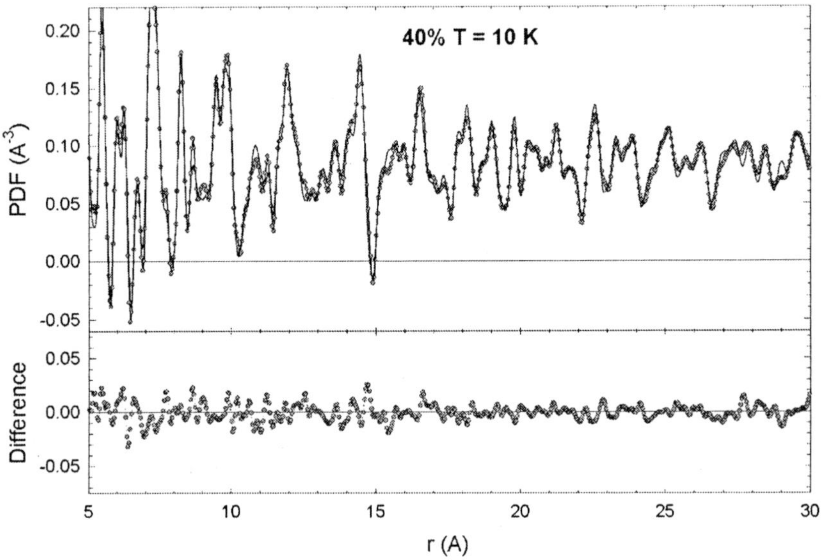

Fig. 3. PDF of $La_{0.6}Sr_{0.4}MnO_3$ in the medium-range obtained by the pulsed neutron diffraction (circles) compared to the one calculated to the crystal structure (solid line) [11]. The lower figure is the difference, which is within the noise.

3.2 Long-range strain field around a polaron

Cubic perovskite manganite, $A_{1-x}A'_xMnO_3$, where A is a trivalent ion such as rare earths and A' is a divalent ion such as alkaline earth and Pb, is known to show a remarkable ionic-size effect. When the average ionic size of the A site ions (A and A') is smaller than about $\langle r_A \rangle = 1.2$ Å in the 9-coordination ionic radius [13] the system is insulating since the charges are localized as polarons, and when it is larger charges are delocalized rendering the system metallic. The stability of polarons is known to be determined through the balance between the localizing forces, electron-lattice coupling (g) and antiferromagnetic spin ordering, and the delocalizing forces, electron kinetic energy represented by the transfer integral (t) and the elastic energy (K). The effective coupling constant is given by

$$\lambda = \frac{g^2}{Kt\langle\cos(\theta/2)\rangle} \tag{9}$$

where θ is the angle two neighboring spins make [14,15]. If the magnitude of λ is greater than a certain value electron-lattice coupling dominates and polarons are formed, while if it less polarons are unstable and the system remains metallic. Since when the ionic size is small the Mn-O-Mn bond becomes buckled, the current popular view is that t is reduced due to buckling and this increases λ [12]. However, the expected amount of change in λ is very small, making this scenario unlikely [16].

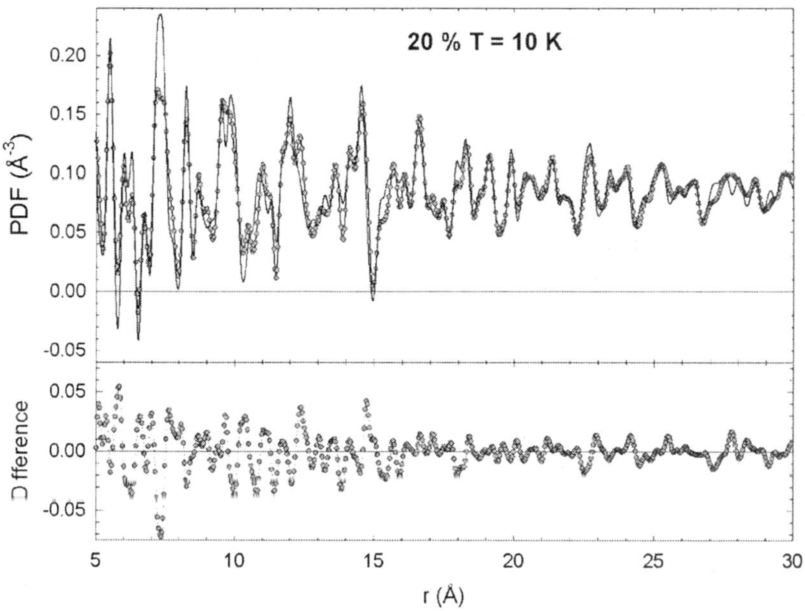

Fig. 4. PDF of $La_{0.8}Sr_{0.2}MnO_3$ in the medium-range obtained by the pulsed neutron diffraction (circles) compared to the one calculated to the crystal structure (solid line) [11]. The lower figure is the difference. The difference is beyond the noise for $r < 15$ Å, suggesting that the local structure is correlated up to 15 Å

We proposed an alternative explanation based upon the local structure [17,18]. For a hole doped system ($x < 0.5$) a polaron is created by eliminating the local JT distortion on the site, by reducing two long Mn-O bonds to short ones. When the A-site ionic size is large and the Mn-O-Mn bond is straight, this will create a long-range strain field. The effect of the long-range strain field is to increase the polaron energy and thus reduce the effective coupling constant λ. This effect was evaluated by using the continuum approximation of Eshelby [19], and was found to reduce the value of λ by nearly a factor of two [20]. However, if the ionic size is small and the Mn-O-Mn bond is buckled, the reduction in the Mn-O distance can be accommodated locally by unbuckling the bond, thus without creating the long-range strain field. Fig. 5 shows the calculated value of λ as a function of the ionic radius [20]. By increasing the ionic size, the value of λ is reduced by a factor of two, and this must be the reason why polaron becomes unstable as the ionic size is increased. This mechanism also explains why the crossover occurs at a specific value of the ionic size, when the Mn-O-Mn within the polaron becomes straight.

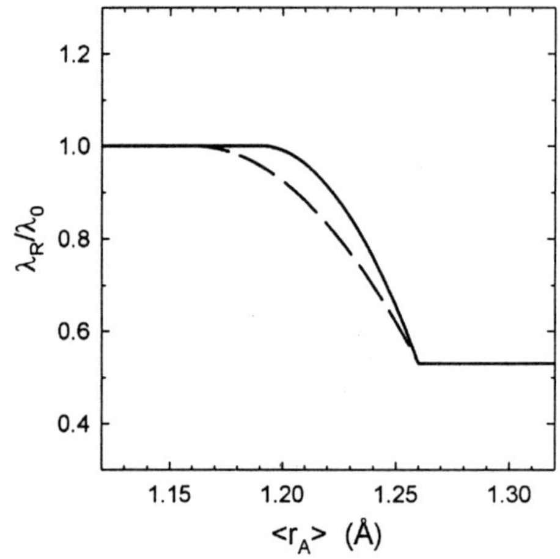

Fig. 5. The effective electron-lattice coupling constant, λ, as a function of the A-site ionic radius, $<r_A>$, normalized by λ_0, the constant without the long-range strain field [20].

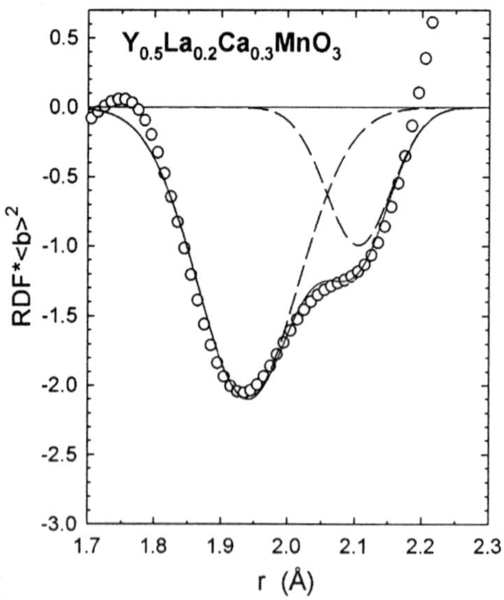

Fig. 6. The Mn-O peak of the RDF, $4\pi r^2 <b>^2 \rho_0 g(r)$; for $Y_{0.5}La_{0.2}Ca_{0.3}MnO_3$ fitted with two gaussian peaks describing local JT distortion without the long-range strain field (a) and for $Y_{0.5}La_{0.2}Ca_{0.3}MnO_3$ fitted with three gaussian peaks describing the effect of long-range strain field (b) [20].

Proving the presence of the long-range strain field directly by the medium-range PDF is difficult, since it involves extensive modeling, and yet the results may not be convincing. Instead, the details of the short-range structure provided clear evidence. To explain how this is done it is useful to review the Eshelby theory. In this theory, a spherical hole is created in an elastic body, and is fitted with an object with a different size and shape. First, the object is deformed so that is fits the hole. At this moment deformation occurs only in the inserted object. Then the system is relaxed so that the matrix is also deformed, producing a long-range strain, to accommodate the object. As a result the strain in the inserted object is significantly reduced, roughly to a half of the original amount. Thus in the case of the manganites if the strain is accommodate locally the strain exists only at the polaron site, but if the long-range strain is there the matrix is also deformed.

Modeling of the local structure of $Y_{0.5}La_{0.2}Ca_{0.3}MnO_3$ with respect to the local JT distortion clarified which the case is for this compound. As shown in Fig. 6 the first Mn-O peak of the pulsed neutron PDF obtained with the GLAD at the IPNS, which is negative because of the negative scattering length of Mn, can be modeled well by assuming that the local JT distortion is removed only from the polaron sites (a), but not if one assumes also the matrix is deformed (b) [20]. Thus in this compound the polaron is not accompanied by the long-range strain field. Indeed the A-site ionic radius of this compound (1.14 Å for the 9-coordination [21]) is well below the critical value (1.20 Å), so that charges are localized as polarons.

4. CONCLUSIONS

Structural characterization usually means carrying out an x-ray diffraction measurement and determining the crystal structure. For complex materials that have local deviations from the average structure, however, such a conventional structural analysis does not provide essential information necessary for understanding the properties. Since the properties are more sensitive to the local structure than the global structure, the knowledge of the local structure is essential in characterizing complex materials. Among the local probes the PDF analysis is unique in its capability to determine the medium-range structure as well as the short-range structure. In this paper we discussed the accuracy of the PDF method in comparison to other methods, using the results on CMR manganites as an example. The unique capability of the PDF method to describe the MRO is one of the major advantages of the PDF method.

ACKNOWLEDGMENTS

The author is thankful to the collaborators of the works sited here, D. Louca, Th. Proffen and S. J. L. Billinge for their contributions. The work was supported by the National Science Foundation through DMR01-02565.

REFERENCES

1. T. Egami and S. J. L. Billinge, *Underneath the Bragg peaks: Structural analysis of complex materials* (Elsevier, Oxford) to be published.
2. B. E. Warren, *X-ray Diffraction* (Addison-Wesley, Reading, 1969; Dover, New York, 1990).

3. Th. Proffen, T. Egami, S. J. L. Billinge, A. K. Cheetham, D. Louca and J. B. Parise, *Appl. Phys. A*, to be published.
4. B. H. Toby and T. Egami, *Acta Cryst., A* **48**, 336 (1992).
5. T. Egami, *Mater. Trans. JIM,* **31**, 163 (1990).
6. B. E. Warren, H. Krutter and O. Morningstar, *J. Amer. Cer. Soc.*, **19**, 202 (1936).
7. T. Egami, in *Structure and Bonding*, **vol. 98**, ed. J. B. Goodenough (Springer-Verlag, Berlin, 2001) p. 115.
8. S. J. L. Billinge, *et al.*, *Phys. Rev. Lett.* **77**, 715 (1996).
9. D. Louca, T. Egami, E. L. Brosha, H. Röder, and A. R. Bishop, *Phys. Rev. B* **56**, R8475 (1997).
10. D. Louca and T. Egami, *Phys. Rev. B* **59**, 6193 (1999).
11. T. Egami and D. Louca, *J. Superconductivity* **13**, 247 (2000).
12. C. H. Chen and S.-W. Cheong, *Phys. Rev. lett.*, **76**, 4042 (1996).
13. H. Y. Hwang *et al., Phys. Rev. Lett.* **75**, 914 (1995).
14. A. J. Mills, P. B. Littlewood, and B. I. Shraiman, *Phys. Rev. Lett.* **74**, 5144 (1995); A. J. Millis, B. I. Shraiman and R. Mueller, *Phys. Rev. Lett.* **77**, 175 (1996).
15. H. Röder, J. Zang and A. R. Bishop, *Phys. Rev. Lett.* **76**, 1356 (1996).
16. P. G. Radaelli *et al.*, *Phys. Rev. B*, **56**, 8265 (1997).
17. T. Egami and D. Louca, *J. Superconductivity* **12**, 23 (1999).
18. D. Louca, T. Egami, W. Dmowski and J. F. Mitchell, *Phys. Rev. B*, in press.
19. J. D. Eshelby, *Proc. Roy. Soc. A* **241**, 376 (1957).
20. T. Egami and D. Louca, unpublished.
21. R. D. Shannon, *Acta Cryst.* **A 32**, 751 (1976).

X-RAY TRUNCATION ROD ANALYSIS OF THE REVERSIBLE TEMPERATURE DEPENDENT SURFACE STRUCTURE OF $LAALO_3$

R.J. Francis, S.C. Moss, A.J. Jacobson, and R. Barabash*

1. INTRODUCTION

The surface structures, compositions, and terminations of ABO_3 oxides with the perovskite structure are of major importance due to their extensive use as substrates for thin film growth, in particular for the growth of superconducting films of materials such as $YBa_2Cu_3O_7$. $LaAlO_3$ is particularly attractive in this regard since its lattice mismatch with $YBa_2Cu_3O_7$ is small (<2%), and its high melting point (2453K) minimizes interfacial dislocations and chemical reactions with the films.[1] To facilitate its use as a growth substrate the surface structure and composition of $LaAlO_3$ should be thoroughly characterized. $LaAlO_3$ has a bulk rhombohedral perovskite structure at room temperature and undergoes a transformation to a simple cubic structure at about 710K,[2] although several groups have reported that the change occurs gradually between room temperature and 770K.[2-4] This transformation involves changes in bond angles of only tenths of a degree and microtwinning is not observed.

The [001] surface of $LaAlO_3$ can terminate in either LaO_x or AlO_x-type structures at either the La or Al layers respectively. Wang and Shapiro have studied the surface of $LaAlO_3$ using reflection electron microscopy (REM),[5] and due to the lack of contrast seen in the REM images suggested that the surface of $LaAlO_3$ is terminated exclusively by either a La-O or Al-O layer and not a mixture of both. More recently Rabalais and co-workers have performed a more detailed study of the surface of $LaAlO_3$ using time-of-flight scattering and recoiling spectrometry (TOF-SARS), AFM and PES.[6,7] They found the surprising result that the surface termination of $LaAlO_3$ depends on temperature. At temperatures less than 423K, the surface is exclusively terminated by an Al-O layer, while at temperatures above 523K the surface seems to be terminated by an La-O layer. Between 423K and 523K mixed terminations were conjectured and it was proposed that the surface stoichiometry changed from AlO_x to LaO_x. This change was found to be fully reversible.

* R.J. Francis and S.C. Moss, Physics Department, University of Houston, Houston, Texas 77204-5506. A.J. Jacobson, Chemistry Department, University of Houston, Houston, Texas 77204-5641. R. Barabash, Metals and Ceramics Division, Oak Ridge National Laboratory, Oak Ridge, TN 37831-6118.

Such a complete change in the surface termination of an oxide material is a unique observation and clearly has major implications for the use of $LaAlO_3$ as a thin film substrate. Rabalais et al.[6] suggested that the surface termination change was caused by the formation of surface oxygen vacancies at high temperature which drives the migration of the La atom to the surface and the Al atom into the bulk. However, despite the range of techniques that have been used to study the surface of $LaAlO_3$,[5,6] to date none have been capable of providing direct structural information about the surface. X-ray crystal truncation rod (CTR) analysis is capable of providing such information since the shape of the CTR profile depends sensitively on the precise atomic arrangement of the surface (and near-surface) atoms.[8,9] A recent example of the utility of this technique was provided by a study the structure of the surface of hydrated Al_2O_3 (0001) crystals, in which it was found that the hydrated surface consisted of a semi-ordered layer of water above an oxygen terminated surface, in contrast to the non-hydrated surface which is aluminum terminated.[10]

We have utilized high brilliance synchrotron X-ray radiation to study the surface structure of [001] oriented single crystals of $LaAlO_3$ at both room temperature and ca. 670K (well above the surface structural change observed previously) in order to determine the detailed atomic movements associated with the surface structural change. These results have implications for the use of $LaAlO_3$ as a substrate for thin film growth.

2. THEORY/CALCULATION

Several detailed discussions of the calculation of X-ray scattering by surfaces have been published previously.[8,9,11-14] However, a brief discussion is included here. For the case of an *infinite* crystal lattice it is well known that the diffraction pattern consists of narrow δ-like functions (broadened somewhat by instrumentation and sample effects) at the points in reciprocal space at which the Laue conditions are satisfied (the Bragg points). However, if finite size effects are included then the diffraction peaks are broadened by an amount which is inversely related to the dimension of the diffracting region of the crystal. One consequence of these finite-size broadening effects is that if a crystal is truncated in any direction, for example by a surface, a significant (decreasing) amount of intensity is scattered continuously away from the Bragg peaks. The diffraction pattern from the surface of a crystal which is extended in two dimensions but truncated in one therefore consists of scattering which is sharp in two directions but continuous in the third (a "rod" of scattering). This rod-shaped scattering extends in the direction normal to the crystal surface and is referred to as crystal truncation rod (CTR) scattering.

For a perfectly terminated, perfectly smooth, [001] crystal surface the structure factor amplitude of the CTR scattering is given by:[9]

$$F_{hkl}(l) = F_{u,\,hkl}(l)\sum_{n=0}^{N} e^{i2\pi nl}\,, \tag{1}$$

where N is the number of unit cells in the z direction and $F_{u,hkl(l)}$ is the bulk unit cell structure factor. Since N is very large this reduces to the limit:

$$F_{hkl}(l) = F_{u,hkl}(l)\frac{1}{1-e^{i2\pi l}}, \tag{2}$$

and the intensity is given by:

$$I_{hkl}(l) \alpha \mid F_{u,hkl}(l)\mid^2 \frac{1}{2\sin^2(\pi l)}. \tag{3}$$

For an ideal surface the CTR profiles therefore consist of $1/\sin^2(\pi l)$ tails around each Bragg peak. However, the intensity distribution along the rod is strongly influenced by any deviation from ideal termination (such as surface relaxation, the presence of vacancies or interstitials etc.), or surface roughness. For surfaces that deviate from ideal termination, the total CTR scattering amplitude is calculated by considering the interference between the scattering arising from the bulk structure and that arising from the surface structure. The total CTR scattering *amplitude* is given by the linear sum of the scattering arising from the bulk and that arising from the surface layer:

$$F_{hkl}^{CTR}(l) = F_{hkl}^{Bulk}(l) + F_{hkl}^{Surf}(l). \tag{4}$$

The bulk structure factor amplitude is exactly as given above for the case of an ideally terminated surface. The surface structure factor amplitude is, in general, given by:

$$F_{hkl}^{Surf}(l) = \sum_j \delta_j f_j \exp[2\pi i(x_j h + y_j k + z_j l)], \tag{5}$$

where δ_j is the occupancy of the jth surface atom, f_j is the form factor multiplied by the Debye-Waller factor of the jth surface atom, and x_j, y_j, and z_j are the absolute positional parameters of the surface atoms in units of the bulk unit cell. The number of terms depends on how complex the surface model is and how many surface layers are assumed to participate in the surface relaxation. In principle, the model can be made arbitrarily complex. The total CTR amplitude is therefore given by:

$$F_{hkl}^{CTR}(l) = F_{u,hkl}^{Bulk}(l)\frac{1}{1-e^{i2\pi l}} + \sum_j \delta_j f_j \exp[2\pi i(x_j h + y_j k + z_j l)], \tag{6}$$

and hence the CTR intensity for a relaxed but perfectly smooth surface is given by:

$$I_{hkl}^{CTR}(l) = A(l)\mid F_{u,hkl}^{Bulk}(l)\frac{1}{1-e^{-i2\pi l}} + \sum_i \delta_j f_j \exp[2\pi i(x_j h + y_j k + z_j l)]\mid^2, \tag{7}$$

where A(l) includes correction terms for incident X-ray flux, polarisation, Lorentz and resolution factors, etc.

In general, surface roughness will have the effect of causing the CTR intensity around Bragg peaks to fall off faster than for the case of a perfectly smooth surface. Various models of surface roughness have been proposed including those by Robinson,[13] Andrews and Cowley,[11] and Harada.[15] The one applied here (although others were considered, see below) is the Robinson model in which a single parameter β ($0 \leq \beta \leq 1$) is introduced which describes the fractional occupancy of the first surface layer. Successive layers then have an occupancy of β^n where n is the number of the layer. [It should be noted that the layers here refer to *entire unit cells*, not single atomic layers.] This has the effect of reducing the diffraction intensity by a factor:

$$\frac{(1-\beta)^2}{[1+\beta^2-2\beta\cos(2\pi l)]}. \tag{8}$$

The overall CTR intensity is thus given by:

$$I_{hkl}^{CTR}(l) = A(l)\frac{(1-\beta)^2}{[1+\beta^2-2\beta\cos(2\pi l)]}\left| F_{u,hkl}^{Bulk}(l)\frac{1}{1-e^{i2\pi l}} + \sum_j \delta_j f_j \, j\exp[2\pi i(x_j h + y_j k + z_j l)]\right|^2 . \tag{9}$$

The parameters of the model are therefore the scale factor, the roughness parameter, the atomic occupancy factors, and the atomic positional parameters.

3. EXPERIMENTAL

The truncation rod profiles reported in this paper were recorded on station X22C of the National Synchrotron Light Source, Brookhaven National Laboratory. This beamline is equipped with a bent, cylindrical, platinum coated focusing mirror (spot size ~ 1 mm) and a Ge(111) double crystal monochromator.[16] For these experiments monochromatic X-rays with a wavelength of 1.2399 Å were used. A highly polished, circular [001] oriented single crystal of $LaAlO_3$ of diameter 10 mm and thickness 5 mm was mounted within the ultrahigh vacuum (UHV) surface diffractometer on X22C, which is operated at a base pressure of 1 x 10^{-10} Torr. The experimental arrangement has been described in detail previously.[16] Prior to mounting, the crystal had been annealed in flowing oxygen at 1473K. Atomic force microscopy (AFM) images of the surface following this annealing revealed a smooth high quality surface, consisting of mostly unit cell, or multiple unit cell, steps with an rms oscillation of ~0.8 nm. Prior to collection of the CTR data the crystal surface was cleaned by heating to ca. 900K under UHV for several hours. Although some small degree of contamination may remain following this treatment, surface contaminants not in registry with the surface structure will have no effect on the CTR analysis since background diffuse scattering arising from such contaminants is subtracted as part of the analysis procedure.

CTR data was collected along the 00L, 10L and 20L rods. A z-axis geometry was used in which incident and diffraction angles were the same. For the 00L rod, 83 CTR data points were collected between [0,0,0.2] and [0,0,4.3]. For the 10L and 20L rods 43 CTR data points were collected between L = 0.2 and L = 2.3. For each data point transverse rocking scans were performed through the CTR and the structure factors were obtained by integration of these rocking scans, after subtraction of the diffuse scattering that peaks under the rod but which is spread over several degrees. Figure 1 shows a typical experimental transverse rocking scan taken at [0,0,0.8] revealing a sharp truncation rod peak and the much broader diffuse scattering beneath it which is related both to background contamination and lateral correlation of the roughness. The data were corrected for polarization, Lorentz, sample area, and detector resolution effects.

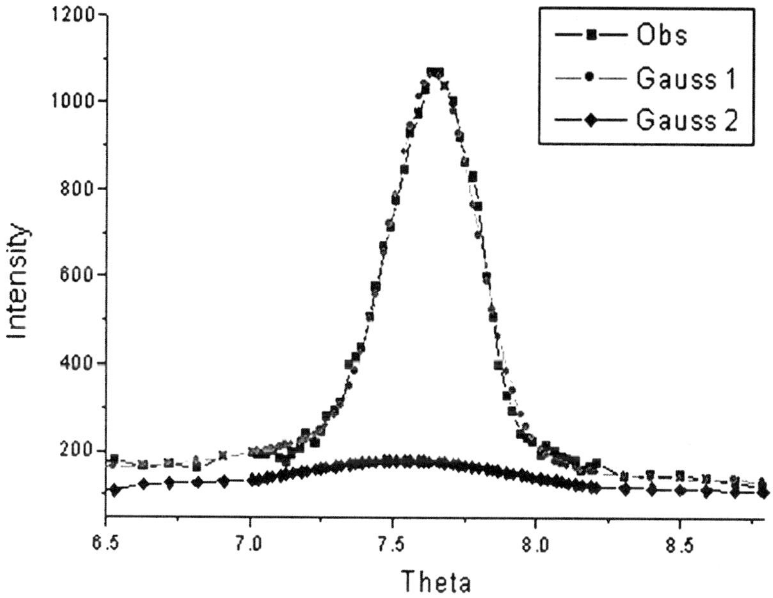

Figure 1. Example of a typical transverse rocking scan showing the sharp truncation rod peak and the much broader diffuse scattering peak. This scan was taken at room temperature on the 00L rod at [0,0,0.8].

For the high temperature data set, the crystal was heated using a tungsten filament located at the base of the crystal. The temperature of the crystal was monitored using a thermocouple attached to a groove cut along middle of the crystal circumference. The filament current was adjusted until the reading was 710K+/-5K. Since there is expected to be some difference between this temperature and the temperature of the surface, the exact surface temperature is not known. However, based on previous experience, the surface temperature is expected to be about 670K. Certainly, this is well above the temperature of the surface structural change observed previously.[6]

Least squares fitting of the data was performed using the ROD program[17] written by Elias Vlieg which is freely available for download from the European Synchrotron Radiation Facility (ESRF) web site. This program allows the input of an arbitrarily complicated surface structure, and will calculate the individual bulk and surface scattering contributions, plus the total scattering. Least squares fitting of the calculated scattering to the observed data can then be performed using the atomic positional parameters, occupancy parameters and Debye-Waller factors of each of the atoms, either individually or as a group. A simulated annealing procedure was also used for the top unit cell but the results did not improve the fits using the ROD routine which included more layers (2 unit cells).

Another important aspect of the fitting procedure was the use of bond distance constraints to prevent chemically unreasonable solutions to the least squares fitting results. These were particularly important given that only x0L data were collected and it is therefore impossible to refine independently the x and y positional parameters of each atom. However, the chemical bond distance and valency requirements[18] of each atom place severe restrictions on the possible structural solutions. These constraints were incorporated into the least squares fitting procedure by calculating an energy penalty[19] associated with deviations from the ideal bond lengths[18] between the appropriate atoms, and adding this to the residuals calculated for the CTR data. By adjusting the relative weights of the CTR residuals and the bond distance terms it was possible to derive structural models which both fitted the CTR data well and in which all bond distances and atom valancies were chemically sensible. Furthermore, given the overall near cubic symmetry of the underlying bulk lattice it was assumed that each atomic movement in **a** was either accompanied by zero movement in **b** or a movement equal to that in **a,** as appropriate. For example, it is clear that for the aluminum atom in the 1st surface layer a substantial lateral movement in **a** cannot be accompanied by a equal lateral movement in **b** since this would bring it unreasonably close to the lanthanum atom, whereas no such restriction applies for the oxygen atoms in the 1st surface. Because the bulk rhombohedral structure is so nearly cubic, when averaged over the 4 <111> directions, we assumed an equal probability, using cubic notation, that a displacement that occurs along the cubic **a**-axis was equally likely to occur along the **b**-axis in another domain. Our data is thus taken along h0L rods but these could also be read as 0kL rods. Given the relatively small amount of lateral data and the necessary assumptions made about the atomic movements in the **b** lattice direction for an [h0L] rod, it is clear that the uncertainties in the lateral positions of the atoms are much greater than the uncertainties in the vertical positions. However, the combination of least-squares fitting to the CTR data, together with the bond distance and valency constraints, places severe restrictions on the number of possible solutions, and the atomic positions of the atoms can be determined with some confidence.

4. RESULTS

The room temperature CTR plots for the 00L, 10L, and 20L rods are shown in figure 2 a-c respectively. The high temperature CTR plots are shown in figure 3 a-c respectively. Small, but significant, changes between the room temperature and high temperature data are seen, indicative of a surface structure change between the two temperatures. Least squares fitting (based on the minimization of the residual R(log10[I])) (where I is the integrated intensity of the CTR peaks) of the complete 00L, 10L and 20L data set was performed simultaneously using a model consisting of a fixed bulk structure and an adjustable surface region. As mentioned previously, interference between the scattering amplitudes from the surface and the undisturbed bulk structures gives rise to alterations in the diffraction intensities along the CTR. The CTR profile is found to be very sensitive to the precise atomic positions of the surface atoms. Since $LaAlO_3$ has the possibility of terminating in either LaO_x or AlO_x surfaces, both terminations were tested in the fitting procedure. The aluminum-terminated surface was clearly found to fit the data better, and this was used as the basic starting model in the fitting procedure.

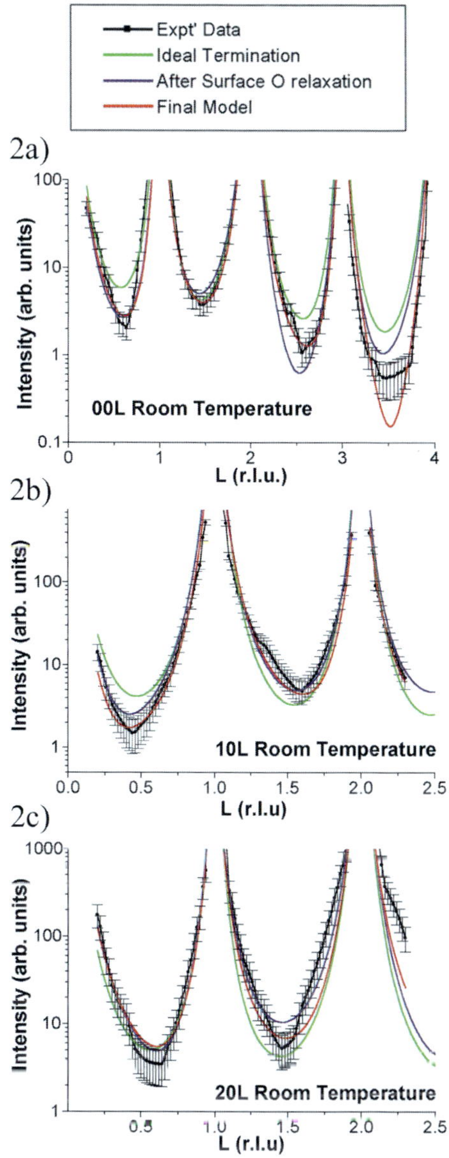

Figure 2. (Color) Experimental and calculated CTR profiles recorded at room temperature for a) the 00L rod, b) the 10L rod, and c) the 20L rod. In each figure the experimental data points are plotted in black together with estimated errors (for clarity in the 00L profile only every other points is plotted), and the final best fit calculated profile is shown as a solid red line. Also shown are calculate profiles for an ideally terminated surface (the solid green line) and the calculated profiles after surface oxygen relaxation (the solid blue line).

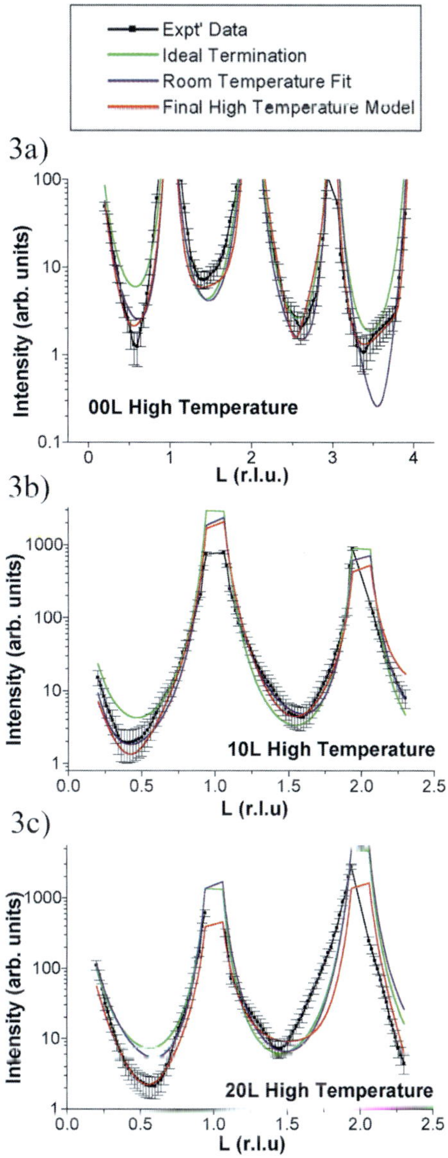

Figure 3. (Color) Experimental and calculated CTR profiles recorded at ca. 670K for a) 00L rod, b) 10L rod, and c) 20L rod. In each figure the experimental data points are plotted in black together with estimated errors (for clarity in the 00L profile only every other points is plotted), and the final best fit calculated profile is shown as a solid red line. Also shown are the calculated profiles for an ideally terminated surface (the solid green line) and the calculated profiles for the best fit room temperature structure (the Solid blue line).

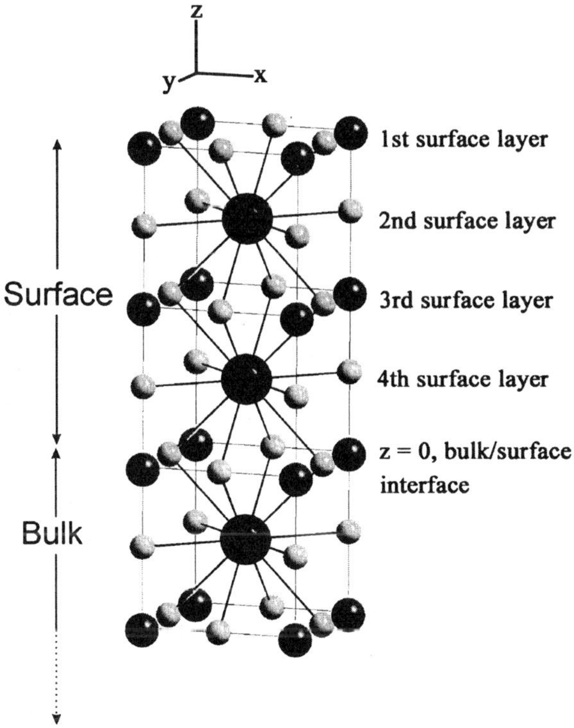

Figure 4. Base aluminum terminated model used in the surface structure refinement showing the four topmost atomic layers (referred to in the text as the "surface" atoms) which were allowed to relax during the refinement. La atoms are shown as black spheres, Al atoms as medium gray spheres, and O atoms as light gray spheres.

4.1 Model/Fitting Details

The base model (see figure 4) for both the room and high temperature structures consisted of an aluminum terminated surface in which the topmost four surface layers (that is, two AlO layers and two LaO layers) were allowed to relax. From this point the atoms in these four layers will be referred to as the "surface" atoms. Each of the atoms in these top four layers was allowed to move independently in all three lattice directions, with the constraints described in the experimental section above. The fitting procedure was performed using all 169 data points in all three rods simultaneously. In addition to the 30 atomic positional parameters, two scale parameters for the specular and non-specular data, a roughness parameter according the model of Robinson, [13] and an overall temperature factor were refined. (Of course, individual atoms will have separate Debye-Waller factors, but this distinction was small compared with the static displacements reported here.) During the fitting procedure a weighted residual was minimized with respect to the logarithm of the measured intensity to avoid the dominance of high intensity CTR points very close to the Bragg reflections. The calculated CTR's for the unrelaxed aluminum terminated surface (shown in figures 2 and 3 as green lines) show very significant deviations from the observed data for both the room temperature and high temperature data sets in all three rods, indicating the presence of a significant structural relaxation from the ideally terminated surface.

4.2 Room Temperature Structure

For the room temperature data set, allowing the surface oxygen atoms to relax away from the ideal bulk values (i.e. a flat Al-terminated surface) resulted in a significant improvement in the overall fit for each of the CTR profiles. This fit is shown in figure 2 as a blue line. In particular, the quality of the fit in the low L region of the 00L profiles was greatly improved. However, the calculated 00L profiles still showed substantial deviations at high L, and the 10L and 20L profiles also showed some deviations. These mismatches were found to diminish greatly when the surface aluminum atoms were also allowed to relax. Allowing the other surface atoms to relax resulted in slight but important improvements to the overall fit. The final best fit of the calculated patterns to the observed data is shown in figure 2 as a red line. The coordinates of the best-fit model and the ideally terminated surface are given in table 1. The most significant structural change from the ideally terminated surface involves a considerable relaxation of the surface oxygen atoms away from the surface (by ~0.14c, where c is the z-axis lattice parameter) together with a concomitant relaxation of the aluminum atom into the surface (by ~0.05c) and the oxygen in the second layer towards the surface (by ~0.08c). Small lateral movements (0.08-0.1c) of the oxygen atoms in the first two layers are also found to occur. The vertical and lateral movements of the remaining surface atoms are found to be much smaller. In particular the La atom in the second layer is found to move an insignificant distance from its bulk position, and the movements of the atoms in the third and fourth layers are negligible. The fairly large surface relaxation has the primary effect of changing the surface aluminum atom coordination from a pseudo-octahedral (i.e. octahedral coordination in which one of the oxygen vertices is missing), to a distorted square pyramidal coordination. This is reasonable, since it would be expected that a 5-coordinate pseudo-octahedral configuration would be a highly unfavorable coordination geometry for aluminum, and that the surface would relax in such a way as to alter this coordination. The final refined surface structure is shown viewed near parallel to the b-axis in figure 5, with an inset view of the change in Al coordination from the ideally-terminated case to the refined room temperature structure.

The final best fit CTR profile shown in figure 2 is, in general, excellent. No R-factor was calculated, however. The reason for this lies, for example, in the fact that the primary deviations from the observed 00L data are due not to a difference in calculated intensity at the anti-Bragg points, but rather to how quickly the calculated and observed intensities rise from these points towards the Bragg reflection points. At low L the observed intensities rise more rapidly than the calculated intensities on both sides of the anti-Bragg points, whereas at high L the reverse is true. The rapidity with which CTR intensities rise on either side of the anti-Bragg point is intimately connected with the roughness of the surface, with a rougher surface giving rise to a more rapidly varying intensity.[13] The observed data, therefore, are strongly suggestive of an L-dependent roughness parameter, β, which has a high value at low L, but a somewhat lower value at high L. Such an L-dependence of the roughness parameter is not predicted by any of the proposed models for surface roughness,[13,15] however, and the origin of this effect is unclear at this time. This feature of our results remains very interesting and lies, we believe, in the details of

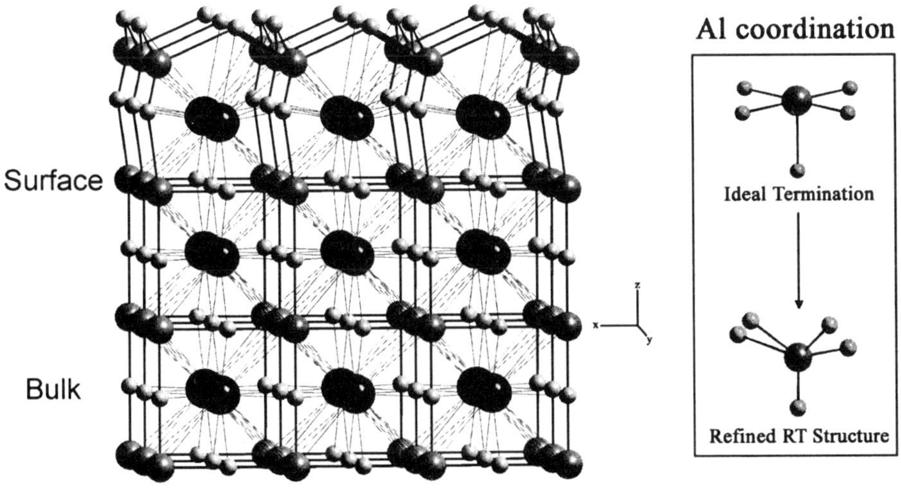

Figure 5. View of the refined room temperature surface structure of $LaAlO_3$ shown viewed near parallel to the b-axis. La atoms are shown as black spheres, Al atoms as medium gray spheres, and O atoms as light gray spheres. *Inset:* Diagram showing the Al coordination change of the topmost aluminum atom from the ideally terminated case (pseudo-octahedral) to the refined room temperature structure (distorted square-pyramidal).

the individual surface atoms Debye-Waller factors which may be both large and asymmetric and scales different from the bulk values. This aspect of our fitting is currently being considered more carefully with Dr. R. Barabash. Nevertheless, the overall refined value of β for all of the data is in excellent agreement with the value of the RMS roughness obtained from AFM studies of the crystal used in this experiment. The RMS roughness of the surface can be calculated from the refined value of β using the relation $\sigma rms = \beta^{0.5} * d/(1-\beta)$, where d is the lattice spacing perpendicular to the surface.[13] Putting the refined value of $\beta = 0.589$ into this equation gives a value for the RMS roughness of 0.707 nm. The value obtained using AFM was 0.803 nm.

We therefore conclude that at room temperature, under the conditions of our study, the surface of $LaAlO_3$ is aluminum-terminated with a large relaxation of the surface oxygen atoms away from the bulk and the surface aluminum atom into the bulk causing a change in aluminum coordination from 5-coordinate pseudo-octahedral to distorted square-pyramidal.

4.3 High Temperature Structure

When the calculated CTR profiles for the derived room temperature structure (shown as green lines in figure 3) are compared with the observed high temperature data, small, but significant, discrepancies are seen, indicating a structural change upon heating. The most obvious differences are seen between the relative intensities at the L = 0.5 and 1.5 points of the 00L and 20L profiles, and between L = 3 and 4 in the 00L profile. Whereas for the room temperature data an excellent fit between the observed and calculated data could be obtained using only the first two atomic layers, it was found that for the high temperature data refinement of the positional parameters of the entire first four layers were necessary to minimize the deviations from the observed data.

Table 1: Final refined fractional unit cell coordinates of the best-fit surface model for the room temperature structure, together with the values for an ideally terminated Al-O surface. The Δx and Δz values express the change in surface position in the final refined model with respect to the positions in the ideally terminated surface. †Refinement of the atomic positions in the 3rd and 4th surface layers gave values negligibly different from the bulk values. They were therefore fixed in the final refinement.

Layer		Bulk Termination			Relaxed Surface				
		x	y	z	x	y	z	Δx (nm)	Δz (nm)
1	Al	0	0	2	-0.005	0	1.95	-0.0019	-0.0189
	O	0.5	0	2	0.58	0.08	2.14	0.0303	0.0530
	O	0	0.5	2	-0.08	0.58	2.14	-0.0303	0.0530
2	La	0.5	0.5	1.5	0.505	0.505	1.485	0.0019	-0.0057
	O	0	0	1.5	-0.08	0	1.56	-0.0303	0.0227
3	Al	0	0	1	0†	0†	1†	-	-
	O	0.5	0	1	0.5†	0†	1†	-	-
	O	0	0.5	1	0†	0.5†	1†	-	-
4	La	0.5	0.5	0.5	0.5†	0.5†	0.5†	-	-
	O	0	0	0.5	0†	0†	0.5†	-	-
B	Al	0	0	0	0	0	0	-	-
	O	0.5	0	0	0.5	0	0	-	-
	O	0	0.5	0	0	0.5	0	-	-
	La	0.5	0.5	-0.5	0.5	0.5	-0.5	-	-
	O	0	0	-0.5	0	0	-0.5	-	-

The final best-fit CTR profiles are shown as a red line in figure 3. Although residual discrepancies remain, in general the fit between the calculated and observed profiles is good. The final refined positional parameters of all of the surface atoms are compared with the bulk in table 2. As can be appreciated from table 2 at high temperature a much more substantial structural change occurs. The most significant changes involve large vertical relaxations of the surface oxygens away from the bulk (~0.26c) together with a significant vertical relaxation of the surface aluminum atom (~0.16c) and the oxygen atom in the third atom layer (~0.08c). These are accompanied by substantial lateral shifts of the oxygen atoms in the top two surface layers. Small, but much less significant, movements of the lanthanum atom and the other surface atoms are also observed. The final refined surface structure is shown viewed near parallel to the b-axis in figure 6.

The atomic movements associated with the structural change between room and high temperature are shown in detail in figure 7 using shade atoms and arrows. The primary effect of the surface transformation is to change the coordination of the surface aluminum atom from distorted square pyramidal at room temperature to a pseudo-tetrahedral one at high temperature. The observed change in aluminum termination from room to high temperature is clear figure 7. Interestingly, this is the change proposed by Yao et al. in their original paper.[6] However, in marked contrast to their proposed model, no evidence is found for the formation of any surface vacancies. Attempts to introduce surface oxygen

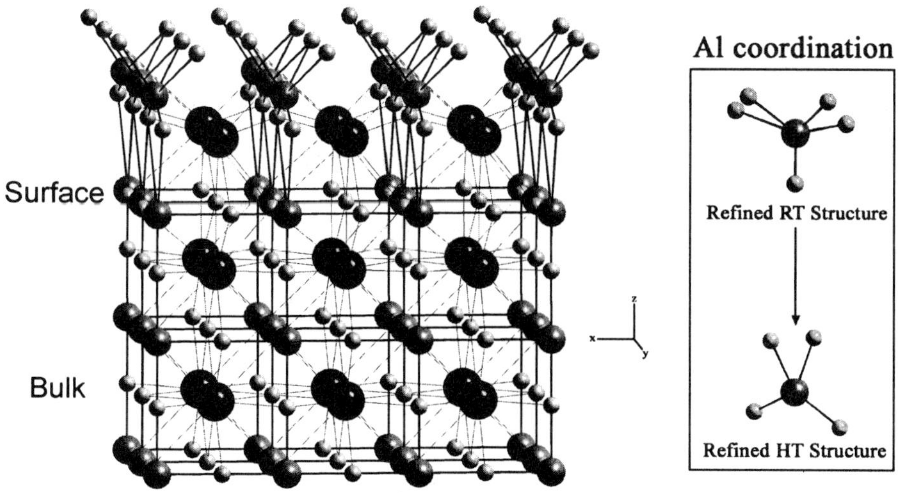

Figure 6. View of the refined high temperature (ca. 670K) surface structure of $LaAlO_3$ shown viewed near parallel to the b-axis. La atoms are shown as black spheres, Al atoms as medium gray spheres, and O atoms as light gray spheres. *Inset:* Diagram showing the Al coordination change of the topmost aluminum atom from the refined room temperature structure (distorted square-pyramidal) to the refined high temperature structure (distorted tetrahedral).

vacancies into the model gave significantly worse fits, and led to chemically unrealistic refined structures. In fact, refinement of the surface oxygen occupancies actually led to values greater than 1 rather than less than 1 (see below). More dramatically, there was no evidence for any significant movement of the lanthanum atom towards the surface of the crystal. The position of the lanthanum atom is highly constrained by the data due to its very strong effect on the CTR profiles. Figure 8 shows the effect on the CTR profiles of progressively moving the La atom by 0.025c and 0.05c from its bulk position. As is clear in figure 8, even these small shifts produce dramatic changes in the CTR profile, especially at high L. The pronounced kink in the observed high temperature 00L CTR profile at ~ L = 3.5 can be very well reproduced by a very small shift (0.012c) of the La atom from its bulk position (see figure 3a). Larger La atom shifts produce significantly greater kinks and can be completely discounted. For this reason the position of the lanthanum atom can be determined with high accuracy (see table 2).

4.4 Discussion

The CTR data clearly rules out a reversal of metal atom surface termination, as conjectured by Yao et al.[6] A consideration of the high temperature structure, however, readily accounts for their findings. As can be appreciated from figure 7 the lateral movement of the surface oxygen atoms has the effect of greatly increasing the *access* to the lanthanum atom by ions impinging on the surface of the crystal. Thus, despite the very small movement of the lanthanum atom, this lateral motion of the oxygen atoms exposes the La atoms and leads to a significant enhancement of the lanthanum atom signature seen in the

Table 2: Final refined fractional unit cell coordinates of the best-fit surface model for the high temperature structure, together with the values for an ideally terminated Al-O surface. The Δx and Δz values express the change in surface position in the final refined model with respect to the positions in the ideally terminated surface.

Layer		Bulk Termination			Relaxed Surface				
		x	y	z	x	y	z	Δx (nm)	Δz (nm)
1	Al	0	0	2	-0.23	0	1.833	-0.0871	-0.0633
	O	0.5	0	2	0.72	0.22	2.24	0.0833	0.0909
	O	0	0.5	2	-0.23	0.73	2.24	-0.0871	0.0909
2	La	0.5	0.5	1.5	0.524	0.524	1.488	0.0091	-0.0045
	O	0	0	1.5	0.32	0	1.63	0.1212	0.0492
3	Al	0	0	1	0.00	0	0.96	0.000	-0.0152
	O	0.5	0	1	0.44	-0.06	0.95	-0.0227	-0.0189
	O	0	0.5	1	-0.07	0.57	0.95	-0.0265	-0.0189
4	La	0.5	0.5	0.5	0.492	0.492	0.489	-0.0030	-0.0042
	O	0	0	0.5	-0.02	0	0.49	-0.0076	-0.0038
B	Al	0	0	0	0	0	0	-	-
	O	0.5	0	0	0.5	0	0	-	-
	O	0	0.5	0	0	0.5	0	-	-
	La	0.5	0.5	-0.5	0.5	0.5	-0.5	-	-
	O	0	0	-0.5	0	0	-0.5	-	-

ion-scattering experiments.[6] As regards the apparent L-dependent DWF noted earlier, we suggest the following possible explanation.

In the first place we must consider the static DWF *exp*(-2*W*), usually given as $W = 8\pi^2 \overline{\mu_{st}^2} \sin^2\theta / \lambda^2$, where $\overline{\mu_{st}^2}$ is the mean-square static displacement for the surface atoms. Near the surface we in principle should take into account the following:

a) the displacement fields around all atoms, j, will change as they are no longer centrally symmetric due simply to the surface and their new positions;

b) the displacement of atoms becomes inhomogeneous due to the increased probability for certain atoms to alter their vertical (and lateral) positions. In such cases, *W* depends not only on the type of atom under consideration but, naturally, on the number, *n*, of the plane relative to the surface (*n*=1 is the surface layer, *n*=2 the next, etc.). The general expression for the DWF (see Ref. 20) contains both real and imaginary terms. Inside the bulk crystal – away from the surface – crystals with an inversion center lose their imaginary contribution, the DWF is the same as for simple roughness, and *W* maybe written as proportional to L^2, where all constant factors of π^2, etc. are incorporated into *L* and the displacements, u_j, are projected onto the z-axis. (Lateral terms can also, of course, be included.)

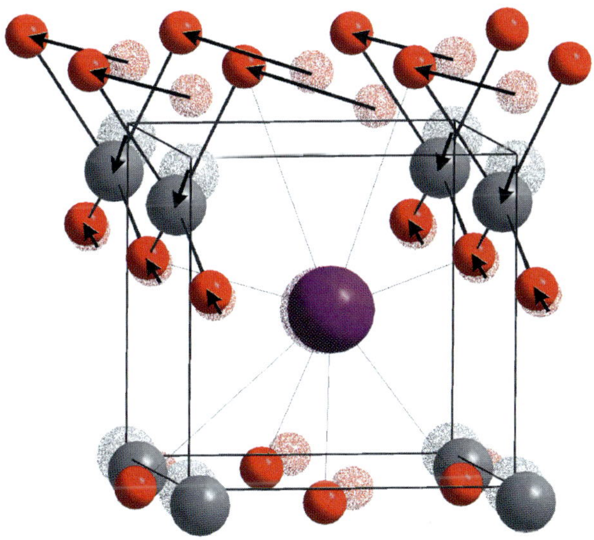

Figure 7. (Color) Diagram showing the atomic movements associated with the structural change from room to high (ca. 670K) temperature. The refined high temperature positions are shown as solid spheres, with La atoms shown in purple, Al atoms in gray, and O atoms in red. The refined room temperature positions are shown as partially grayed out spheres. The black arrows represent the individual atom movements for the two topmost surface layers. The black cube corresponds to the bulk unit cell.

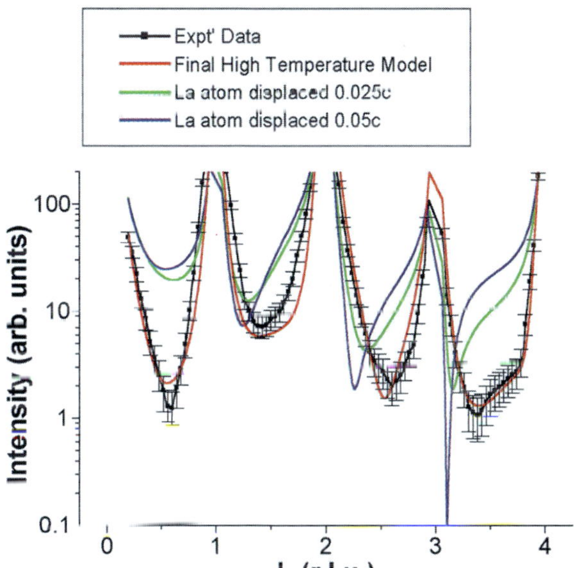

Figure 8. (Color) Observed and calculated CTR profiles showing the dramatic effect on the CTR curves of very small movements of the surface La atoms away from their bulk positions. The black curve with error bars is the observed high temperature 00L CTR data. The red curve is the final best fit calculated profile. The green and blue curves show the calculated CTR profiles after moving the La atom just 0.025c and 0.05c respectively away from the best fit position. As can be appreciated even these very small shifts can be completely discounted.

Near the surface, however, especially in our case of (selected) large movements, the displacement fields around all atoms are not centrally symmetric. The expression for W – for the individual atoms as commonly incorporated into f_j – now contains both real and imaginary terms:[20,21]

$$W_j = W_j^{'} + iW_j^{''}, \tag{10}$$

where

$$W_j^{'} = \sum_{j,n} c_n (1 - \cos Lu_{jn}); \tag{11}$$

$$W_j^{''} = \sum_{j,n} c_n (Lu_{jn} - \sin Lu_{jn}), \tag{12}$$

and c_n is the coordination number in the n-th plane. If we consider for simplicity the case where the displacements are large only in the first planar coordination shell, the expression above can be written:

$$W_j^{'} = c_1 \sum_j (1 - \cos Lu_j); \tag{13}$$

$$W_j^{''} = c_1 \sum (Lu_j - \sin Lu_j). \tag{14}$$

These expressions contain a limited number of terms and clearly oscillate with L, as observed experimentally. Moreover, they yield the correct sign. For larger L, there appears a seemingly low value for the surface roughness. If we now take into account the thermal DWF, it is clear that the displaced surface atoms have somewhat weaker bonds with the layer below than the normal bulk value, especially for movements perpendicular to the surafce. The thermal DWF, W_{fth}, also contains, to low order, both real and imaginary terms:[21]

$$W_{fth} = W_{fth}^{'} + iW_{fth}^{''}, \tag{15}$$

where,

$$W_{fth}^{'} = \frac{1}{2}\left\langle (Lu_{jn})^3 \right\rangle - \frac{1}{24}\left[\left\langle (Lu_{jn})^4 \right\rangle - 3\left\langle (Lu_{jn})^2 \right\rangle^2\right] + \cdots \tag{16}$$

$$W_{fth}^{''} - \frac{1}{6}\left\langle (Lu_{jn})^3 \right\rangle + \cdots \tag{17}$$

The above equations may also result in a non-linear dependence of W on L^2. Such anharmonic corrections become essential at high temperatures. While the application of these general ideas to our data is underway, we note here only that they may in principle be responsible for the unusual oscillatory damping of our rod scans. Results on this will be presented at a later date, in which we will indicate more than the uppermost layer in our evaluation.

5. SUMMARY

The [001] surface structure of $LaAlO_3$ has been studied using X-ray truncation rod analysis at both room temperature and ca. 670K. Modeling of 00L, 10L, and 20L CTR profiles has enabled a detailed picture of the surface structure of $LaAlO_3$ at both room and higher temperatures to be determined. At room temperature the surface is found to be aluminum terminated, and a fairly minor structural rearrangement relative to the ideally terminated surface is found to occur, mainly involving a relaxation of the surface oxygen atoms away from the surface and a small relaxation of the aluminum atoms into the surface. This has the primary effect of changing the surface aluminum atom coordination from a pseudo-octahedral to a distorted square pyramidal coordination. At the higher temperature a much more significant structural transformation occurs involving the top four atomic surface layers. This is primarily characterized by a considerable relaxation of the top surface oxygen atoms and a concomitant relaxation of the aluminum atom into the surface, together with substantial lateral movements of the oxygen atoms in the top two layers. No evidence is found for a reversal of surface termination or for the formation of surface oxygen vacancies. These results are consistent with the previous studies of the surface of $LaAlO_3$. The surface structural change is found to be completely reversible, with the structure reverting to the low temperature form on cooling.

ACKNOWLEDGEMENT

The authors would like to thank Scott Perry and Paul van der Heide of the University of Houston for performing AFM and PES measurements respectively on the $LaAlO_3$ crystals. We would also like to thank Hubert Zajonz for assistance in the data analysis, and Doon Gibbs for assistance with the data collection. The assistance of J.H. Li is also gratefully acknowledged. This work was supported by the NSF, MRSEC on DMR-9632667.

REFERENCES

1. R. K. Simon, C. E. Platt, K. P. Daly, A. E. Lee, and M. K. Wagner, Appl. Phys. Lett. **53**, 2677 (1988).
2. S. Geller and V. B. Bala, Acta Crystallogr. **9**, 1019 (1956).
3. H. Fay and C. D. Brandle, *Crystal Growth* (Pergamon, Oxford, 1967).
4. H. M. O'Bryan, P. K. Gallager, G. W. Berkstresser, and C. D. Brandle, J. Mater. Res. **5**, 183 (1990).
5. Z. L. Wang and A. J. Shapiro, Surf. Sci. **328**, 141 (1995).
6. J. Yao, P. B. Merril, S. S. Perry, D. Marton, and J. W. Rabalais, J. Chem. Phys. **108**, 1645 (1998).
7. P. A. W. van der Heide and J. W. Rabalais, Chem. Phys. Lett. **297**, 350 (1998).
8. I. K. Robinson, R. T. Tung, and R. Feidenhans'l, Phys. Rev. B **38**, 3632 (1988).
9. I. K. Robinson and D. J. Tweet, Rep. Prog. Phys. **55**, 599 (1992).
10. P. J. Eng, T. P. Trainor, G. E. B. Jr., G. A. Waychunas, M. Newville, S. R. Sutton, and M. L. Rivers, Science **288**, 1029 (2000).

11. S. R. Andrews and R. A. Cowley, J. Phys. C: Solid State Phys. **18**, 6427 (1985).
12. R. Feidenhans'l, Surf. Sci. Rep. **10**, 105 (1989).
13. I. K. Robinson, Phys. Rev. B **33**, 3830 (1986).
14. H. You, in: *2nd International Conference on Surface X-ray and Neutron Scattering*, Bad Honnef, Germany, 1991).
15. J. Harada, Acta. Cryst. **A48**, 764 (1992).
16. D. Gibbs, B. M. Ocko, D. M. Zehner, and S. G. J. Mochrie, Phys. Rev. B **42**, 7330 (1990).
17. E. Vlieg, J. Appl. Cryst. **33**, 401 (2000).
18. N. E. Brese and M. O'Keeffe, Acta. Crystallogr. **B47**, 192 (1991).
19. J. S. Pederson, Surf. Sci. **210**, 238 (1989).
20. M.A. Krivogalz, *X-ray and Neutrons Diffraction in Non-ideal Crystals*, (Springer, NY, 1996)
21. R. Barabash, unpublished work.

LOCAL DISLOCATION STRUCTURE FROM LAUE DIFFRACTION

R.I. Barabash, G.E. Ice, B.C. Larson, W. Yang*

1. INTRODUCTION

A general kinematic treatment of x-ray scattering by crystals with dislocations was developed by Darwin [1], Krivoglaz *et al.* [2], Warren [3], and Wilkens [4]. This approach is widely used for the analysis of dislocation substructure and local rotations in single crystals measured by means of rocking curves [5-13]. The main disadvantage of the rocking curve technique is the need to rotate the sample; rotations introduce uncertainties in the real space coordinates of the scattering volume [14]. Laue white beam measurements however are performed at a fixed orientation of the sample. Special instrumentation [15] makes it possible to correlate different points of the Laue intensity with diffraction from regions within the crystal (Fig.1a).

Here, we apply the general kinematic treatment of x-ray scattering from dislocations to determine the dislocation structure of crystals from white beam microdiffraction Laue measurements. Although white-beam Laue diffraction is a classic method widely used to determine crystal symmetry and orientation, unit cell volumes cannot be determined with standard Laue measurements, and the typical precision is not high enough for measurement of strain or dislocation structure. The development of ultra-brilliant third-generation synchrotron x-ray sources [16,17] and recent progress in x-ray optics [18] has led to intense x-ray beams with submicron spatial resolution. Together with new instrumentation, it is now possible to use the Laue method to quantitatively determine elastic strain and local orientation (texture) distribution in individual grains or subgrains [19,20]. Also one can extract information about the local dislocation structure of the subgrain [21]. Below we expand the general kinematic approach for X-ray (or neutron) scattering by distorted crystals to analyze dislocation arrays and local lattice rotations by white-beam Laue diffraction.

It is known that plastic deformation involves several structural levels [22]. At the first level, there are individual dislocations in the volume with typical size l_1: $\sqrt{n^{-1}} > l_1 > b$ (where n is dislocation density, b is the Burgers vector). As a result of strong interaction between dislocations, random dislocation distributions become

*Metals and Ceramics and Solid State Divisions, Oak Ridge National Laboratory Oak Ridge TN 378316118.

From Semiconductors to Protein, edited by S.J.L. Billinge and M.F. Thorpe
Kluwer Academic/Plenum Publishers, 2002

unstable, and they form correlated dislocation arrangements with screening radius l_2 due to grouping of dislocations into different kinds of dislocation walls. Some part of the dislocations remains distributed randomly and the rest may form different kinds of non random dislocation and disclination arrangements. The size of the second structural level is defined by equation: $D_{fr} > l_2 > \sqrt{n^{-1}}$. Here D_{fr} is a typical size of a fragment or cell with approximately the same crystallographic orientation. These dislocation arrangements are usually interpreted as mesoscale defects. For crystals with an excess of tilt dislocation walls, the transverse broadening of a reflection is proportional to the number of excess dislocations inside the wall and to the total number of excess walls. The next structural level l_3 is related to the typical grain size D_{gr}: $D_{gr} > l_3 > D_{fr}$.

At the macroscopic level l_4 with typical size $l_4 > 10D_{gr}$, the statistical fluctuations of individual defects average towards a mean value, and solids can be approximated as continuous media with some elastic-plastic properties. Classic X-ray techniques usually provide averaging over a large number of grains and give average information related to the 3rd or 4th structure level. With white-beam microdiffraction it is possible to quantitatively analyze all the above structure levels in individual grains or subgrains. In particular, it is now possible to analyze details of structure corresponding to the 2nd structural level.

It should be noted that white x-ray microdiffraction makes it possible to distinguish the intensity distributions related to different grains (or subgrains) within a size ~ 1-100 μm. This scale corresponds to the so-called "mesoscopic" level of dislocation structure which can not be probed by classic x-ray methods. For example, typical rocking curves provide averaging over much larger regions while TEM gives very local information. The observed local lattice rotations at high deformation result from the dislocation structure within the above "mesoscopic " level. For the first time these local material properties can be quantitatively characterized using broad-bandpass x-ray microbeams.

2. GENERAL ANALYSIS OF INTENSITY NEAR MICROBEAM LAUE SPOTS

2.1. Experimental Geometry

In traditional white-beam Laue diffraction, a continuum spectrum is diffracted by a single crystal (or single grain). The crystal scatters the beam into a characteristic Laue pattern that depends on the crystal space lattice, its orientation and the incident-beam energy distribution. Each characteristic reflection *hkl* is scattered specularly from the crystal planes *hkl*, which allows the planar orientations with respect to the incident beam to be determined directly. For microdiffraction it is most convenient to operate with the detector near 2θ~90° (Fig. 1a). As shown in Fig. 1b, the intensity of the Laue pattern from a single grain is formally proportional to a radial integral through reciprocal space weighted by the incident-beam spectral distribution. Because x-rays are penetrating, an x-ray microbeam experiment produces a Laue pattern from every fragment or cell that it intercepts along its path. These overlapping Laue patterns can be separated by triangulation or other techniques [23].

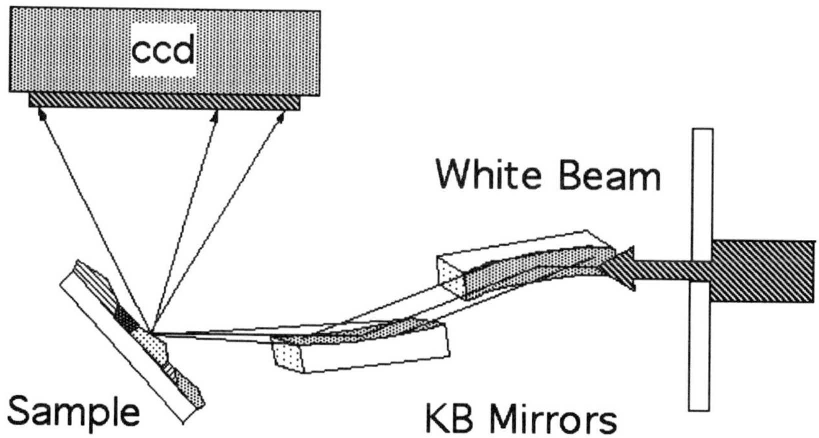

a

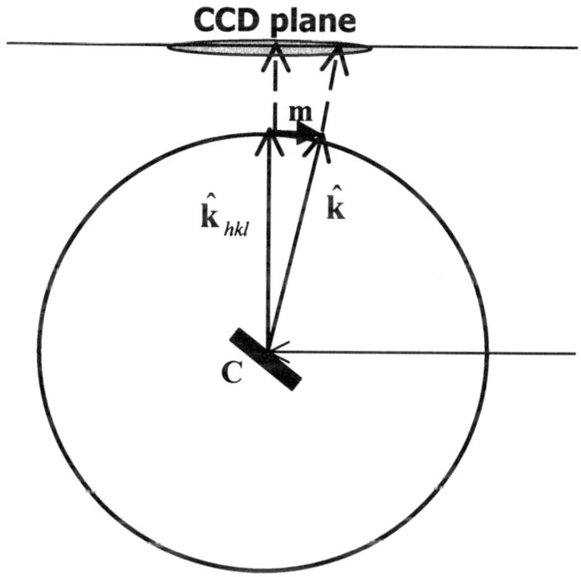

b

Figure 1. Experimental setup for white beam Laue diffraction (a) and formation of Laue image as a function of vector ***m*** (b).

2.2. Formation of Laue Image from a Distorted Crystal

If we ignore temporarily the complications of overlapping Laue patterns and overlapping orders, then the precise shape of each Laue spot depends formally on the orientation of the grain/subgrain, the incident beam spectral distribution near the Bragg energy of the reflection, and the distribution of elastic and plastic deformation in the grain/subgrain. The initial spectral distribution can be described by $I_0(k)$ where $(k = |\mathbf{k}_0|)$. In reciprocal space the exact positions of regular reflections (*hkl*) are related to the orientation, space lattice and of the real-space unit cell size. The momentum transfer corresponding to a Bragg/Laue reflection $\mathbf{G}_{hkl}$ we define as, $\mathbf{k}_{hkl}$ - $\mathbf{k}_0$, where $(|\mathbf{k}_0| = |\mathbf{k}_{hkl}| = k)$. Here $\mathbf{k}_0$ is the incident wave vector and $\mathbf{k}_{hkl}$ is the scattered wave vector that satisfies the Bragg/Laue conditions. The diffuse scattering intensity depends on the deviation $\mathbf{q} = \mathbf{Q} - \mathbf{G}_{\mathbf{hkl}} = \mathbf{k} - \mathbf{k}_{hkl}$ between the diffraction vector $\mathbf{Q}$=$\mathbf{k}$ - $\mathbf{k}_0$, and the momentum transfer $\mathbf{G}_{hkl}$ for a Laue reflection. For each wavelength the center of the Ewald sphere (C) and the origin in reciprocal space (O) are separated by the distance $1/\lambda$. This distance is different for each reflection *hkl* and corresponds to different Ewald spheres, which complicates the analysis for polychromatic beams. We write the Laue equation in a manner that takes into account the spectrum of λ in the beam. Then the region of high intensity around each reciprocal lattice vector $\mathbf{G}_{hkl}$in Laue geometry I_L is a function of a vector $\mathbf{m} = \hat{\mathbf{k}} - \hat{\mathbf{k}}_{hkl}$. The vector $\mathbf{m}$ is a difference between unit vectors $\hat{\mathbf{k}} = \mathbf{k}/k$ and $\hat{\mathbf{k}}_{hkl} = \mathbf{k}_{hkl}/k_{hkl}$, that is defined by the direction of propagation of the scattered radiation (Fig.1b). Here $\mathbf{k}$ is the wave vector of the scattered intensity close to $\mathbf{k}_{hkl}$. Hence (111), (333) and similar reflections are projected simultaneously into the same point in the observed Laue pattern. Near regular reflections (*hkl*) in the Laue pattern, the principal contribution in deformed crystals is made by the intensity distribution $I(\mathbf{q})$, which exhibits characteristic features due to scattering from the dislocation-induced distorted regions in the deformed material. The presence of other defects (vacancies, small dislocation loops, impurities etc.) can also influence the intensity distribution. However, their influence is typically an order of magnitude less than that of excess dislocations. In the present paper we restrict ourselves to the case where the intensity distribution is dominated by an excess of straight edge dislocations of one sign, and consider different possible arrangements of such dislocations.

The Laue intensity in a given direction arises from a radial integral. The integration can generally be performed by numerical integration of $I(\mathbf{q})$ weighted by the intensity distribution $I_0(q)$ where $I_0(q)$ is the intensity distribution of the incident white spectrum. Within this approximation, the resulting intensity distribution by the deformed crystal in the white microbeam method can be written as following:

$$I_L(\mathbf{m}) = A\int I_0(k)I(\mathbf{q})dk \tag{1}$$

Where $\Delta k = |\mathbf{k}| - |\mathbf{k}_{hkl}|$ and $\mathbf{q} = |\mathbf{k}_{hkl}|\mathbf{m} + \Delta k\mathbf{G}/|\mathbf{k}_{hkl}|$.

Thus

$$\mathbf{q} = |\mathbf{k}_{hkl}|\mathbf{m}_{\perp} + (|\mathbf{k}_{hkl}|\mathbf{m}_{rad} + \Delta k\mathbf{G}/|\mathbf{k}_{hkl}|). \quad (2)$$

Here $|\mathbf{k}_{hkl}|$ is the radius of the Ewald sphere that passes through $\mathbf{G}_{hkl}$., and A is a constant. Equation (2) is valid when $\frac{\Delta k}{|k_0|} << 1$. In the first approximation, the vector **m** is perpendicular to $\mathbf{k}_{hkl}$.

The intensity distribution I (*q*) of X-ray (or neutron) scattering due to defects can be computed from the expression

$$I(\mathbf{Q}) = \left|\sum_i f_i \exp(i\mathbf{Q} \bullet (\mathbf{R}_i^0 + \mathbf{u}_i))\right|^2. \quad (3)$$

Here f_i is the scattering factor from an individual atom i, with relaxed co-ordinates $\mathbf{R}_i = \mathbf{R}_i^0 + \mathbf{u}_i$ due to the presence of defects. Here u_i is the displacement from the equilibrium positions $\mathbf{R}_i^0$ corresponding to the crystal before defomation. We calculate the $\mathbf{u}_i$ using continuum elastic theory.

2.3. Laue Diffraction by Crystals with Dislocations

Let us consider a set of edge dislocations illustrated in Fig.2a. All dislocations from this set have the same Burger's vector, b_i, parallel to the X-axis with dislocations lines parallel to the Z-axis. In a plane that cuts the crystal perpendicular to a set of dislocation lines we can numerate the possible positions of dislocations with an index t. To describe the distribution of dislocations belonging to the set in this plane we introduce the number c_t. It can take two values: $c_t = \begin{cases} 1,(+) \\ 0,(-) \end{cases}$. If there is a dislocation in the position t, the corresponding number $c_t = 1$, and in positions without dislocations $c_t = 0$ (Fig.3). Each edge dislocation creates a displacement $\boldsymbol{u}_{it}$ of the i-th scattering cell in the plane XOY perpendicular to its line. The projections of the displacement vector in this plane for the simple case of elastically isotropic crystals, are described by the following equations [24]:

$$u_x = \frac{b}{2\pi}\left[\tan^{-1}\frac{y}{x} + \frac{xy}{2(1-\mu)(x^2+y^2)}\right]; \quad (4a)$$

$$u_y = -\frac{b}{2\pi}\left[\frac{1-2\mu}{2(1-\mu)}\ln(x^2+y^2) + \frac{(x^2-y^2)}{4(1-\mu)(x^2+y^2)}\right]. \quad (4b)$$

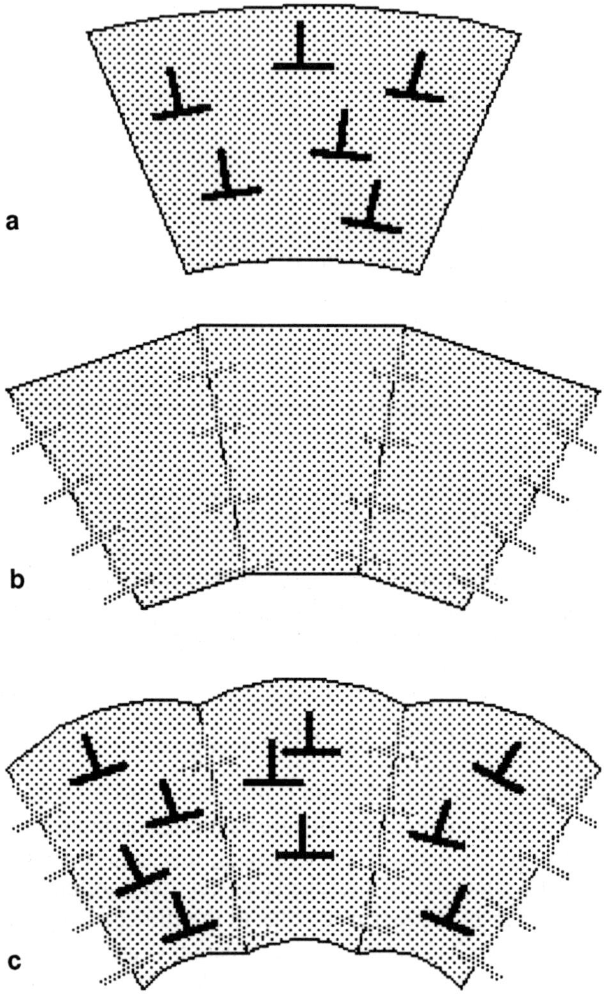

Figure 2. Scheme of crystals with different arrangement of unpaired dislocations:

a - randomly distributed individual unpaired dislocations;

b - randomly distributed tilt dislocation walls;

c - randomly distributed tilt dislocation walls with randomly distributed individual unpaired dislocations in inner regions.

Here μ is Poisson's ratio. The total displacement of the i-th cell $\boldsymbol{u}_i$ is due to all dislocations and is defined by the equation

$$\mathbf{u}_i = \sum_t c_t \mathbf{u}_{it} \,. \tag{5}$$

The intensity distribution of a crystal with dislocations may be written as follows [2 - 4]:

$$I(\mathbf{q}) = f^2 \sum_{i,j} e^{i\mathbf{q}\bullet\boldsymbol{\Delta}} e^{-T} \tag{6}$$

Where f is the average scattering factor of the matrix atoms; $\boldsymbol{\Delta} = \boldsymbol{R}_i^0 - \boldsymbol{R}_j^0$ is the undistorted distance vector between the lattice cells i, j, and

$$T = nS \sum_t 1 - e^{\left(i\mathbf{Q}\bullet(\mathbf{u}_{it} - \mathbf{u}_{jt}\right)} \,. \tag{7}$$

Here S is the area of one dislocation in a transverse plane; n is the total dislocation density, and nS is a dimensionless quantity that indicates the fraction of lattice sites covered with dislocations. The correlation function T differs for different dislocation arrangements. In the most general case it contains both real and imaginary parts. If all dislocations are paired the imaginary part is equal to zero and the correlation function T is real. If there are unpaired dislocations the function T contains both real and imaginary parts. These two limiting cases give rise to distinctly different intensity distributions as discussed below.

2.4. Influence of Paired Dislocations on Scattering

For an equal number of random "+**b**" and "-**b**" dislocations the average deformation tensor is negligible and broadening of the diffuse scattering is induced by random local fluctuations in the unit cell orientations and d spacing that tend to cancel out over long length scales. Due to the character of the displacement field around edge dislocations, displacements occur only in planes perpendicular to the direction of dislocation lines τ. As a result, coherence is not changed along the direction τ, and the diffuse intensity in this direction is the same as for crystals without dislocations. Perpendicular to τ, the diffuse distribution is roughly symmetric with a characteristic full width at half maximum ($FWHM$) dependent on the total dislocation density n: $FWHM \propto \sqrt{n}$ [2,4,10].

2.5. Influence of Geometrically Necessary Dislocations on Scattering

2.5.1. Geometrically Necessary Randomly Distributed Individual Dislocations

This model corresponds to the dislocation arrangement shown in Fig.2a. For excess dislocations, the diffuse scattering distribution relates the distortion tensor, ω_{ij}, due to excess dislocations, n^+, to the dislocation density and system (direction, and Burgers

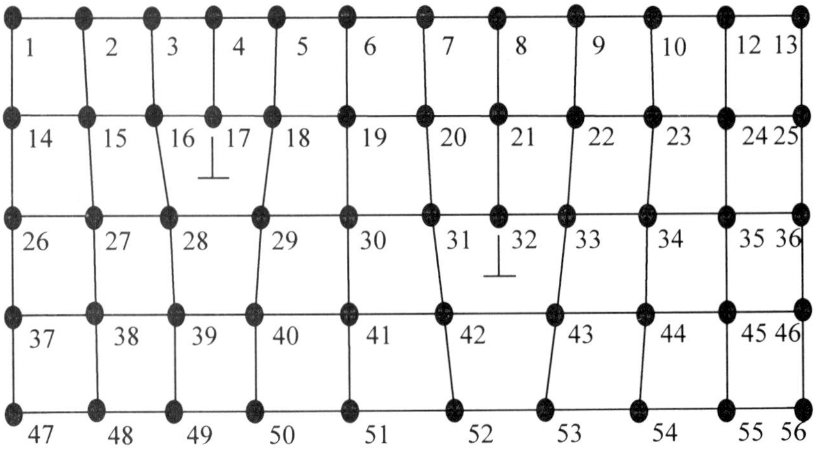

Figure 3. Random numbers c_l describe the dislocation distribution in the transverse plane: $c_{17}= c_{32}=1$, for other lattice sites $c_l=0$.

vector). The dislocation density tensor can be written in terms of the anti-symmetric Levi-Civita tensor of third rank $\varepsilon_{\tau lm}$ and a dislocation density tensor of second rank $\rho_{\tau n}$. [2, 24-26]

$$\varepsilon_{\tau lm}\left(\frac{\partial\omega_{mn}}{\partial x_l}\right) = -\rho_{\tau n} \quad , \tag{8}$$

Here the Levi-Civita tensor has elements $\varepsilon_{123}=\varepsilon_{231}=\varepsilon_{312}=1$; $\varepsilon_{213}=\varepsilon_{132}=\varepsilon_{321}= -1$; and all other $\varepsilon_{ijk}=0$. The tensor of dislocation density $\rho_{\tau n}$ has indices τ that specify the crystallographic direction of the line, and the indices n that specify the Burgers vector direction. For a general set of dislocations with dislocation density ρ_0, line directions parallel to the plane normal *lmn*, and Burgess vector *l'm'n'*, we write the tensor of dislocation density as,

$$\rho_{\tau n} = \begin{pmatrix} \rho_o ll' & \rho_o lm' & \rho_o lm' \\ \rho_o ml' & \rho_o mm' & \rho_o mn' \\ \rho_o nl' & \rho_o nm' & \rho_o nn' \end{pmatrix}. \tag{9}$$

For example, with edge dislocations as shown in Fig. 3a, τ=(001), n=(100) and there is only one non-zero component, $\rho_{zx}=n^+b$. From Eqs. (8,9) it follows that there are only two non-zero components of the mean deformation tensor: $\omega_{xy}=-\omega_{yx}=n^+bx$. We note that this distortion field is antisymmetric and represents a pure rotation about the Z axis that increases with displacement in x. Arbitrary systems of edge or screw dislocation can be similarly treated, but for the following discussion we restrict ourselves to edge dislocations.

The correlation function T, can be expanded with respect to small displacements and has both real and imaginary parts. The first term is imaginary and is linear with respect to x and n^+:

$$T_1 = -iC_1 \sum_{\lambda} n_{\lambda}^{+}(\mathbf{R}_i \bullet \mathbf{b}_{\lambda})([\mathbf{Q} * \mathbf{R}_{ij}]\boldsymbol{\tau}_{\lambda}) \quad (10)$$

From the structure of this equation it follows that dislocations parallel to the diffraction vector do not contribute to the function T_1 and do not influence the intensity of scattering (orientation factor for these dislocations is zero). T_1 describes the influence of mean distortions due to geometrically necessary randomly distributed individual dislocations. This is in agreement with the result of Chu *et al.* [27] who treated the net of dislocations taking into account their mean distortion field. The real part of the correlation function is the same as for paired dislocations and can be written as

$$T_2 = C_2 (Qb)^2 \ln L \sum_{\lambda} n_{\lambda} \varphi_{\lambda} . \quad (11)$$

Here C_1, C_2 are the proportionality coefficients, λ is the number of the dislocation slip system, L is the size of the subgrain (or the cut off radius), φ_λ is the orientation factor for each dislocation system

2.5.2. Geometrically Necessary Randomly Distributed Tilt Dislocation Walls

This model corresponds to the structure shown at Fig. 2b. Each wall provides a rotation between two neighboring mosaic blocks around the line of the wall. The unit vector $\boldsymbol{\omega}$ parallel to the rotation axis of each wall coincides with the unit vector $\boldsymbol{\tau}$ along the dislocation lines in the case of a tilt boundary.

We consider pure tilt boundaries formed by equidistant edge dislocations (so called "thin walls"). Such boundary does not produce any long-range strain but only rotations. The remaining stresses in the boundary region are periodic with a wavelength $b^* = b\csc\Theta/2$. These stresses are only appreciable at the distances less then b^* from the boundary [24]. If h is the distance between these dislocations in the wall, one can consider the boundary as a single defect producing the local rotation field Θ. The disorientation angle Θ due to the boundary is defined by the equation $\frac{b}{h} = 2\sin\frac{\Theta}{2}$, where $\frac{b}{h} \approx \Theta$ for small angle boundaries.

To characterize this model quantitatively we define the average distance between the walls D and write the number of walls per unit length as $\frac{1}{D}$. The total density of unpaired dislocations grouped in the walls is denoted by $n^+ = \frac{1}{Dh}$. We define an axis X perpendicular to the plane of the wall, and an axis Z parallel to the direction of dislocation lines in the wall. For this coordinate system, (Fig.2b) two non-zero components of the mean deformation tensor are equal: $\omega_{xy}=-\omega_{yx}=\Theta x/D$. Mean deformation tensor also results in pure rotations about the Z axis that increase with displacement in x. The number of unpaired walls per unit length and the length of the crystal in the X direction determines the total rotation of the lattice.

2.5.3. *Geometrically Necessary Tilt Dislocation Walls and Individual Dislocations*

The previous two models present two limiting cases. Generally, more complicated structures with a hierarchy of dislocation structures is formed. Real crystals usually contain an array of different types of dislocation structures. Some of the dislocations are distributed randomly, and the rest may form different kinds of non-random arrangements (Fig.2c). The results of our simulated scattering by this general dislocation structure (when some part of dislocations are grouped into the walls and some of them are randomly distributed in the inner regions) are shown in the next section. As discussed above, rotations introduced by a system of unpaired edge dislocations must be perpendicular to the direction of the dislocation line (τ) for both unpaired individual dislocations as for unpaired dislocation walls. As a result, the diffuse scattering along τ remains the same as for the crystal without excess dislocations. Along the direction of dislocation lines, the width of the reciprocal space intensity distribution is again of the order $\delta q_{rad} \sim 2\pi/L$, where L is the size of the crystal or sub-grain in the direction of the dislocation line.

2.6. Natural Co-ordinates for Laue Measurements

Near a Laue spot, the diffuse scattering induced by unpaired-dislocations can be qualitatively understood along three characteristic directions. In the radial direction, x-ray scattering is primarily sensitive to dilatational changes in the lattice. Local strain fields around dislocations can introduce such changes, but because Laue diffraction is inherently insensitive to diffuse scattering in the radial direction, additional measurements are needed. In the transverse direction ($\perp$ $\mathbf{g}=\mathbf{G}_{hkl}/G_{hkl}$) we can define two natural axes ξ and ν:

$$\xi = \frac{\tau \times \mathbf{g}}{|\tau \times \mathbf{g}|}, \qquad \nu = \frac{\xi \times \mathbf{g}}{|\xi \times \mathbf{g}|} \tag{12}$$

With this co-ordinate system, the diffuse scattering is strongly elongated in the ξ direction, and the $FWHM_\xi$ depends on the density of unpaired dislocations n^+, mutual orientation between the active dislocation system, the momentum transfer unit vector $\mathbf{g}$, and the penetration depth L (Fig. 4):

$$FWHM_\xi \propto n^+ L f_\xi(\hat{\tau} \bullet \hat{G}_{hkl}) \tag{13}$$

In the direction ν, the half width is more complicated and depends on the total dislocation density, n and the grain size depending on the orientation of τ to $\mathbf{g}$:

$$FWHM_\nu \propto b\sqrt{nl}\, f_\nu(\hat{\tau} \bullet \hat{G}_{hkl}) \tag{14}$$

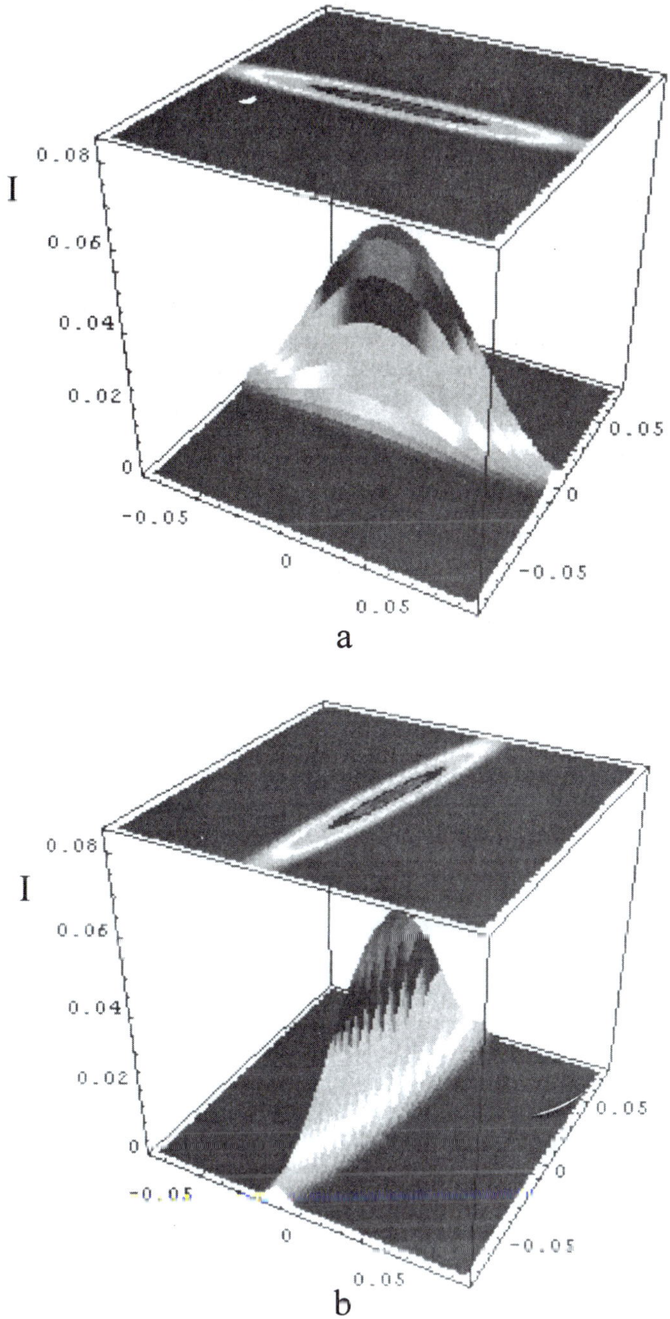

Figure 4. Orientation of dislocation line defines the streak axes (Burger's vector **b** is parallel to $[110]$): a - dislocation line $\hat{\tau}$ is parallel to $[\bar{1}1\bar{2}]$ direction; $\xi = [3\bar{1}\bar{2}]$; $\nu = [1\bar{5}4]$; b - dislocation line $\hat{\tau}$ is parallel to $[1\bar{1}\bar{2}]$ direction; $\xi = [1\bar{3}2]$; $\nu = [\bar{5}14]$.

Here $f_\xi(\hat{\tau}\bullet\hat{G}_{hkl})$ and $f_\nu(\hat{\tau}\bullet\hat{G}_{hkl})$ are orientation functions for ξ and ν axes, $l \approx \ln(L\sqrt{nl})$.

2.7. Shape of Laue Image

An analysis of the *FWHM* along the above two axes enables us to formulate the following criterion:

- If $n^{+}L >> 0.1\sqrt{nl}$, the intensity of scattering with white microbeam is mainly influenced by the geometrically necessary dislocations (related to correlated deformations of the lattice). For example this condition is valid if L=1μm, n=10^{11}cm^{-2}, n^{+}=n. In this case $FWHM_\xi >> FWHM_\nu$ and the inherent energy weighted integral is representative of the key features of $I(\mathbf{q})$. It is also possible to analyse the main features of the unpaired dislocation structure of the material.
- In the opposite case, $FWHM_\xi \sim FWHM_\nu$ and the intensity distribution is almost isotropic. Here the intensity of scattering measured in a white microbeam experiment is significantly influenced by total dislocation density in both directions. In this case, a detailed knowledge of intensity distribution in 3-dimentioned reciprocal space is necessary to adequately understand dislocation structure. For this case the white x-ray microbeam intensity should be differentiated with respect to $|\mathbf{k}_0|$.

3. SIMULATIONS

3.1 Splitting of Laue Reflections

3.1.1. Reconstruction of Laue Image due to Grouping of Individual Dislocations into the Wall

Naturally the models of random unpaired individual dislocations or dislocation walls present two limiting cases. In reality crystals have features of both models and correspond more nearly to a model with fragments and with unpaired randomly distributed individual dislocations between the dislocation walls (Fig. 2c). In this case the correlation function T (Eq.7) is a sum of two functions describing the first and second simple models.

To understand the main features of white beam scattering from crystals with various dislocation arrangements, we simulated the intensity of scattering by crystals with two model fragments (Fig. 5). Each fragment had a size of 2500 interatomic spacings. The total value of excess dislocations was kept constant. A total dislocation density of n^{+}=10^{12}cm^{-2} was chosen. This value is typical of highly deformed crystals. To initiate the simulation, all dislocations are randomly distributed with a dislocation density n^{+}=10^{12}cm^{-2} (Fig. 5a). The deformation fields around each dislocation are superimposed and the intensity was calculated from Eq. 6. The corresponding contour map has a huge broadening along the ξ axis with a typical "flat-top" shape (Fig 5a). As dislocations are removed from the fragments and added to the common dislocation wall, a correlated

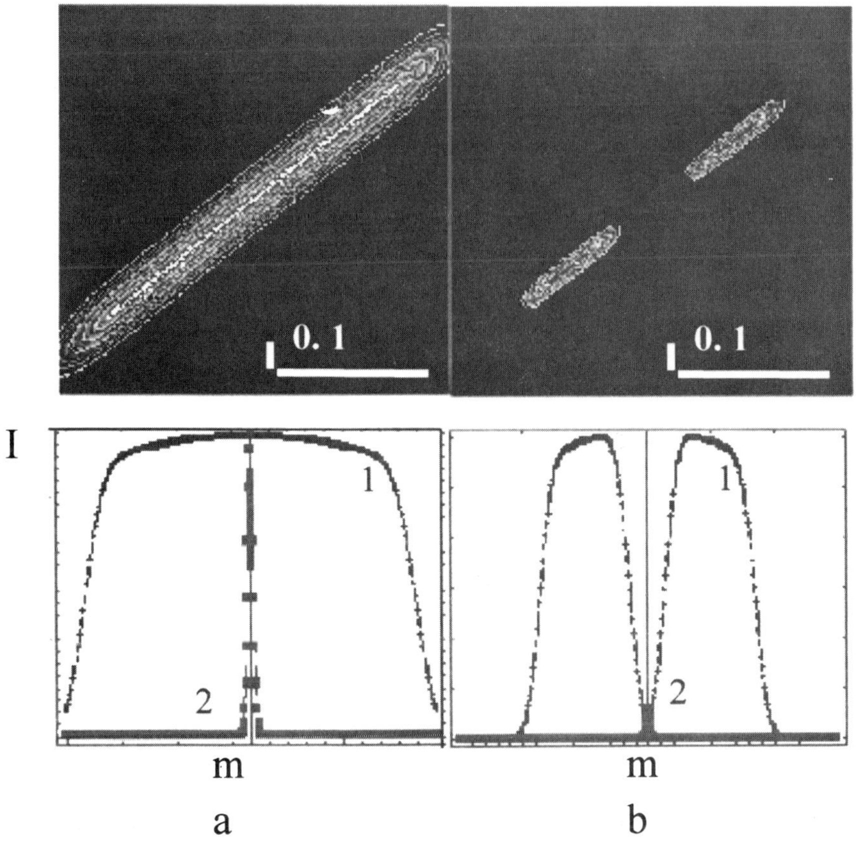

Figure 5. Simulated contour maps and intensity distributions for white beam Laue scattering by mosaic crystal along ξ (1) and ν (2) axes for (222) reflection from grouping of dislocations in the wall (total number of dislocations in the boundary and randomly distributed inside the fragments is constant). Size of the crystal L~2500*d*: a-individual dislocations; b- formation of two separate spots.

misorientation develops between the two parts of the crystal. When the distance between dislocations inside the wall is equal to approximately 21 interatomic spacings, this wall creates a rotation of about $\Theta \approx 2^0$ between the two fragments. This reduces the density of random dislocations inside each fragment to $n=2*10^{11}cm^{-2}$. The intensity distribution changes dramatically as shown in the "flat-top" intensity distribution along ξ axis.

Further grouping of the excess dislocations into the wall results in splitting of the reflection into two separate spots (Fig.5b). This splitting is observed when $n=4*10^{10}cm^{-2}$ and $\Theta \approx 2.12^0$. When $n=2.5*10^{10}cm^{-2}$ and $\Theta \approx 2.14^0$ the intensity distribution corresponds to separate spots. This simulation illustrates that splitting of white beam reflection (*hkl*) into spots depends on the following parameters: density of excess dislocation inside the wall, misorientation angle created by dislocation sub-boundary, size of a fragment, size of scattering volume.

3.1.2. Condition for Splitting of Laue Image

To understand this dependence, let us assume that along the penetration depth L the X-ray beam intersects several fragments with average size D_{fr}. Each fragment contributes to the diffraction. The number of such contributions is equal L/D. Each boundary produces average misorientation Θ. The average distance between the Laue maximums, formed by two adjacent fragments is $\Delta = \frac{Q}{k_0}\Theta$. If this distance exceeds the average $FWHM_{fr}$ of the Laue image for each fragment along the ξ axis, splitting of the Laue spot will occur. If almost all dislocation walls are unpaired, the following criterion can be used:

$$K = \left(\frac{\Theta Q}{k_0 FWHM_{fr}} \right). \tag{15}$$

If $K<1$ the intensity distribution of white beam reflection is continuous, if $K>1$ the white beam reflection is split into separate spots.

3.1.3. Different influence of unpaired individual and boundary dislocations on Laue scattering

Unpaired boundary and individual dislocations result in the same total macrosopic lattice rotation of the crystal. Both of them result in streaking of Laue reflections. Due to local strains, individual dislocations influence the length of the streak more than the same number of dislocations in a boundary. Moreover the *FWHM*, in the narrow direction of the streak is most strongly influenced by individual dislocations and can be used to separate boundary dislocations from the ones inside the fragment. For better separation of unpaired boundary and individual dislocations, the white x-ray microbeam intensity should be differentiated with respect to $|\mathbf{k}_0|$. This can be done by scanning the incident x-ray energy with an incident beam monochromator.

3.2. Size Dependence of Laue Intensity Distribution

To show the dependence of the Laue intensity distribution on the penetration depth (L) covered by the beam passing through the sample (L is a number of interatomic distances), we simulated Laue contour maps for different values of L by keeping a constant total value ($L*n^{+}$=constant) of unpaired dislocations in the whole scattering region (Fig. 6). The orientation of dislocation system in this calculation corresponds to Burgers vector parallel to [110], dislocation line parallel to $[\bar{1}1\bar{2}]$ and the gliding plane $(1\bar{1}\bar{1})$. The simulations show that such a dislocation system results in a rod like intensity distribution (Fig.6) around the reflection (222). The axis of the largest elongation of this intensity rod is located in the plane (111) parallel to the direction ξ=$[3\bar{1}\bar{2}]$. In the plane perpendicular to ξ the contour map is almost a circle with a very small diameter. The character of the intensity distributions along the ξ and the ν axes is absolutely different as expected (Fig. 6).

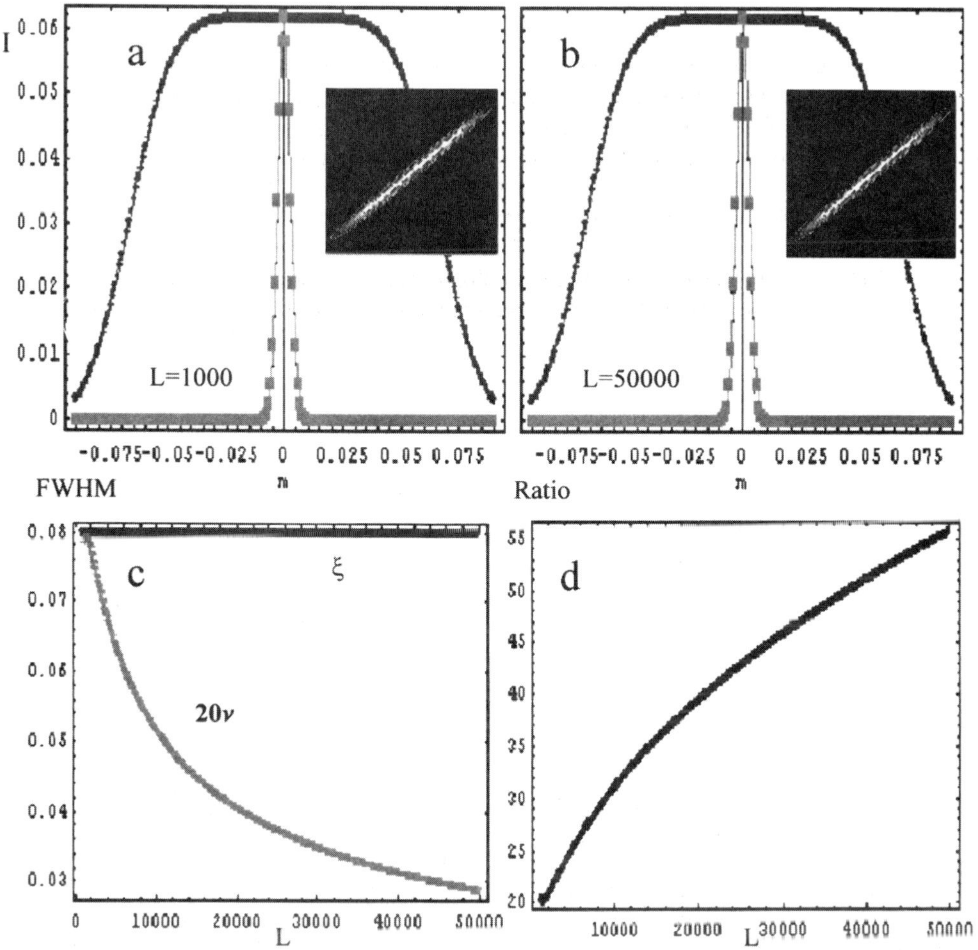

Figure 6. Dependence of Laue intensity distribution on the size of scattering region L (a, b); $FWHM_{\xi}$ in the long (ξ) and 20 $FWHM_{\nu}$ in the narrow (ν) directions of the streak (presented with a coefficient 20 to fit the scale) (c) and ratio of $FWHM_{\xi}/FWHM_{\nu}$ in these two directions with L.

The intensity distribution along the ξ axis has flat-top shape with large FWHM. Along the *ν* axis it is very narrow and is best described with a Gaussian function that is very sensitive to dislocation density. FWHM along the ξ directions (Fig. 6c) does not change with L (as total number of excess dislocation along the penetration depth remaines the same), while along the *ν* direction (Fig. 6c) it decreases (as dislocation density decreases). This means that the total dislocation density can be determined from the intensity distribution along the *ν* direction, while the total number of unpaired (geometrically necessary) dislocations, providing local rotation along the penetration depth of the beam, can be obtained from the intensity distribution along the ξ direction.

3.3. Analysis of Heterogeneous Dislocation Network Around Indentation in a Cu Single Crystal

This microbeam-Laue technique was applied to a complicated heterogeneous dislocation structure arising from plastic deformation in a nanoindented Cu single crystal.

The orientation of the Cu single crystal surface normal was parallel to [111], and the upper side of the pyramidal nanoindent is almost parallel to $[1\,\bar{1}\,0]$. The corresponding coordinate axes around the nanoindentation are shown in Fig. 7. Laue images were taken at various positions in and near the pyramidal nanoindent crater. .Some experimental images are presented together with simulated images. Both the experimental data and the model images reveal long streaks with distinctive orientations that can abruptly change direction. These streaks have been qualitatively discussed previously [15, 21].

To understand the shape of experimental Laue images and to check the sensitivity of the Laue image to different possible orientations of dislocations, we have simulated the Laue images corresponding to different arrangements of the primary unpaired dislocations. In FCC crystals, typical edge dislocation lines are parallel to the directions <112> with Burgers vectors being parallel to the directions <110> and corresponding glide planes {111} [24]. The plastic response of the material subjected to indentation can be understood by formation of excess or "geometrically necessary" dislocations (GND) that appear in the material to accommodate the indentor [15, 29]. Usually only a few of these dislocation systems are activated depending on the conditions of deformation. The change of orientation of the primary dislocation set surrounding the nanoindentation results in a reorientation of the Laue images (Fig.7). A simple image (uniaxial elongated streak) is obtained for the position (A). The streak can be described by the dislocation system with Burgers vector parallel to $[0\,\bar{1}\,\bar{1}]$, a dislocation line parallel to $[2\,\bar{1}\,1]$ and total dislocation density n=2*10^{11}cm^{-2} with an excess dislocation density n^{+}=0.5n. To model the vertical part of the image, a second slip system with Burgers vector parallel to $[\bar{1}\,0\,\bar{1}]$ and dislocation line parallel to $[1\bar{2}\,\bar{1}]$ was required. Images taken at the positions (B) and (C) (Fig. 7) are more complicated. At these positions, the penetrating beam intersects several fragments with different rotation axes relatively to the indentation. Plastic deformation during indentation activates different slip systems of the dislocations in these fragments. New instrumentation makes it possible to observe the dramatic change of activated dislocation networks along the beam penetration depth. Depth resolved experimental and simulated diffraction images for positions B and C are shown at Figures 7 b, c. The orientation of the activated dislocation system is shown at the figure. With depth, the beam intersects several regions with different GND dislocation network.

The transition from one dislocation network to another and corresponding Laue images around the indentation are shown in the Figures 7 b, c. These dislocation sets form a dislocation network surrounding the whole indentation and can be used to explain the elastoplastic response of the crystal during nanoindetation.

4. CONCLUSION

X-ray microdiffraction with broad-bandpass x-ray beams provides a powerful new tool for the study of plastic deformation in materials. Plastic deformation with a primary set of excess (GND) dislocations results in elongated streaks in the Laue image that can be used to determine the dislocation density. The orientation and density of the primary dislocation sets are determined for a nanoindented Cu single crystal.

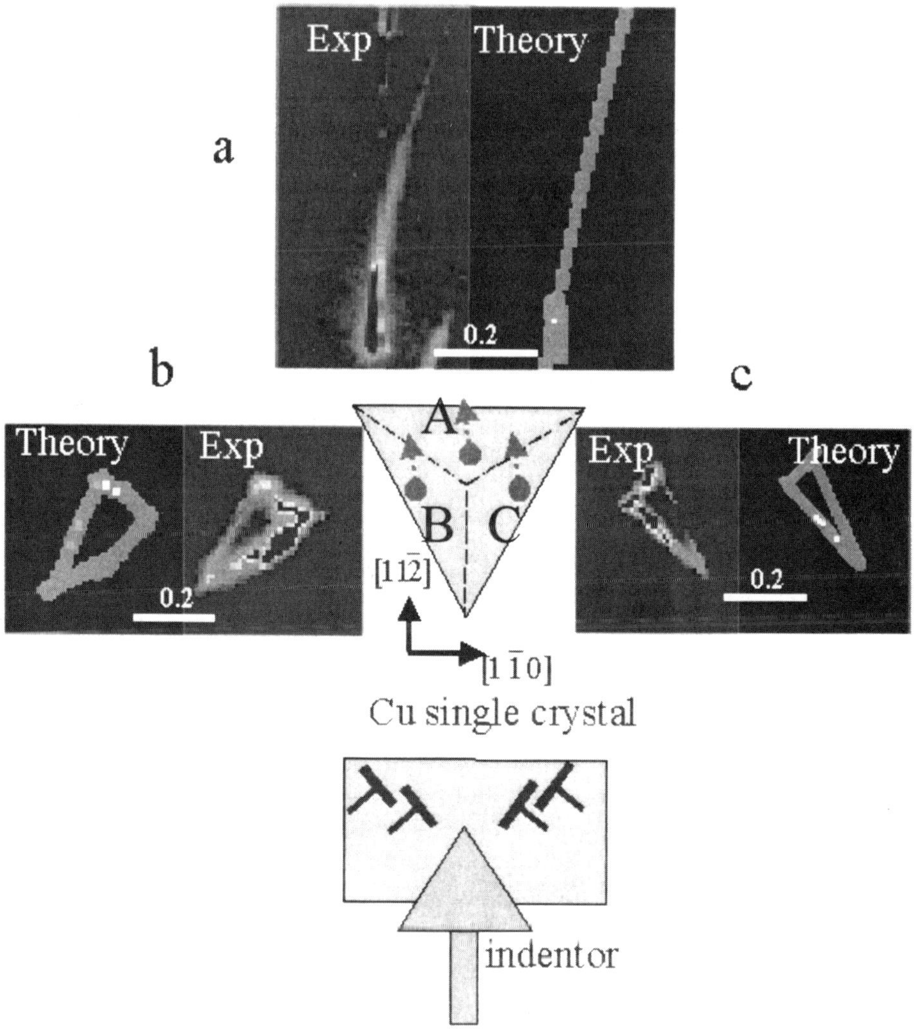

Figure 7. Laue images reveal which dislocations are operative around the indentation. The following dislocations systems are used to model Laue images for A, B and C positions of the indentation: a-Two activated dislocation system provide total rotation around $[1\bar{1}0]$ direction. 1) $\hat{\tau} \parallel [2\bar{1}1]$, $\hat{b} \parallel [011]$; 2) $\hat{\tau} \parallel [1\bar{2}\bar{1}]$, $\hat{b} \parallel [101]$; b-Beam intersects five regions with different orientations of geometrically necessary dislocations: 1) $\hat{\tau} \parallel [101]$, $\hat{b} \parallel [10\bar{1}]$; 2) $\hat{\tau} \parallel [\bar{1}\bar{2}1]$, $\hat{b} \parallel [101]$; 3) $\hat{\tau} \parallel [\bar{1}1\bar{2}]$, $\hat{b} \parallel [110]$; 4) $\hat{\tau} \parallel [1\bar{1}\bar{2}]$, $\hat{b} \parallel [110]$; 5) $\hat{\tau} \parallel [\bar{2}\bar{1}1]$, $\hat{b} \parallel [011]$; c-Beam intersects four regions with different actual dislocation systems and different rotation axes: 1) $\hat{\tau} \parallel [011]$, $\hat{b} \parallel [01\bar{1}]$; 2) $\hat{\tau} \parallel [1\bar{1}\bar{2}]$, $\hat{b} \parallel [110]$; 3) $\hat{\tau} \parallel [\bar{1}1\bar{2}]$, $\hat{b} \parallel [110]$; 4) $\hat{\tau} \parallel [101]$, $\hat{b} \parallel [10\bar{1}]$. Scattering by dislocation system with $\hat{\tau} \parallel [101]$, found for B and C positions is equivalent to the scattering by two octahedral systems ($\hat{\tau} \parallel [121]$ + $\hat{\tau} \parallel [1\bar{2}1]$) with $\hat{b} \parallel [10\bar{1}]$ and equal dislocation density. To distinguish those cases, simultaneous analysis of several Laue spots is usually performed.

ACKNOWLEDGEMENT

Research sponsored by the Division of Materials Sciences and Engineering, Office of Basic Energy Sciences, U.S. Department of Energy, under Contract DE-AC05-00OR22725 with UT-Battelle, LLC[29].

REFERENCES

1. C.G. Darwin, Phil.Mag., **27**, 315,657 (1914); **43,** 800 (1922)
2. Krivoglaz, M.A., *Theory of X-Ray and Thermal Neutron Scattering by Real Crystals*, (Plenum Press, New York, 1969); Krivoglaz, M.A., *X-ray and Neutron Diffraction in Nonideal Crystals*, (Springer, Berlin, 1996).
3. Warren, B.E., *X-ray diffraction*, (Dover Publications, INC., New York, 1969)
4. Wilkens, M., Phys. Stat. Sol. A. **2**, 359-370,(1970); *Fundamental Aspects of Dislocation Theory* II,p. 1195-1221, Edited by J.A. Simmons, R. de Wit and R. Bullough, Washington: Nat.Bur.Stand. Special Publ.317, **II, pp. K1**-K7,(1970).
5. M. Wilkens, T. Ungar, H. Mughrabi, Phys.Stat.Sol.(a) (1987) **104,**157-170.
6. C.H. Lang, W. Schneider, H. Mughrabi, Acta Metall. Mater.(1995),**43,** 5,1751-1764
7. Thomson, R. and Levine, L.E., Acta Cryst. **A53** (1997), 590-602.
8. Barabash, R. and Klimanek, P., J. Appl. Cryst. , **32,** (1999), 1050-1059.
9. Breuer, D., Klimanek, P., Pantleon W., J. Appl. Cryst. ,**33,** (2000), 1284-1294.
10. Barabash R., Klimanek P. Z.Metallkunde,**92**, (2001), 1, 70-75.
11. Castelnau, O., Chauveau, T., Drakopoulos, M., Snigirev, A. Snigireva, I. Schroer, C., Ungar, T.: Mater. Sci. Forum, Vols347-349, (2000) [ISSN 0255-5476], 297-302.
12. E.M. Lauridsen, D. Juul Jensen, H.F. Poulsen, U. Lienert, Scr. Mat.(2000)**43**, 561-566
13. L.Margulies, G.Winther, H.F. Poulsen, Science, (2001)**291,** 2392-2394
14. J.S.Chung, G.Ice, J.Appl.Phys.,(1999) **86,** 9, 5249-5255.
15. B.C.Larson, N. Tamura, J.S.Chung, G.Ice, J.D.Budai, J.Z.Tischler, W.Yang, H.Weiland, W.P.Lowe, Proc.Mat.Res.Soc.Symp. Fall, **590**, 247, (2000).
16. B. Buras and S. Tazzari, European Synchrotron Radiation Facility Report of European Synchrotron Radiation Project, Cern LEP Division, Geneva, Switzerland, (1984).
17. G.K. Shenoy, P.J. Viccaro, and D.M.Mills, *Characteristics of the 7 GeV Advanced Photon Source:A Guide for Users* (Argonne National Laboratory, Argonne, IL, 1988), No.**ANL-88-9.**
18. G.E.Ice, X-Ray Spectrum. **26**, 315 (1997).
19. M.A.Marcus, A.A.Macdowell, E.D.Isaacs, K.Evans-Lutterodt, G.Ice, Mat. Res. Soc. Symp. Proc. vol. **428**, (1996), 545-556.
20. J.S.Chung, N. Tamura, G.Ice, B.C.Larson, J.D.Budai, Mat. Res. Soc. Symp. Proc. Vol.**563**, (1999), 169-174.
21. R. Barabash, G.E. Ice, B.C. Larson, G.M. Pharr, K.-S. Chung, W. Yang, APL, **79**, 4, (2001)
22. Rybin, V.V., *High Plastic Deformations and Fracture of Metals,* (Moscow, Metallurgia, in Russian, 1986.), p.223.
23. S.R.Stock, A.Guvenilir, D.P.Piotrovski, Z.U.Bek, Mat.Res.Soc.Sympos.Proc., vol.**375** (1995), 275-280.
24. Nabarro,F.R.N.(1987), *Theory of Crystal Dislocations.* New York: Dover Publications, INC.
25. L. Landau and E. Lifshitz, *Theory of Elasticity,* (Pergamon, Oxford, 1959).
26. G. Arfken, *Mathematical Methods For Physicists*, (Academic Press, 2nd ed, New York, 1970), pg. 132.
27. S.N.G.Chu, A.T. Macrander, K.E.Strege, W.D.Johnston, I.Appl.Phys., **57**, 249,(1985)
28. A.Gouldstone,H.-J.Koh,A.E. Giannakopulus, S.Suresh, Acta Mater.**48**,(2000) 2277-2295.
29. The submitted manuscript has been authored by a contractor of the U.S. Government under contract No. DE-AC05-00OR22725. Accordingly, the U.S. Government retains a nonexclusive, royalty-free license to publish or reproduce the published or reproduce the published form of this contribution, or allow others to do so, for U.S. Government purposes

PROCESSES OF SELF-ORGANIZATION DURING EPITAXIAL GROWTH OF SEMICONDUCTOR SUPERLATTICES - AN X-RAY SCATTERING STUDY

Václav Holý, Günther Bauer, Julian Stangl, and Gunther Springholz *

1. INTRODUCTION

Self-organization processes taking place at a surface during epitaxial growth of semiconductor layered systems represent a promising way for the fabrication of low dimensional objects.[1] Two types of self-organization have been observed during growth. In the so called layer-by layer growth mode, the growth can be described as a lateral movement of monolayer steps at a growing vicinal surface. During the growth, the monolayer steps agglomerate creating a nearly periodic distribution of step bunches divided by nearly atomically flat terraces. In a multilayer, this process creates a template for a nearly periodic sequence of one-dimensional quantum wires. The reason for this bunching tendency is so far not completely clear. In a series of papers[2–4] the bunching of monolayer steps has been explained as a consequence of an inhomogeneous strain at the growing surface caused by a rough interface buried underneath and by a lattice mismatch of the growing layer with respect to the substrate. An inhomogeneous surface strain distribution affects the surface diffusion of adatoms and consequently the mobility of migrating monolayer steps. Then, a nucleation of a step bunch is more probable in a region with a large strain gradient. This description of the bunching process was able to explain the major features observed, namely the improving periodicity of the bunches with increasing number of interfaces, and a oblique replication of the position of the bunches at subsequent interfaces.[3]

The bunching of the monolayer steps has been studied also during homoepitaxial growth,[5] where no lattice mismatch is present. This observation cannot be explained by an inhomogeneous local strain and another mechanism must be active in this case. Monte-Carlo simulations of a homoepitaxial growth taking into account energy barriers for the hopping of an adatom "downstairs" from a monolayer step (Schwoebel barrier[6]) and "upstairs" on the step showed that a bunching of the steps can occur in homoepitaxy indeed.

The second type of the processes leading to a self-organization of the growing surface is based on an elastic relief of internal stresses due to the creation of nearly periodic sequence of small islands (quantum dots) at the surface (so called Stranski-Krastanow growth

*Václav Holý, Masaryk University Brno, Czech Republic CZ-611 37. Günther Bauer, Julian Stangl and Gunther Springholz, J. Kepler University Linz, Austria A-4040.

From Semiconductors to Protein, edited by S.J.L. Billinge and M.F. Thorpe
Kluwer Academic/Plenum Publishers, 2002

mode). In SiGe/Si superlattices, two families of islands are observed.[7,8] Small islands ("hut clusters") have pyramidal shapes with crystallographically equivalent facets, their twofold or fourfold symmetry is induced by the crystallographic symmetry of the growing surface. The larger islands ("domes") have complicated shapes with various types of facets. The population of these families depends on the growth conditions (temperature, flux of the adatoms etc.).

In a superlattice, the positions of the quantum dots at different interfaces are correlated. In most cases (for instance SiGe quantum dots in a SiGe/Si superlattice on a (001) Si substrate), the dots are correlated vertically creating columns. Laterally, the dot columns create a disordered square array along $\langle 100 \rangle$, however the lateral correlation of the dots is much weaker than the vertical one. In ZnCdSe/ZnSe systems, the ZnCdSe quantum dots are vertically anticorrelated.[9] It is generally believed now that the positions of quantum dots in a superlattice are determined by elastic properties of the host crystal. A quantum dot buried below the surface causes a local strain field that propagates towards the free surface. In an elastically soft direction this strain exhibits a maximum which increases the probability of the nucleation of a dot at the corresponding position at the surface.[11–13] This effect is well pronounced in elastically anisotropic crystals such as PbTe. In a PbSe/PbEuTe superlattice the arrangement of the PbSe dots is periodic in all three dimensions and the dots create a nearly perfect three-dimensional "dot lattice".[13,14]

The optical properties of low-dimensional objects strongly depend on the homogeneity of their sizes and chemical composition. Therefore, it is an important task to characterize the structure of both free standing and buried objects below the surface. Several methods have been employed for this purpose (atomic force microscopy – AFM, transmission electron microscopy – TEM) but only x-ray scattering makes it possible to determine *non-destructively* the shape of the objects, their positions and also their chemical composition. Specular and non-specular x-ray reflection (XRR) probes the neighborhood of the origin of the reciprocal lattice and thus is sensitive to the shape of the objects, to their positions but *not* to the strains around them. Wide angle x-ray diffraction (XRD) probes the region of the reciprocal lattice around a non-zero point, thus it is extremely sensitive to the strains and consequently to the lattice mismatch and chemical composition of the objects.

The main difficulty in the application of x-ray scattering from low dimensional objects consists in their extremely small volumes. For most objects, the signal from the objects would completely disappear in the tails of the substrate scattering in a usual coplanar experimental geometry. In order to enhance the signal from the objects, a grazing-incidence geometry can be used, both in the small-angle version (Grazing-incidence small angle scattering – GISAXS)[10] and in wide-angle diffraction (Grazing-incidence diffraction – GID). In the coplanar XRR geometry the measurable space frequencies of the surface/interface morphology are limited by very small scattering angles. This limitation can be overcome in the non-coplanar GISAXS arrangement.

In this paper we present some recent examples of the application of these methods as well as a basic theoretical description of the scattering from low-dimensional objects.

2. GRAZING-INCIDENCE SMALL ANGLE SCATTERING

The experimental arrangement of the GISAXS method is sketched in Fig. 1. The primary beam with the wave vector $\mathbf{K}_i$ hits the sample surface under a small angle α_i close to the critical angle α_c of total external reflection. Usually, the scattered radiation with the wave vector $\mathbf{K}_f$ is measured by a position-sensitive detector (PSD) oriented parallel to the sample surface so that we obtain the intensity I as a function of the in-plane angle ψ for

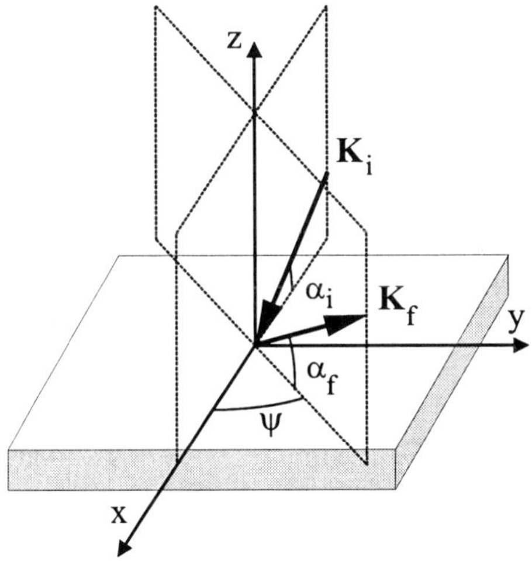

Figure 1. Sketch of the GISAXS geometry

constant take-off angle α_f. The intensity is expressed as a function of the scattering vector

$$\mathbf{Q} = \mathbf{K}_f - \mathbf{K}_i. \tag{1}$$

Performing the measurements for various azimuthal directions of the incidence plane we obtain an intensity map $I(Q_x, Q_y, Q_z = \text{const.})$. In order to enhance the signal from the interface objects, it is useful to avoid strong truncation rod scattering at $\mathbf{Q}_{\|} = 0$. This can be performed by a slight difference between the angles $\alpha_{i,f}$.

The main advantage of GISAXS lies in its depth sensitivity. Changing α_i we can tune the depth in the sample, where the scattering comes from (the information depth Λ) from few nm for α_i and/or α_f close to α_c, to several μm for larger α's.

The cross-section of diffuse scattering in GISAXS can be obtained by means of distorted-wave Born approximation (DWBA[15]) as

$$\left(\frac{d\sigma}{d\omega}\right)_{\text{diff}} = \frac{1}{16\pi^2} 2K^2 |\Delta\delta|^2 \left|t_i t_f\right|^2 \left|\Omega^{\text{FT}}(\mathbf{Q}_t)\right|^2 G(\mathbf{Q}_t), \tag{2}$$

where $\Delta\delta$ is the contrast of the refractive indices of the quantum dot and its neighborhood, $\Omega^{\text{FT}}(\mathbf{q})$ is the Fourier transformation of the shape function $\Omega(\mathbf{r})$ of a dot (unity in the dot volume and zero outside) and

$$G(\mathbf{Q}_t) = \left\langle \sum_m \sum_n e^{-i(\mathbf{Q}_t . \mathbf{R}_m - \mathbf{Q}_t^* . \mathbf{R}_n)} \right\rangle \tag{3}$$

is the geometrical factor depending on the dot positions $\mathbf{R}_m$ averaged over all possible dot configurations. Using DWBA, we assume that the primary radiation is transmitted through the sample surface, the transmitted wave with the amplitude t_i is scattered by the dots and the scattered wave is transmitted through the surface again (with the transmission coefficient t_f). This concept is valid for the dots buried below the surface; in the case of free standing objects one has to take into account also the scattering of the reflected wave and/or the reflection of the scattered wave.[16,17] Due to the refraction at the sample surface, the scattering vector $\mathbf{Q}$ is changed, its corrected value is denoted $\mathbf{Q}_t$.

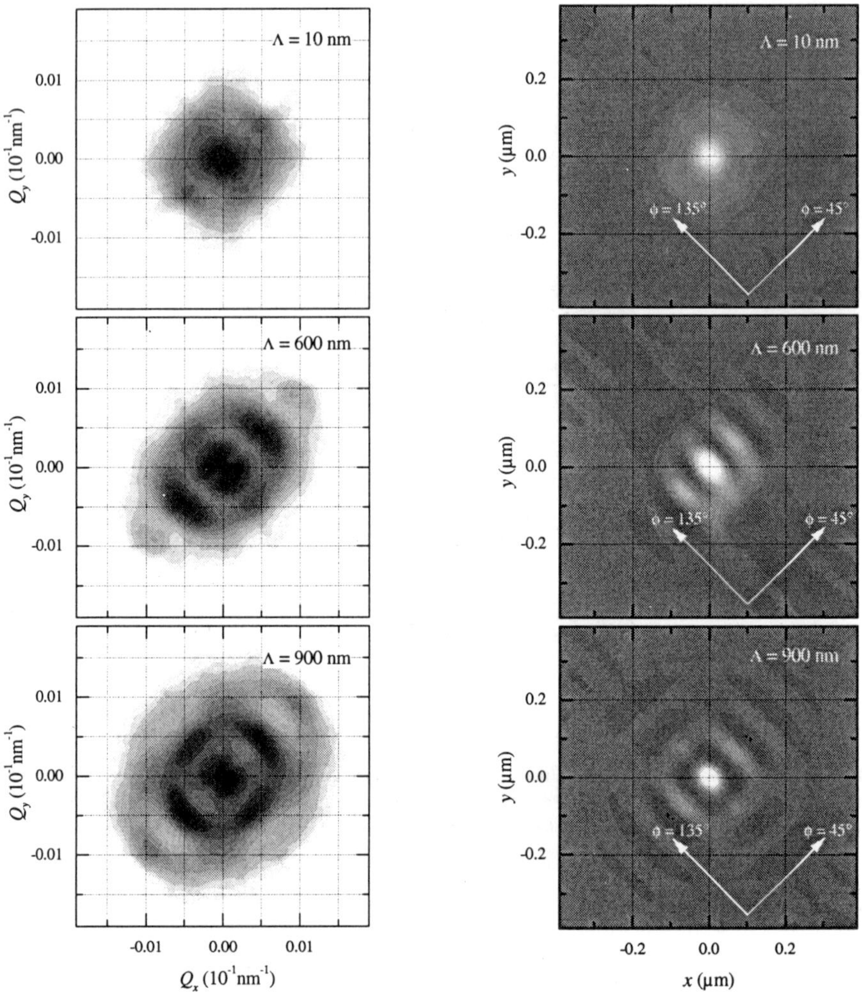

Figure 2. The GISAXS intensity maps of a SiGe/Si superlattice measured for three different information depths Λ (left panels) and their Fourier transformations (right panels).

The main problem in the interpretation of the GISAXS data is to distinguish between the influence of the dot shape expressed by $\Omega^{\mathrm{FT}}(\mathbf{Q}_t)$ in Eq. (2), and the dot positions contained in the correlation function $G(\mathbf{Q}_t)$. If the dots are much smaller than their mean lateral distance, it is suitable to perform the two-dimensional Fourier transformation of the measured intensity distribution $I(\mathbf{Q}_\parallel, Q_z = \text{const.})$

$$J(\mathbf{x}; Q_z) = \frac{1}{4\pi^2}\int \mathrm{d}^2\mathbf{Q}_\parallel I(\mathbf{Q}_\parallel, Q_z)\mathrm{e}^{\mathrm{i}\mathbf{Q}_\parallel\cdot\mathbf{x}} = \text{const.}\,[\delta(\mathbf{x}) + p(\mathbf{x})] \otimes \Phi(\mathbf{x}; Q_{tz}) \tag{4}$$

where $\otimes$ means convolution, $p(\mathbf{x})$ is the pair correlation function of the lateral dot positions and

$$\Phi(\mathbf{x}; Q_{tz}) = \frac{1}{4\pi^2}\int \mathrm{d}^2\mathbf{Q}_\parallel \left|\Omega^{\mathrm{FT}}(\mathbf{Q}_\parallel, Q_{tz})\right|^2 \mathrm{e}^{\mathrm{i}\mathbf{Q}_\parallel\cdot\mathbf{x}} \tag{5}$$

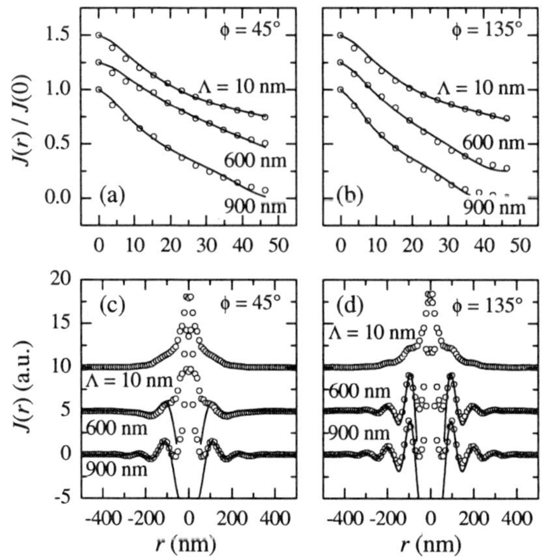

Figure 3. (a,b) measured (points) values of J in the region for small r for two different azimuthal directions ϕ and three different information depths Λ and their fits using the model of ellipsoidal dots (thin lines). (c,d) for large r the experimental data for J (circles) have been fitted using the short-range-order model of the dot arrangement (lines).

For small dots, the function $J(\mathbf{x}; Q_z)$ for small $|\mathbf{x}|$ is determined only by the dot shape, since the pair correlation function is constant in this range. For $|\mathbf{x}|$ larger than the dot diameter, the function $\Phi(\mathbf{x}; Q_{tz})$ is zero and J is influenced only by the dot positions.

2.1. SiGe quantum dots in a SiGe/Si superlattice

As a demonstration of the method we show here our results on SiGe quantum dots in a SiGe/Si superlattice.[18] The investigated Si/SiGe multilayer sample was grown by solid source molecular beam epitaxy on a (001) Si substrate with a miscut of 2°. The multilayer consisted of 20 periods containing nominally 2.5 nm thick $Si_{1-x}Ge_x$ layer ($x_{Ge} = 0.45$) and a 10 nm thick Si. On top of the last Si spacer layer 5 monolayeres of pure Ge were deposited at 500°C.[19] Under these growth conditions, the step bunches are formed at the growth surface, which are accompanied by strain fields influencing the nucleation of SiGe islands. It was intended to use these step bunches in order to obtain a more regular island arrangement and hence a narrower distribution of the island sizes.

The GISAXS experiments have been performed at the TROÏKA II undulator beamline at the ESRF, Grenoble, using an x-ray wavelength of $\lambda = 0.15545$ nm. The diffusely scattered intensity from the islands was measured by a PSD oriented parallel to the sample surface. Series of GISAXS scans were recorded for different $\alpha_{i,f}$, (i.e. for different Q_z's) and for azimuthal positions of the sample ranging from 0° to 180° in steps of 5°. From the scans measured at given Q_z we constructed a two-dimensional intensity distribution in the Q_x-Q_y plane. The resulting in-plane reciprocal space maps are shown in Fig. 2. Performing the numerical two-dimensional Fourier transformations of these distributions after Eq. (4), we obtained the functions $J(\mathbf{x}; Q_z)$ shown in Fig. 2 as well.

For small $|\mathbf{x}|$ we have compared the values of J with the calculation based on a simple

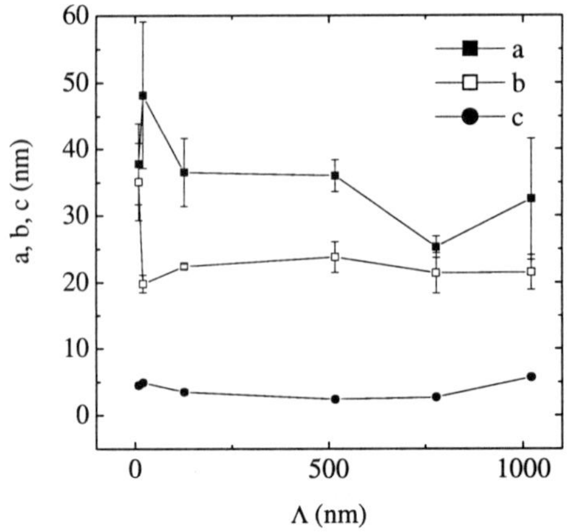

Figure 4. The dependence of the parameters of the dot shape (see text) on the information depth Λ

assumption that the dots have the shape of an upper half of an ellipsoid. We determined the lateral half-axes of the ellipsoid a and b ($a > b$), the vertical half-axis c (the dot height) and the azimuthal angle of the longer axis of the ellipsoid with respect to [110] direction. In Fig. 3 we have plotted radial cuts from the two-dimensional distribution of J shown in Fig. 2 in the azimuthal directions $\phi = 45°$ and 135° with respect to [110] (denoted by arrows) and their theoretical fits.

The pair correlation function of the dot positions was obtained from the radial cuts of J for larger $|\mathbf{x}|$. These cuts were compared with the one-dimensional pair-correlation function assuming a short-range-order model of the dot distribution[20] assuming that the distances L of the neighboring dots are random with a given mean value. From the fits of the experimental cuts to the theory shown in Fig. 3(c,d), we have determined the mean dot distance $\langle L \rangle$ and its root mean square (r.m.s.) deviation as functions of the azimuthal angle ϕ (see our previous paper[18] for more details).

From the fits it follows that the dots are arranged approximately in a square array along ⟨100⟩ directions, however, the dispersion of the dot distances in the [100] direction parallel to the miscut is smaller than in the other one, due to the step bunching during growth of the multilayer.[19] The positions of the dots obey very well short-range-order model.

The parameters a, b and c of the dot shape depend on Q_z. This can be attributed to the dependence of the dot shape on the depth below the multilayer surface, since changing Q_z we change also the information depth Λ. The depth dependence of these parameters is plotted in Fig. 4. In spite of relatively large errors of the parameters, the tendency of growing the parameter a towards the surface is obvious.

2.2. PbSe quantum dots in a PbSe/PbEuTe (111) superlattice

The SiGe quantum dots shown in the previous section were correlated vertically and the lateral periodicity of the dot positions was rather poor. A completely other situation was observed in a PbSe/PbEuTe superlattice. The arrangement of the PbSe quantum dots in the superlattice depends on the thickness D of the PbEuTe spacer layers. For layer thicknesses smaller than approx. 35 nm, the dots at different interfaces are arranged vertically and their lateral positions are nearly random. For D between 35 and 60 nm, the dots create a nearly perfect trigonal lattice, the lattice parameter being proportional to D. Therefore, in this

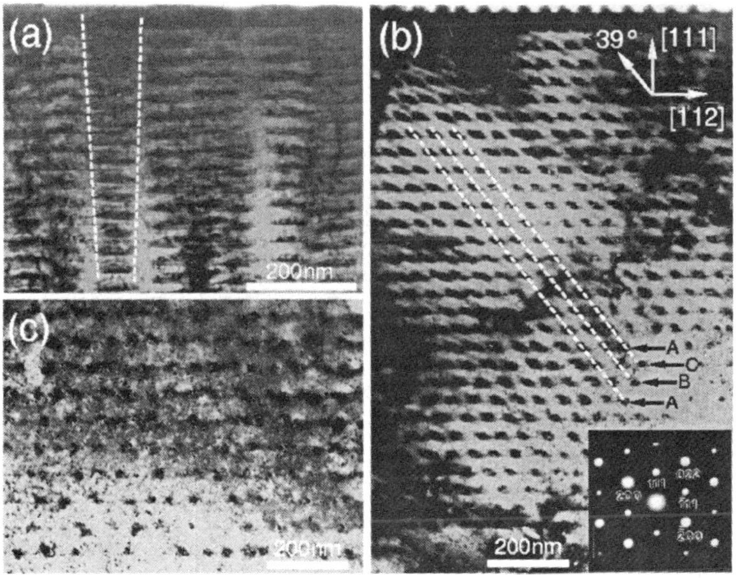

Figure 5. Cross-section transmission electron microscopy of PbSe/PbEuTe superlattices #966 (a), #925 (b) and #906 (c) with various types of dot stacking, namely vertical stacking (a), trigonal stacking (b) and completely random arrangement of the dots (c). The inset shows the electron diffraction pattern of the sample. The TEM pictures have been obtained by H. H. Kang and L. Salamanca-Riba at the University of Maryland.

range of thicknesses, the dot distance can be tuned by changing D. Finally, for $D > 60$ nm, both the vertical and horizontal positions of the dots are completely random. This strange behavior has been explained by a calculation of the strain distribution at a growing surface due to a buried dot[13] and a good agreement with the experimental data was achieved.

We have investigated three samples denoted #966, #925 and #906, with the spacer thicknesses $D = 35.5$, 47 and 66 nm, respectively, covering all thee types of the dot stacking. The details of the growth procedure can be found elsewhere.[13,14] The cross-section transmission electron micrographs (TEM) of the samples are in Fig. 5

A similar conclusion can be drawn from coplanar x-ray diffraction maps in symmetrical 111 diffraction measured at the TROÏKA II and ROBL beamlines at ESRF, Grenoble with a wavelength $\lambda = 1.54$ Å (Fig. 6). The first stacking type (only vertical stacking) gives rise to horizontal intensity sheets in reciprocal space crossing the superlattice satellites in Fig. 6 (a). The trigonal stacking of the dots produces a sequence of satellite intensity maxima shown in Fig. 6 (b) corresponding to the lattice points of the reciprocal "dot lattice". If the dot positions are not correlated, no satellite maxima can be observed and the intensity distribution in Fig. 6 (c) corresponds to that of a single dot.

The widths of the intensity maxima in reciprocal space is inversely proportional to the degree of periodicity of the dot positions. Within the short-range order model, this width should also be proportional to the square of the order of the satellite maximum. On the other hand, in a long-range order model, the dots are randomly displaced from ideal periodic positions, the displacements of different dots are not correlated. In this model, the width of the intensity maxima is independent of its order. Therefore, from the measured widths of the maxima we can decide whether the dot arrangement follows a short-range or long-range order model. This is shown in Fig. 7, where we have plotted the horizontal

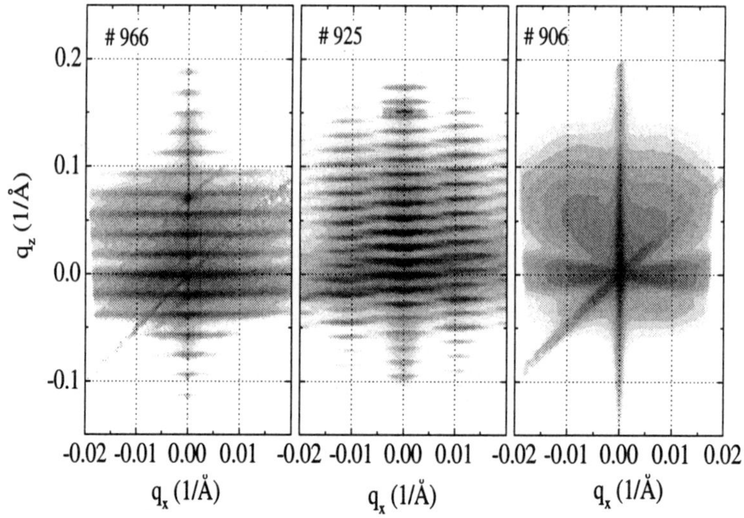

Figure 6. Reciprocal space maps of diffracted intensity of PbSe/PbEuTe, coplanar symmetrical diffraction 111 of PbSe/PbEuTe superlattices.

and vertical widths $\delta q_{\|}, \delta q_{\perp}$ of the maxima of samples # 966 and # 925 as functions of the horizontal and vertical components of the reduced scattering vector

$$\mathbf{q} = \mathbf{Q} - \mathbf{h},$$

where **h** is the diffraction vector. For sample #966 we found a parabolic dependence of $\delta q_{\perp}$ on $q_{\|}$. Therefore, the lateral dot positions are vertically correlated in this sample with r.m.s. deviation of about 5 nm, which is much smaller than the dot size. Thus, the dots are perfectly vertically correlated in this sample. The vertical width $\delta_{\perp}$ does not depend on $q_{\perp}$ indicating that the vertical correlation of the vertical dot positions is perfect and it is only limited by the growth stability.

In sample #925 the vertical correlation of the vertical dot position is nearly perfect with the r.m.s. deviation of about 0.2 nm. The lateral correlation of the lateral dot position obeys the short-range order model with r.m.s. deviation of 6 nm. Thus, the dots are periodically arranged in the lateral direction approximately up to the 7th neighbor. The vertical

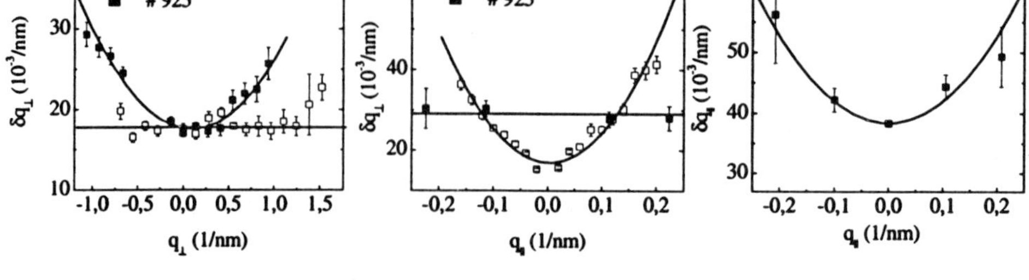

Figure 7. The horizontal and vertical width of the intensity maxima obtained from the intensity maps in Fig. 6 and their fits using the short-rand and log-range order models.

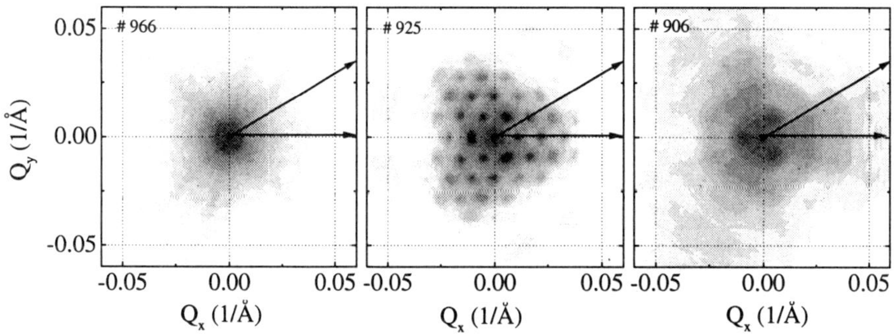

Figure 8. Intensity distributions $I(Q_x, Q_y, Q_z = 0.1Å^{-1})$ on PbSe/PbEuTe superlattices. The arrows denote the directions, where the linear scans have been extracted.

correlation of the lateral dot positions is nearly perfect in this sample. This means that the lateral disorder of the dots propagates during the growth without any significant change.[21]

We have also investigated the PbSe/PbEuTe samples by GISAXS using the same experimental arrangement as for the SiGe dots in the previous section. Fig. 8 shows the measured intensity distributions for constant $Q_z = 0.1$ Å^{-1}. In the map of sample #925 the lateral satellites can be seen corresponding to the trigonal stacking of the dots. For the other samples no intensity maxima can be observed, since in these samples the dots are not arranged laterally.

From the intensity maps we have extracted linear scans along the directions denoted by arrows. We have simulated these scans using Eq. (2) assuming that the dots have the form of a truncated pyramid with C_{3v} symmetry and with rounded edges. The results are summarized in Fig. 9 showing also the form of the dot base in the insets. From the figure it is obvious that for smallest spacer thickness D (sample #966), where the vertical stacking occurs, the dots have the largest lateral dimensions. For intermediate thicknesses the dot stacking is trigonal and the dots are slightly smaller, the smallest dots have been observed for largest D, where no dot arrangement occurs.

2.3. Self-assembled SiGe wires in a SiGe/Si superlattice

We have studied a 20 period SiGe/Si superlattice having the thickness 10.0 nm of the Si space layers and nominally 2.5 nm of $Si_{0.55}Ge_{0.45}$ layers. The superlattice has been grown by molecular beam epitaxy on Si (001) substrate with a large miscut of 3.5° towards [100]. The details of the growth were described previously.[22] Using atomic force microscopy (AFM) we have observed a periodic ripple structure on the surface with the period L of about 90 nm along the miscut direction and a peak-to-valley amplitude of about 1 nm.

The GISAXS experiments have been performed at the TROÏKA II beamline at ESRF, Grenoble using the same arrangement as in the previous sections. The measurements of the intensity distribution along the Q_y axis (see Fig. 1) have been carried out in two azimuthal directions of the incidence plane. In arrangement (a), the ripples were parallel to the incidence plane so that the Q_y axis in reciprocal space was perpendicular to the ripples. In arrangement (b) the ripples were perpendicular to the incidence plane and parallel to the scan trajectory in reciprocal space.

Fig. 10 shows the GISAXS scans measured in both arrangements for various values of Q_z. The angles $\alpha_{i,f}$ were adjusted so that the constant value $Q_x = 2.5 \times 10^{-3}$ Å was

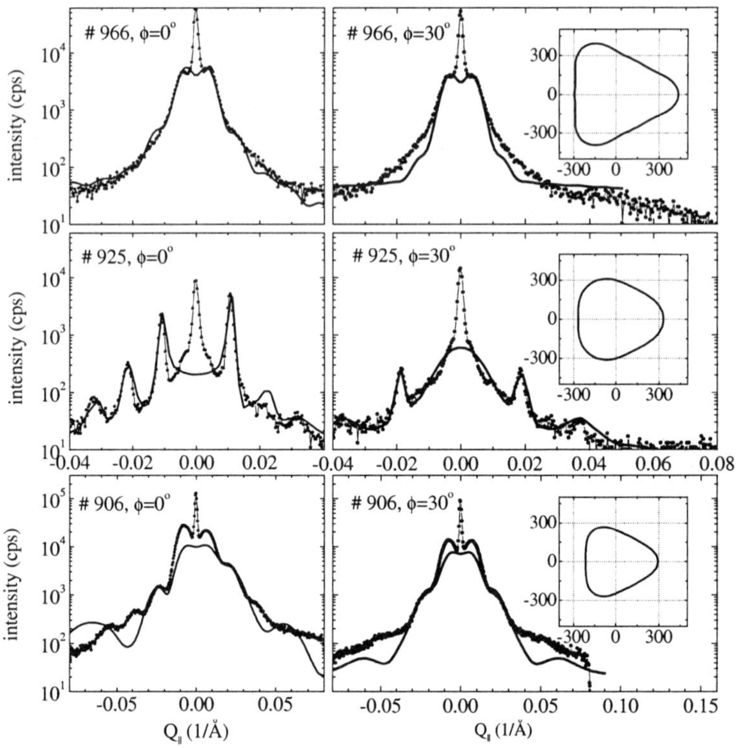

Figure 9. The linear GISAXS scans extracted in two azimuths $\phi = 0$ and 30° from the measured intensity maps in Fig. 8 (dots) and their fits assuming a pyramidal shape of the dots (full lines). In the insets we have plotted the shapes of the dot bases following from the fits, the axes of the insets are in Å.

achieved in all scans. For all the scans, the information depth Λ was larger than the thickness of the whole superlattice, and thus the measured signal corresponds rather to the structure of the buried interfaces than that of the surface.

In the arrangement (a) we observe a sequence of lateral satellites corresponding to a periodic sequence of buried quantum wires with the same period $L = 90$ nm as that of the surface ripples. The widths of the satellites are connected to the experimental resolution function and to a disorder of the wire positions. In geometry (b) no satellites appear.

We have analyzed the measured data using Eq. (2). Assuming a short-range order model of the wire positions and the geometry (a), this formula can be re-written in the following way

$$\left(\frac{d\sigma}{d\omega}\right)_{\text{diff}} = \text{const.} \left|t_i t_f\right|^2 \sum_P \mathcal{R}_P(Q_y - P) \left|\Omega^{\text{FT}}(P, Q_{tz})\right|^2, \tag{6}$$

where $P = 2p\pi/L$ (p is integer) are the vectors of the one-dimensional lattice reciprocal to the wire positions, Ω^{FT} is the Fourier transformation of the two-dimensional shape function of the wire cross-section, and $\mathcal{R}_P$ is the function describing the shape of the P-th lateral satellites assumed Lorentzian with the FWHM proportional to P^2. In Fig. 10 we show also the fits of the measured scans using this formula, where the satellite heights $F_P \equiv \left|\Omega^{\text{FT}}(P, Q_{tz})\right|^2$ have been taken as free parameters. The very good quality of the fit reflects the fact that the wires are arranged according the short-range order model.

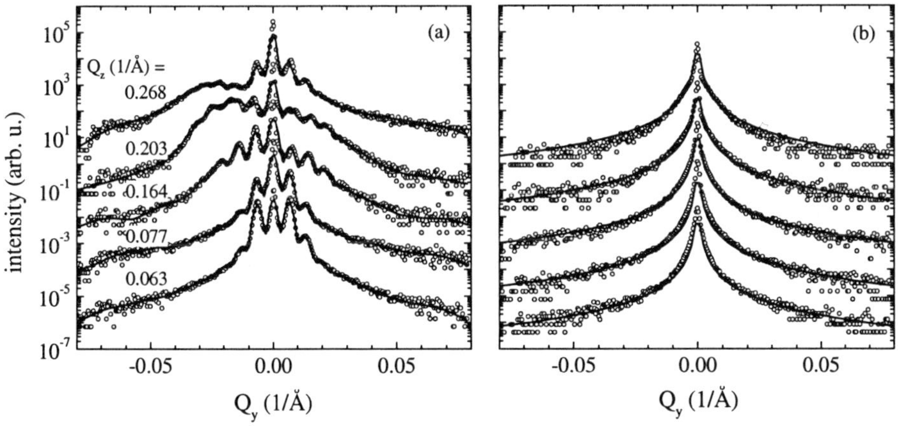

Figure 10. The GISAXS scans of a SiGe/Si superlattice with self-assembled quantum wires measured in two geometries (a) and (b) for various Q_z (points) and their fits using the theory described in the text (lines).

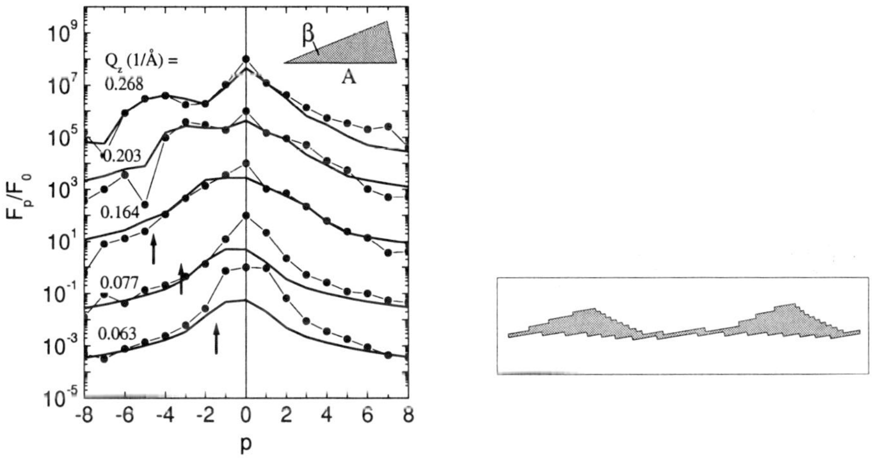

Figure 11. In the left panel, the heights F_P of the lateral satellites are plotted (points) along with their fits (lines) assuming the wire cross-section in the inset. The meaning of the arrows is explained in the text. The right panel shows the shape of the wire cross-section.

The curves in (b) have been fitted to a single Lorentzian peak, from which the mean length 250 nm of the wires follows. This value is in a good agreement with AFM data, therefore the geometry of the buried wires corresponds approximately to that of the surface ripples.

Knowing the heights F_P we were able to determine the mean shape of the wire cross-section fitting them to theoretical values assuming a triangular wire cross-section shown in the inset of Fig. 11. In this figure we compare the experimental heights with their best fits. The heights of the satellites are determined by the Fourier transformation of the triangular cross-section of the wires. This transformation exhibits maxima lying in reciprocal space along the normals to the sides of this triangle. Thus, from the Q_z dependence of the positions of the side maxima of the envelope curves of the satellite heights (denoted by arrows in Fig. 11) we can determine the slope β of the sidewall of the wire. We found $\beta \approx 6°$

which is *larger* than the miscut of the substrate. Therefore, the sidewall of the wire must contain monolayer steps "upstairs" (see the right panel in Fig. 11). This structure of the wires cannot be created only by bunching of monolayer steps and another mechanism must be active moving the adatoms "upstairs".

3. GRAZING INCIDENCE DIFFRACTION

In contrast to GISAXS, grazing incidence diffraction (GID) is sensitive both to the shape of the objects and to the strain fields in the objects and around them (if they are buried).

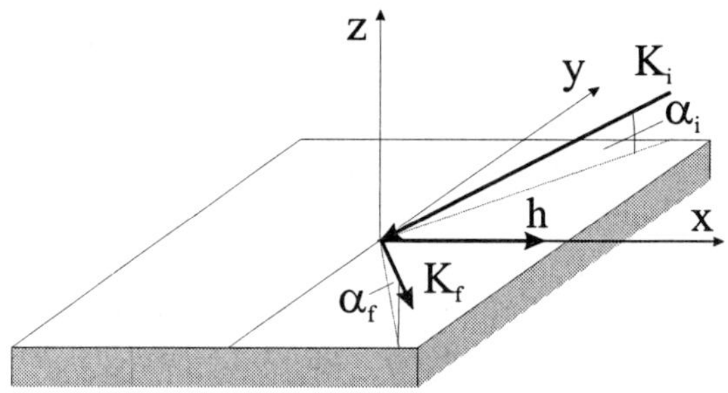

Figure 12. Sketch of the grazing incidence diffraction method.

A sketch of the GID method is shown in Fig. 12. The primary radiation collimated in two orthogonal directions hits the sample under a small angle α_i close to the critical angle α_c. The transmitted wave diffracts; in the symmetrical case (shown in the figure) the diffracting net planes are perpendicular to the sample surface and the diffraction vector **h** is parallel to it. The diffracted beam exits the sample under a small angle α_f. Usually, the diffracted radiation is measured by a PSD oriented perpendicularly to the surface so that the PSD spectrum corresponds to an α_f scan in reciprocal space.

The main problem in the interpretation of GID data is to distinguish between the scattering from the objects and from the strain fields around them. Both factors give rise to diffuse scattering accompanying the coherent diffraction maxima in reciprocal space. Under certain simplifying conditions (large lattice mismatch between these object and the host crystal, large size of the objects) these two factors can be distinguished.[23] However, in many cases a simulation of the diffuse scattering and a detailed comparison with experimental data are necessary.

The simulation of diffuse scattering in GID is based on the DWBA approach. Using this method, we divide the scattering potential $\hat{\mathbf{V}}$ of the sample into two parts, the first one $\hat{\mathbf{V}}_A$ corresponds to the perfect system without the interface objects, the second $\hat{\mathbf{V}}_A$ accounts for the objects. The scattering due to $\hat{\mathbf{V}}_B$ is calculated using two eigensolutions $|E_A^{(i,f)}\rangle$ of the wave equation with $\hat{\mathbf{V}}_A$.

In GISAXS, the undisturbed system is assumed amorphous, and these solutions are the transmitted waves with the wave vectors $\mathbf{K}_{ti,f}$; these waves are scattered from the objects. In GID the situation is more complicated. Since the undisturbed system (sample without the objects) is crystalline, the undisturbed wavefields $|E_A^{(i,f)}\rangle$ should also contain the waves diffracted in the undisturbed system.[24] The calculation of these wavefields is not trivial, since the undisturbed system is usually semiinfinite and the diffraction process is dynamic.

Including diffraction into the undisturbed wavefields we obtain additional scattering channels (scattering of the transmitted wave by the object and subsequent diffraction of the scattered wave, for instance). Since the diffraction condition is fulfilled in a very narrow region of reciprocal space, these additional channels would modify the intensity distribution only in certain points in reciprocal space. Therefore, this complicated ansatz is not necessary, in most cases. We can restrict ourselves to an *amorphous* undisturbed system and include its crystalline structure into $\hat{\mathbf{V}}_B$. Then, close to the reciprocal lattice point of the substrate, the differential cross-section of diffuse scattering can be described in a way similar to GISAXS:

$$\left(\frac{d\sigma}{d\omega}\right)_{\text{diff}} = \text{const.} \left|t_i t_f\right|^2 F(\mathbf{q_t}) G(\mathbf{q}_t), \tag{7}$$

where $\mathbf{q}_t$ is the reduced scattering vector $\mathbf{q} = \mathbf{Q} - \mathbf{h}$ corrected to refraction and the satellite intensity is now given by[25,26]

$$F(\mathbf{q}_t) = \left| \int_V d^3\mathbf{r}\, e^{-i\mathbf{q}_t.\mathbf{r}} \left[\chi_{\mathbf{h}}(\mathbf{r}) e^{-i\mathbf{h}.\mathbf{u}(\mathbf{r})} - \chi_{\mathbf{h}}^{\text{host}} \right] \right|^2. \tag{8}$$

Here we have denoted $\mathbf{u}(\mathbf{r})$ the displacement field caused by a single object, $\chi_{\mathbf{h}}(\mathbf{r})$ is the $\mathbf{h}$-th coefficient of the polarizability of the crystal with the objects, and $\chi_{\mathbf{h}}^{\text{host}}$ ist the constant polarizability coefficient of the host crystal. The shape of the objects enters the function $\chi_{\mathbf{h}}(\mathbf{r})$ since we assume that the polarizabilities of the objects and the host crystal are different. Equation (8) was derived under the assumption that only buried and not free standing objects are present in the sample and the displacement fields caused by all objects are same. The latter assumption might not be valid, especially if we take into account a surface relaxation of internal stresses caused by buried objects close to the surface. In this case, Eq. (8) should be slightly modified.[25]

For free standing quantum dots, a "brute force method"[27] has been developed for the simulation of diffuse scattering in GID. In this method, the scattered x-ray intensity was obtained by a numerical Fourier transformation of $\exp(-i\mathbf{h}.\mathbf{u}(\mathbf{r}))$. This approach made it possible to estimate the gradient of the chemical composition in a free standing dot from GID measurements.

The displacement field $\mathbf{u}(\mathbf{r})$ can be calculated using a numerical finite element method. If the objects are buried below a flat surface, the field can also be obtained by an analytical solution of the equilibrium equations[13,28,29]

$$\frac{\partial \sigma_{jk}}{\partial x_k} + f_k = 0, \quad j,k = x,y,z \tag{9}$$

under the boundary conditions

$$\sigma_{jk} n_k \big|_{\text{surface}} = 0. \tag{10}$$

Here we have denoted σ_{jk} the elements of the stress tensor, $\mathbf{n}$ is the surface normal vector, and

$$f_k = \text{const.} \frac{\partial}{\partial x_k} \left(\Delta(\mathbf{r}) \Omega(\mathbf{r}) \right)$$

is the density of the volume force caused by the lattice mismatch $\Delta(\mathbf{r})$ in the object with the shape function $\Omega(\mathbf{r})$. The analytical solution of the elasticity problem is much simpler than the finite element method and makes it possible to change easily the parameters of the objects during the simulation or fitting the experimental data.

3.1. Grazing incidence diffraction from SiGe quantum dots in a SiGe/Si superlattice

The GID method was used for the investigation of SiGe quantum dots in a 10 period SiGe/Si superlattice, each period consisted of an 2.1 nm $Si_{0.5}Ge_{0.5}$ layer and 20 nm thick Si spacer. TEM investigations revealed vertically arranged islands (dots) at the SiGe/Si interfaces, the diameter of the islands was abot 80 nm and the height 2 nm. The GID experiments were performed at the wiggler beamline BW2 at HASYLAB, Hamburg. The wavelength of the parallel and monochromatic incident beam was $\lambda = 0.13627$ nm, the resolution was increased by using a Si(111) analyzer crystal. The experiments are described in detail elsewhere.[26]

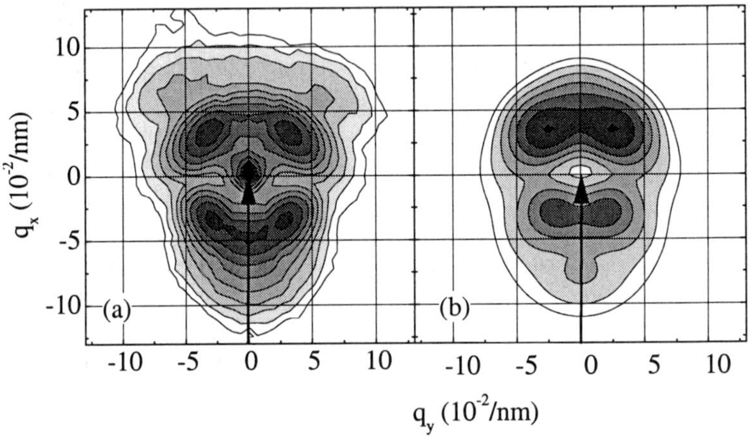

Figure 13. The measured (a) and simulated (b) $q_x q_y$ distribution of the scattered intensity, 220 GID diffraction. The arrows indicate the direction of the diffraction vector.

In Fig. 13 we show the distribution of the diffracted intensity in the $(q_x q_y)$ plane measured in GID diffraction 220 for constant incidence angle $\alpha_i = \alpha_c$. In this geometry the information depth is smaller than the superlattice period so that the primary beam does not irradiate the islands at all. The diffuse scattering takes place at local strain fields propagating from the dots to the surface. Apart from the central coherent maximum (the "nose" of the picture), four lateral satellite maxima appear indicating that the islands are arranged in a disordered $\langle 100 \rangle$ array with the period about 110 nm. From the widths of these maxima the r.m.s. deviation about 30 nm from the island periodicity was deduced. The elongation of the intensity pattern towards the origin of reciprocal space (the "chin" of the face in the negative q_x coordinates) indicates a prevalent tensile tensile deformation of the lattice.

The simulation was performed using the above theory. From the comparison of the simulated and measured maps we estimated the mean Ge content in the dots, we found about 45% of Ge so that the intermixing of Si and Ge int he dots during their overgrowth was rather small.

3.2. Grazing incidence diffraction from self-assembled quantum wires in a SiGe/Si superlattice

The same samples investigated by GISAXS in Sect. 2.3have also been studied by GID with the purpose to estimate the mean chemical composition of the wires. The study is still in progress, here we present our first results.

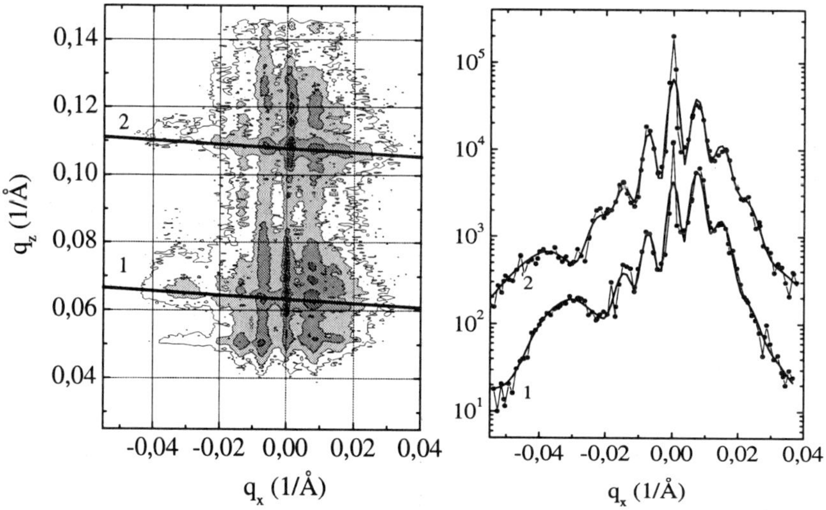

Figure 14. The intensity distribution in the longitudinal $(q_x q_z)$ plane parallel to the diffraction vector measured in 400 GID diffraction at the SiGe/Si superlattice with quantum wires (left panel). From the two dimensional map, the linear scans have been extracted along the lines 1 and 2, the scans are plotted in the right panel (dots) along with their fits (full lines).

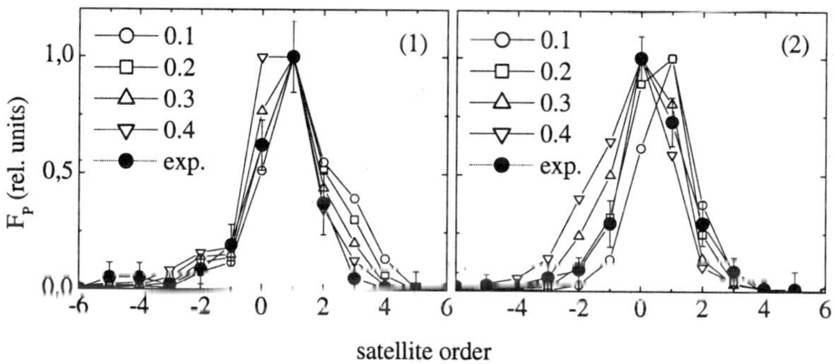

Figure 15. The satelite heights F_P determined from the measured data (full points) and their simulations with various Ge concentrations x_{Ge} in the wires ranging from 0.1 to 0.4. The panels (1) and (2) show the data for the linear scans extracted along the lines 1 and 2, respectively, in Fig. 14.

The GID measurements have been performed at TROÏKA II beamline at ESRF, Grenoble using a wavelength of $\lambda = 1.5545$ Å and a diamond 111 double crystal monochromator. The scattered intensity was recorded using a linear position sensitive detector (PSD) placed in the perpendicular direction to the sample surface. In a single scan we have measured the intensity distribution in the $q_x q_z$ plane parallel to the [400] diffraction vector (so called longitudinal scan, Fig. 14) keeping constant the incidence angle $\alpha_i = 0.5^\circ$. Since this plane was perpendicular to the wires, lateral satellite maxima can be seen similarly to GISAXS. However, the intensity of these maxima is influenced by the strains around the wires.

From the map we have extracted linear scans shown in Fig. 14 as well. For the fitting of these scans we have used the same approach as in GISAXS, namely Eq. (6) and we assumed a short-range order of the wire positions. From the fit we have obtained the heights F_P of the satellites. Then, we have compared these heights with the results of the simulations based on Eq. (8) and on the strain calculation according to Eqs. (9,10). For the shape of the wires we used the results of the GISAXS study (see Sect. 2.3) and the only adjustable parameter was the mean Ge content x_{Ge} in the wires. The results are presented in Fig. 15. It can be seen that the satellites with negative orders are higher due to a prevalent tensile lateral deformation of the wire structure. The relatively best correspondence of the measured heights of the satellite maxima with the simulations was achieved for $x_{Ge} \approx 0.2$ indicating that a considerable intermixing takes place during the overgrowth of the wires.

4. SUMMARY

A combination of grazing incidence small angle x-ray scattering (GISAXS) with grazing incidence diffraction (GID) makes it possible to characterize very small self-assembled wires and islands at interfaces of semiconductor superlattices. GISAXS is sensitive only to the shape and positions of these objects, whereas GID detects mainly the strain fields. Knowing the shape of the objects from GISAXS and using it as a fixed parameter, we are able to estimate the chemical composition of the buried objects from the comparison of GID data with the simulations. Compared to other structure sensitive methods such as AFM or TEM, x-ray scattering is non-destructive and it investigates also buried objects. In contract to these methods, x-ray scattering is non-local, i.e. it yields information on the objects averaged over a macroscopic sample volume.

ACKNOWLEDGEMENT

The work has been supported Ministry of Education of Czech Republic (VS96102), by Grant Agency of Czech Republic (202/00/0354), and by by FWF, Vienna. We acknowledge the assistance of the staff of the TROÏKA beamline at ESRF with the beamline setup, namely O. Konovalov and D. Smilgies. One of us (V. H.) acknowledges the travel support by the University of Houston and Michigan State University.

REFERENCES

1. D. Bimberg, M. Grundmann, and N. N. Ledentsov, *Quantrum Dot Heterostructures*, (Wiley, Chichester 1999).
2. J. Tersoff, Y. H. Phang, Z. Zhang, and M. G. Lagally, *Phys. Rev. Lett.* **75**, 2730–2733 (1995).
3. V. Holý, A. A. Darhuber, J. Stangl, G. Bauer, J. Nutzel, and G. Abstreiter, *Phys. Rev. B* **57**, 12435–12442 (1998).
4. V. Holý, A. A. Darhuber, J. Stangl, G. Bauer, J. Nutzel, and G. Abstreiter, *Semicond. Sci. Technol.* **13**, 590–598 (1998).
5. C. Schelling, G. Springholz, and F. Schäffler, *Phys. Rev. Lett.* **83**, 995–998 (1999).
6. R. L. Schwoebel and E. J. Shipsey, *J. Appl. Phys.* **37**, 3682–3689 (1966).
7. O. G. Schmidt and K. Eberl, *Phys. Rev. B* **61**, 13721–13729 (2000).
8. G. Medeiros-Ribeiro, A. M. Bratkovski, T. L. Kamins, D. A. A. Ohlberg, and R. S. Williams, *Science* **279**, 353–355 (1998).
9. M. Strassburg, V. Kutzer, U. W. Pohl, A. Hoffmann, I. Broser, N. N. Ledentsov, D. Bimberg, A. Rosenauer, U. Fischer, D. Gerthsen, I. L. Krestnikov, M. V. Maximov, P. S. Kop'ev, and Zh. I. Alferov, *Appl. Phys. Lett.* **72**, 942–944 (1998).
10. M. Schmidbauer, T. Wiebach, H. Raidt, M. Hanke, R. Köhler, and H. Wawra, *Phys. Rev. B* **58**, 10523–10531 (1998).
11. J. Tersoff, C. Teichert, and M. G. Lagally, *Phys. Rev. Lett.* **76**, 1675–1678 (1996).
12. V. A. Shchukin, D. Bimberg, V. G. Malyshkin, and N. N. Ledentsov, *Phys. Rev. B* **57**, 12262–12274 (1998).
13. V. Holý, G. Springholz, M. Pinczolits, and G. Bauer, *Phys. Rev. Lett.* **83**, 356–359 (1999).
14. G. Springholz G., V. Holý, M. Pinczolits, and G. Bauer, *Science* **282**, 734–736 (1998).

15. S. K. Sinha, E. B. Sirota, S. Garoff, and H. B. Stanley, *Phys. Rev. B* **38**, 2297–2311 (1988).
16. M. Rauscher, R. Paniago, T. H. Metzger, Z. Kovats, J. Domke, J. Peisl, H. D. Pfanner, J. Schultze, and I. Eisele, *J. Appl. Phys.* **86**, 6763–6769 (1999).
17. Z. Kovacs, M. Rauscher, H. Metzger, J. Peisl, R. Paniago, H.D. Pfannes, J. Schulze, I. Eisele, F. Boscherini, and S. Ferrer, *Phys. Rev. B* **62**, 8223-8231 (2000).
18. J. Stangl, V. Holý, T. Roch, A. Daniel, G. Bauer, J. Zhu, K. Brunner, and G. Abstreiter, *Phys. Rev. B* **62**, 7229–7236 (2000).
19. Jian-hong Zhu, K. Brunner, and G. Abstreiter, *Appl. Phys. Lett.* **73**, 620–622 (1998).
20. P. R. Pukite, C. S. Lent, and P. I. Cohen, *Surf. Sci.* **161**, 39–53 (1985).
21. V. Holý, J. Stangl, G. Springholz, M. Pinczolits, and G. Bauer, *J. Phys. D: Applied Physics* **34**, A1–A5 (2001).
22. V. Holý, T. Roch, J. Stangl, A. Daniel, G. Bauer, T. H. Metzger, Y. H. Zhu, K. Brunner, and G. Abstreiter, *Phys. Rev. B* **63**, 205318–10 (2001).
23. I. Kegel, T. H. Metzger, A. Lorke, J. Peisl, J. Stangl, G. Bauer, J. M. García, and P. M. Petroff, *Phys. Rev. Lett.* **85**, 1694–1697 (2000).
24. A. Daniel, V. Holy, Y. Zhuang, T. Roch, J. Grenzer, Z. Bochníček, and G. Bauer, *J. Phys. D: Applied Physics* **34**, A197–A202 (2001).
25. M. A. Krivoglaz, *X-Ray and Neutron Scattering in Nonideal Crystals*, Vol. I, (Springer, Berlin, 1996).
26. V. Holý, A. A. Darhuber, J. Stangl, S. Zerlauth, F. Schäffler, G. Bauer, N. Darowski, D. Luebbert, U. Pietsch, and I. Vávra, *Phys. Rev. B* **58**, 7934–7943 (1998).
27. T. Wiebach, M. Schmidbauer, M. Hanke, H. Raidt, R. Köhler, and H. Wawra, *Phys. Rev. B* **61**, 5571–5578 (2000).
28. J. W. Flocken and J. R. Hardy, *Phys. Rev. B* **1**, 2447–2455 (1970).
29. R. A. Matsumura and G. Sines, *J. Appl. Phys.* **41**, 3930–3936 (1970).

TOTAL SCATTERING AND REVERSE MONTE CARLO MODELLING OF DISORDERED CRYSTALLINE MATERIALS

Matthew G. Tucker, Martin T. Dove, and David A. Keen*

1. INTRODUCTION

The term disordered crystalline materials may seem a contradiction, since crystalline materials are generally considered as an ordered array of atoms with varying degrees of symmetry. This of course is the correct description of the structure when averaged over all the instantaneous position of the atoms in the material. In a standard Rietveld analysis the variation from these average positions is often accounted for by the use of three-dimensional Gaussian functions, where the widths of these functions are called the atomic displacement parameters and the average atom positions are taken to be the mid points of these distributions. These overall distributions are usually referred to as the temperature factors, since they allows for the effects of thermal motion on the average structure. For most crystalline materials this method describes the overall structure of the material very well. However, if the average structure of the crystalline material can accommodate orientational disorder within these temperature factors, as is often indicated if they are highly anisotropic, then the local structure of the material needs to be determined to obtain a true model of the material. Materials which can be described by coordinated polyhedra, such as silica, can often accommodate rotations of these polyhedra while maintaining the overall average structure. This can lead to disordered crystalline materials and, as will be illustrated later, can be the key to understanding the true form of the high temperature structures of this type of materials.

There are a number of experimental techniques that are used to investigate the local structures of materials and many are discussed in this book. The technique employed here is total scattering. This is a standard powder diffraction experiment but, instead of fitting and subtracting the background that lies under the Bragg scattering, the whole spectrum is carefully corrected and fitted. The Bragg peaks provide information on the average structure and the background, called diffuse scattering, provides information on the local structure of the material. In the case of disordered crystalline materials the diffuse scattering

*Matthew G. Tucker and Martin T. Dove, Mineral Physics Group, Department of Earth Sciences, University of Cambridge, Downing Street, Cambridge, CB2 3EQ, UK, MGT: mtuc98@esc.cam.ac.uk, MTD: martin@esc.cam.ac.uk. David A. Keen, Physics Department, Oxford University, Clarendon Laboratory, Parks Road, Oxford, OX1 3PU, UK and ISIS facility, Rutherford Appleton Laboratory, Chilton, Didcot, Oxfordshire, OX11 0QX, UK, d.a.keen@rl.ac.uk.

is often clearly visible in the scattering spectrum as an oscillating background upon which the Bragg peaks rest. In order to characterize the local structure accurately a high spatial resolution is required. The spatial resolution Δr obtainable from scattering data is given by equation 1, where Q_{max} is the largest value of the scattering vector Q used in the measurement and defined in equation 2. Here θ is the scattering angle and λ is the wavelength of the radiation used.

$$\Delta r = 2\pi/Q_{\max} \quad (1)$$

$$Q = 4\pi \sin\theta/\lambda \quad (2)$$

It is clear from equation 1 and 2 that a large Q_{max} and thus a small λ are required for a high spatial resolution. With a lab based X-ray diffractometer using Cu K_α radiation a $Q_{max} \sim 8Å^{-1}$ and a maximum $\Delta r \sim 0.77$ Å are available. A typical bond length of Si–O in silica is about 1.61 Å, so the resolution would only be $\sim$ 50% of the bond length. Better resolution can be obtained by using Mo anodes or even better using synchrotron radatation. However, since the form factor for X-rays is Q dependent, even when high Q values are obtained, unless great care is taken, the useful data are often so low in amplitude that they are masked by the background. In the past few years total scattering machines have been developed at synchrotrons and they will be a source of useful total scattering data.

An alternative approach to obtain high spatial resolution data is to use neutron radiation. Using a reactor source, neutron wavelengths of 0.5Å are common and this gives a $Q_{\max} \sim 25Å^{-1}$ and a $\Delta r \sim 0.25$Å. This is a reasonable resolution, however, using a spallation source of neutrons, such as at ISIS in the UK, wavelengths of < 0.25 Å are available which give a $Q_{\max} > 50Å^{-1}$ and a $\Delta r > 0.13$Å(or $< 8\%$ of an Si–O bond). At a spallation source neutrons are produced by bombarding a heavy metal target with high energy protons; the collision produces neutrons with a wide range of energies and wavelengths. The disadvantage of this methods is that the overall flux of neutrons produced is much lower than at a reactor source, but this can be overcome by using the time-of-flight (tof) method. Instead of using a monochromator to select the required wavelength, a white beam of neutron is used. The time the neutrons take to reach the detectors is measured, and since the total flight path of the neutrons is known, the energy and wavelength of neutrons reaching the detectors at any particular time is known.

The tof method also differs from more conventional diffraction methods in the way the data is collected. Normally, in order to measure data over the whole range of Q, the detector is moved to vary the diffraction angle θ in equation 2 with a fixed wavelength, λ. However in tof diffraction the wavelength, λ, varies as the energy of the neutron varies and thus each detector "sees" the whole wavelength range from the source and collects data for a range of Q values. This means λ in equation 2 is varying and θ can be held constant. The result of this is that a tof diffractometer has a collection of detectors fixed at a variety of angles. In theory just one detector at each angle would be enough but, to increase the counting statistics, arrays of detectors are used and grouped together into banks around different nominal angles.

When using total scattering to investigate the structure of materials, it is common to concentrate on just one aspect of the data, either the Bragg or the diffuse part of the data, and hence to investigate the average or the local structure. However, since a true model of the material must satisfy both parts of the data simultaneously, it is more satisfactory to fit both aspects of the data with a single model. A very powerful technique for doing this is reverse Monte Carlo (RMC) modelling.

The RMC code used here was first developed by McGreevy and Pusztai[1] as an approach for producing three-dimensional atomic configuration from scattering data. Initially, the technique was applied to the study of liquid and amorphous materials, where previously the only way to obtain models was through the use of interatomic potentials and not from experimental data. It was realized that from a configuration of atoms a radial distribution function can be calculated. In simple terms this involves counting the number of atoms of each type with increasing r in a small range dr, summed over all central atom positions. The radial distribution function is useful since the Fourier transform is directly related to the total scattering measured. Using this information, and the Monte Carlo method to move the atoms in the configuration randomly until the calculated scattering matches the measured data, a three-dimensional atomic configuration that is consistent with the data is obtained. This technique need not be limited to liquid and amorphous materials but, at least in principle, is capable of being used as a powerful tool for modelling of crystalline systems.

We have modified RMC to take account of all the useful information obtainable from total scattering from crystalline materials. This will be described and the application of the new codes will be illustrated using some of our recent results. It is hoped that the reader will see the potential of the method and will be encouraged to try the technique for themselves. If they have seen diffuse scattering in their data or wish to gain some local information where they already have the average structure, then they should find the techniques outlined here very useful.

2. NEUTRON TOTAL SCATTERING

Although the methods outlined in this chapter are applicable to both X-ray and neutron total scattering, the development has concentrated on utilizing neutrons from a spallation source using the time-of-flight method. This is due to the high Q range obtainable and therefore the high spatial resolution as discussed above.

2.1. Total scattering formalism

It is useful at this point to introduce the basic equations of scattering from an isotropic material. Here the formalism reviewed by Keen[2] is followed. The total scattering intensity is given by

$$\frac{1}{N}\frac{\mathrm{d}\sigma}{\mathrm{d}\Omega} = i(Q) + \sum_{j=1}^{n} c_j \overline{b_j^2} \tag{3}$$

where $4\pi \sum_j c_j \overline{b_j^2}$ is the total scattering cross-section of the material and the summation is over the n atom types. There are N atoms in the material and c_j is the proportion of atom type j. $i(Q)$ is related to the total radial distribution function, $G(r)$, by the Fourier transforms

$$i(Q) = \rho_0 \int_0^\infty 4\pi r^2 G(r) \frac{\sin Qr}{Qr} dr \tag{4}$$

$$G(r) = \frac{1}{(2\pi)^3 \rho_0} \int_0^\infty 4\pi Q^2 i(Q) \frac{\sin Qr}{Qr} \mathrm{d}Q \tag{5}$$

with average atom number density $\rho_0 = N/V$ (in atoms/Å^3). $G(r)$ is defined by the partial radial distribution functions $g_{ij}(r)$:

$$G(r) = \sum_{i,j=1}^{n} c_i c_j \bar{b}_i \bar{b}_j (g_{ij}(r) - 1) \tag{6}$$

where

$$g_{ij}(r) = \frac{n_{ij}(r)}{4\pi r^2 \mathrm{d}r\rho_j} \tag{7}$$

$n_{ij}(r)$ are the number of particles of type j between distances r and $r + \mathrm{d}r$ from a particle of type i, and $\rho_j = c_j\rho_0$. Two other correlation functions are commonly used to represent the information contained in $G(r)$:

$$D(r) = 4\pi r\rho_0 G(r) \tag{8}$$

$$T(r) = 4\pi r\rho_0 \left(G(r) + \left[\sum_{i=1}^{n} c_i \bar{b}_i \right]^2 \right) \tag{9}$$

These are often preferred to $G(r)$ because the important quantity in both expressions, $rG(r)$, is the primary quantity involved in the Fourier transform (equations 4 and 5) of the the experimental quantity $Qi(Q)$.

Equations 4 and 5 are at the heart of the RMC method since $G(r)$ is the quantity calculated from the RMC atom configuration and $i(Q)$ is the quantity measured and therefore fitted.

2.2. Experimental procedure

The samples required for total scattering measurements are finely ground powders as in a standard powder diffraction experiment. The sample is usually contained in a cylindrical thin-walled vanadium can, although if the sample reacts with vanadium, silica glass may be used. Vanadium is preferred since it scatters almost completely incoherently, no Bragg peaks appear in the data from the sample container and, since thin cans can be made, the overall scattering from the can is low.

Typically the experiment requires heating or cooling of the sample, so the can is placed in a furnace, closed circuit refrigerator (CCR) or cryostat. In order to obtain a corrected $S(Q)$ from the sample as in equation 3 several of extra data sets need to be collected to characterize the neutron absorption and scattering from the material surrounding the sample. These measurements typically include an empty can within the CCR or furnace, an empty CCR or furnace, an empty instrument and a standard vanadium rod. The latter measurement is required to remove the moderator profile from the data and to normalize the data. Finally, the density and volume of powder in the can must be measured.

These measurements are used to produce a corrected $S(Q)$ for the sample. We use the CORRECT routines produced by Howe et al.[3] Since we can calculate the self scattering (the last term in equation 3), we can use this as an internal check of the corrections at high Q, since $S(Q \to \infty)$ oscillates about this value. Once the corrected $S(Q)$ is produced the self scattering is subtracted to give $i(Q)$ which can then be used in the RMC methods.

3. REVERSE MONTE CARLO

3.1. Standard RMC

As mentioned in the introduction, the RMC code used here was developed by McGreevy and Pusztai and has been described by themselves and others in a number of articles.[1,4–6] Generally the method begins with the production of a three dimensional configuration of atoms with periodic boundary conditions. This can either be the average crystalline structure as is the case here or it could be some other arrangement of atoms, or completely random. An important constraint on this starting configuration is that the starting density and ratio of atoms must be correct, since these will not be changed during the refinement procedure and so must be correct from the outset. As with the standard Monte Carlo technique, an "energy" function is minimized during the refinement process. In RMC a total χ^2 is defined as a sum of various individual χ^2 terms for each of the data being fitted. One such χ^2 is given in equation 10.

$$\chi^2_{i(Q)} = \sum_j [i_{\text{calc}}(Q_j) - i_{\text{expt}}(Q_j)]^2/\sigma^2_{i(Q)} \tag{10}$$

The atoms of the configuration are then moved at random by a random amount (within a set maximum) and provided the move has not resulted in atoms being closer than preset closest approaches the new value of χ^2 is calculated. If χ^2 has decreased the move is accepted. If χ^2 has increased the move is accepted with the probability $\exp(-\Delta\chi^2/2)$, this ensures local minima are avoided. This procedure continues until χ^2 is oscillating about a minimum value for the set of σ's used in the χ^2's. If the fits to the data are not acceptable, the σ's can be reduced and the procedure continued until acceptable fits are obtained.

When fitting $i(Q)$ as in equation 10 it is clear that $i(Q)_{\text{calc}}$ must be calculated from the configuration of atoms. This is done by calculating the total $G(r)$ for the configuration and then Fourier transforming this as in equation 5. However, since the $G(r)$ can only be calculated to $r = L/2$ (where L is the shortest configuration dimension), $i(Q)_{\text{calc}}$ will contain Fourier transform ripples related to this maximum r. This problem can be overcome by convoluting the measured data $i(Q)_{\text{expt}}$ with the Fourier transform of the configuration box function:

$$i^{\text{conv}}_{\text{exp}}(Q) = \frac{1}{\pi}\int_{-\infty}^{+\infty} i_{\text{exp}}(Q')\frac{\sin((Q-Q')L/2)}{Q-Q'}\,dQ' \tag{11}$$

This then facilitates a comparison of the calculated and measured $i(Q)$ data.

The total χ^2 can include any real or reciprocal space function that can be calculated from the configuration and not just $i(Q)$; in fact any known information that can be obtained from the model can be fitted, such as bond lengths and angles. This fact is used to enhance the RMC method for the study of disordered crystalline materials.

3.2. RMCprofile

RMCprofile[7,8] is the name we have given to our version of the RMC program. At its core is the original RMCA program, which has been modified to take advantage of the additional information available when studying crystalline systems. The most notable change is the inclusion of the Bragg data in the fitting process. This has been done in two ways. The Bragg intensities may be extracted from the profile using the Pawley method,[9] and the following χ^2 defined.

$$\chi^2_{\text{Bragg}} = \sum_{hk\ell} |\alpha I^{\text{exp}}_{\text{Bragg}}(hk\ell) - I^{\text{calc}}_{\text{Bragg}}(hk\ell)|^2/\alpha^2\sigma(hk\ell)^2 \tag{12}$$

$$\alpha = \frac{\sum_{hk\ell} I^{\text{calc}}_{\text{Bragg}}(hk\ell) I^{\text{exp}}_{\text{Bragg}}(hk\ell)/\sigma(hk\ell)^2}{\sum_{hk\ell} I^{\text{calc}}_{\text{Bragg}}(hk\ell)^2/\sigma(hk\ell)^2} \tag{13}$$

Where $I^{\text{calc}}_{\text{Bragg}}(hk\ell)$ is calculated from the configuration, $\sigma(hk\ell)^2$ is obtained from the Pawley method and α is a scale factor as defined in equation 13.

An alternative approach is to fit the whole profile during the RMC refinement. This overcomes the difficulty of extracting the bragg intensities of overlapping Bragg peaks, and more faithfully reproduces the Reitveld refinement results. Initially the data is refined using a Reitveld refinement program (for example, GSAS). The lattice parameters, peak shape parameters and the background function are supplied to RMCprofile and the intensities are then calculated from the configuration. The differences between the calculated and experimental profile are then used to define a χ^2 in the normal way:

$$\chi^2_{\text{profile}} = \sum_j \left| I_{\text{expt}}(t_j) - s' I_{\text{calc}}(t_j) \right|^2 / \sigma_I(t_j)^2 \tag{14}$$

Here $I_{\text{expt}}(t)$ is the GSAS data on a time-of-flight scale, t is the neutron flight time, s' is a scale factor. $I_{\text{calc}}(t)$ is defined

$$I_{\text{calc}}(t_j) = \sum_{hk\ell} L_{hk\ell} |F(hk\ell)|^2 R_{hk\ell}(t_j - t_{hk\ell}) + B(t_j) \tag{15}$$

where the structure factor is

$$F_{\text{expt}}(hk\ell) = \frac{1}{N} \sum_j \bar{b}_j \exp(i\mathbf{Q}_{hk\ell} \cdot \mathbf{r}_j) \tag{16}$$

$L_{hk\ell}$ is the Lorentz factor of the $hk\ell$ reflection, $R_{hk\ell}(t_j - t_{hk\ell})$ is the peak shape of the $hk\ell$ reflection arising from the instrumental resolution, and $B(t)$ is the background in the diffraction pattern, which arises from the diffuse scattering component of the total scattering and is treated here in a more conventional way as a polynomial background to be subtracted.

In addition to the inclusion of the Bragg profile, the RMC code has been modified to include bond and angle constraints developed by Keen.[5] These can be used to constrain the system to consist of polyhedra. An example of the additional χ^2 term required, in this case for SiO_4 tetrahedron, is

$$\chi^2_{\text{poly}} = \sum_{\text{Si–O}} |r_{\text{Si–O}} - R_{\text{Si–O}}|^2 / \sigma^2_{\text{Si–O}} + \sum_{\text{O–Si–O}} |\theta_{\text{O–Si–O}} - \Theta_{\text{O–Si–O}}|^2 / \sigma^2_{\text{O–Si–O}} \tag{17}$$

The constraints can be based on data if the bond lengths are taken from the relevant peaks in the $T(r)$. Also the distribution about this mean value, which is determine by $\sigma_{\text{Si–O}}$ in equation 17, can be found from the T(r) peak width.

The final modification to the RMC code that makes up RMCprofile is the inclusion of the instrument peak profile when fitting $i(Q)$ data. As mentioned earlier we have concentrated on the tof method so it is these peak shapes we have used. The calculated $i(Q)$ must be convoluted with the peak shape for a fair comparison the experimental $i(Q)$. This is not a straightforward convolution since the peak shape function is itself a function of Q, so every Q point in the data has its own peak shape function. The convolution has to be performed every atom move, and this can be very time consuming when multiple $i(Q)$ data sets are being fitted. Therefore, it might be desirable to fit the data without initially convolution and only when the fit starts to converge to then begin convoluting.

3.3. MCGRtof

As mentioned in the introduction and shown by equations 4 and 5, the real space information $G(r)$ is obtained by Fourier transforming the measured $i(Q)$. The direct Fourier transform of the $i(Q)$ can introduce transformation ripples into the $G(r)$ if the maximum Q used is not very high, particularly if the $i(Q)$ is not close to zero when Q_{max} is reached.

This can be overcome to some extent by multiplying the $i(Q)$ with a function which comes smoothly to zero at Q_{max}. This has the result of reducing the termination ripples, but with the cost of broadening the features in the $G(r)$. An alternative approach is to use a inverse Monte Carlo technique such as MCGR developed by Pusztai and McGreevy.[10] Here the $G(r)$ is treated as a histogram of points which can be calculated to any value of r. The points are then moved at random by a random amount, the Fourier transform is performed, and the resulting $i(Q)$ is compared to the measured $i(Q)$. In this way, using the RMC algorithm, a $G(r)$ that fits the data is produced, without the need to Fourier transform the experimental $i(Q)$. Since the $G(r)$ can be calculated over any range of r, no trunction ripples are produced. In addition to removing the trunction ripples, MCGR can also help produce a more physically meaningful $G(r)$ since regions can be held at zero, such as at r lengths up to the closest approach distance in the material under investigation, and also kept positive if required.

Often high values of Q are obtained with tof data, so it might seem that the direct Fourier transform method is sufficient for obtaining $G(r)$. However, as mentioned in the introduction, when using the tof method the data are obtained in spectra for each bank of detectors. In the past these different banks were merge into a single data set. This is not very satisfactory, since each bank of data will have a different instrument resolution function. So, for example, features in the low angle banks will be broad whereas the same features in higher angle banks will be sharper. To cope with this problem, it was common to choose parts of each spectra which did not overlap and combine them to produce a single $i(Q)$ spectra. Using MCGR with the instrument resolution function, it is possible to produce a $G(r)$ that fits the data from all the banks of detectors simultaneously. The $G(r)$ is Fourier transformed and then convoluted with the resolution function for a particular detector bank, and then compared to the measured data. This can be done for all the banks simultaneously using the MCGR method, to give a $G(r)$ that fits all the data. We have modified the existing MCGR code to include this feature and have called it MCGRtof.[11]

3.4. RMC methodology

It is useful here to set out how the data is analyzed and RMC used to obtain useful information about disordered crystalline systems. So far we have discussed the method and programs used, but have not made clear how and when each part is brought into play. We will endeavor to briefly explain this now.

Once the data have been collected, they are processed in two parallel ways. Data suitable for Rietveld analysis are produced and refined using GSAS (or any other Rietveld program). The refinement results provide the starting configuration for RMCprofile, the peak shape parameters, the profile background parameters, and the diffraction profile to be fitted. At the same time the data are processed as described above to produce a corrected $i(Q)$. MCGRtof is then used to produce a $G(r)$ using the $i(Q)$ data, and the density from the Rietveld refinement.

Everything is now in place to run RMCprofile. Initially just the polyhedra constraints and Bragg profile are fitted. This moves the starting configuration away from the average structure quickly but without the system becoming unrealistic, since the average structure and polyhedra are being maintained. Then the $i(Q)$ data (from each detector bank used), and the $G(r)$ are then used in the fit. The $G(r)$ is converted to $T(r)$ in RMCprofile and this is fitted, since it is more closely related to the function that is Fourier transformed to produce $i(Q)$. Initially the calculated $i(Q)$ are not convoluted with the instrument resolution functions, but once a minimum in the total χ^2 is approached this option is invoked. By the end of this procedure a three dimensional configuration of atoms that is consistent will all

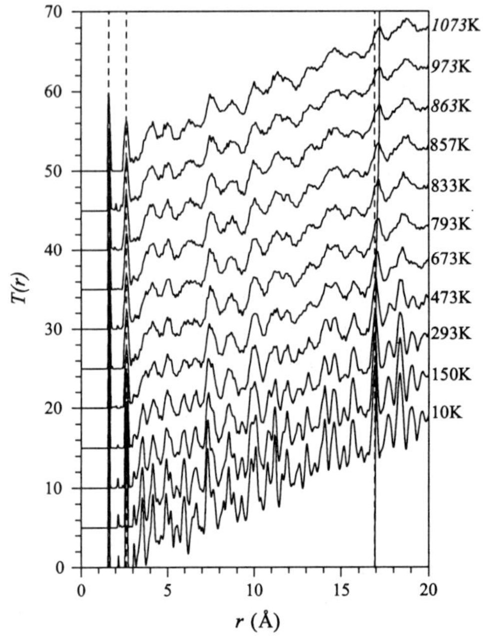

Figure 1. Neutron $T(r)$ functions of quartz over a wide range of temperatures. The dashed lines through the two lowest-r peaks follow the temperature evolution of the Si–O and O–O bond distances, and the full line through the peak at around 17 Å follows an interatomic distance that scales with the average linear thermal expansion. There is a progressive error on our temperature measurements; the nominal measurement of 858 K corresponds to the phase transition.

the data and constraints will be produced. If the fits to particular data sets are not of an acceptable level then it may be necessary to change the relevant σ and run RMCProfile until a new minimum is reached.

The following sections will demonstrate the results that can be obtained using this procedure by looking at three systems we have recently studied. For the first two examples, quartz and cristobalite, the version of RMCprofile which used the extracted Bragg intensities as discussed above was used, and the final example, SF_6, used the version of RMCprofile which fits the Bragg profile.

4. QUARTZ

4.1. Introduction

There were three main objectives for our study of quartz; to investigate the α–β phase transition in quartz from local and long range order perspectives; to produce atomic models that are consistent with the Bragg and diffuse data; and to obtain short-range information from the models. Since this was the first study using this method the aim was also to show the method worked and to develop it where necessary. This study is described in detail elsewhere,[12,13] here we will just give a brief introduction and summary of the results.

We began the study by collecting total scattering data for quartz over the temperature range of 20 to 1073 K. This was done on the now decomissioned LAD diffractometer at

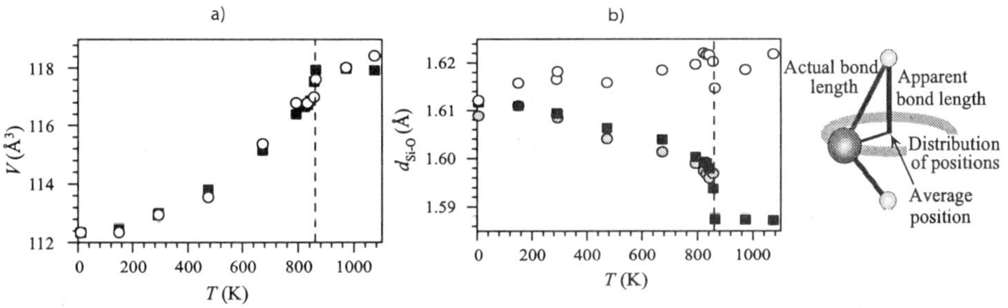

Figure 2. Temperature dependence of the a) unit cell volume and b) Si–O distance. The vertical dashed line shows T_C. In a) data from the Rietveld refinement and the 17 Å peak in $T(r)$ are given by squares and open circles respectively. In b) squares and filled circles represent the distances between average positions, $\langle Si \rangle$–$\langle O \rangle$ given by Rietveld refinement and the RMC respectively, and the open circles represent the average values of the instantaneous bond length, $\langle Si\text{–}O \rangle$ from RMC. The cartoon on the right shows the origin of the shortening of the distance between the mean positions of the atoms compared to the true bond length.

ISIS, UK. The data for each of the 14 measured temperatures were processed as described above and corrected $i(Q)$ obtained. $G(r)$ were produced using MCGR and the data were fitted using the Rietveld method. Finally RMC was run starting from configurations which were $10 \times 10 \times 10$ super cells of the orthorhombic representation of the quartz unit cell obtained from the Rietveld analysis.

4.2. RMC results for Quartz

Figure 1 shows the $T(r)$'s obtained for the various temperatures measured. It is clear from the fact the first few peaks in the $T(r)$'s do not change, apart from a slight broadening with increasing temperature, that the tetrahedra remain relatively unaffected over the whole temperature range despite the structure undergoing a phase transition. The features at a more medium range of r become broader and less intense as the temperature is increased showing the effects of thermal disorder on the medium range order in the structure. At high r there are features that show expansion that is consistent with the volume expansion of the average structure with increasing temperature. One such feature is illustrated with the full line in figure 1 at r 17 Å, where the dashed line indicates the position of the centre of the peak if no expansion was evident with temperature. Probably the most remarkable fact shown by figure 1 is the general lack of influence the phase transition has on the local structure. The top three $T(r)$'s in figure 1 are from quartz in the β-phase, whereas the rest are of α-phase quartz. There is no sharp change when moving from one phase to the other on increasing temperature, instead a gradual increase in disorder with increasing temperature is seen, beginning at temperatures much lower than T_C.

The unit cell volume from the average rietveld structure (filled squares) and calculated from the position of the 17 Å peak in the $T(r)$'s (open circles) are shown in Figure 2a. This clearly shows the average and local structure information are consistent. The graph in figure 2b shows the Si–O bond from the Rietveld structure (filled squares), the average atomic position in the RMC configurations (filled circles) and the average Si–O distance from the actual atom positions (open circle). The Rietveld bond distance appears to contract with increasing temperature, until T_C is reached and above which the bond length remains constant. The bond distance obtained from the average atomic positions in the RMC configuration shows the same contraction. This contraction is not real and indeed is not physical, since the overall system is expanding as shown by the volume of the unit cell.

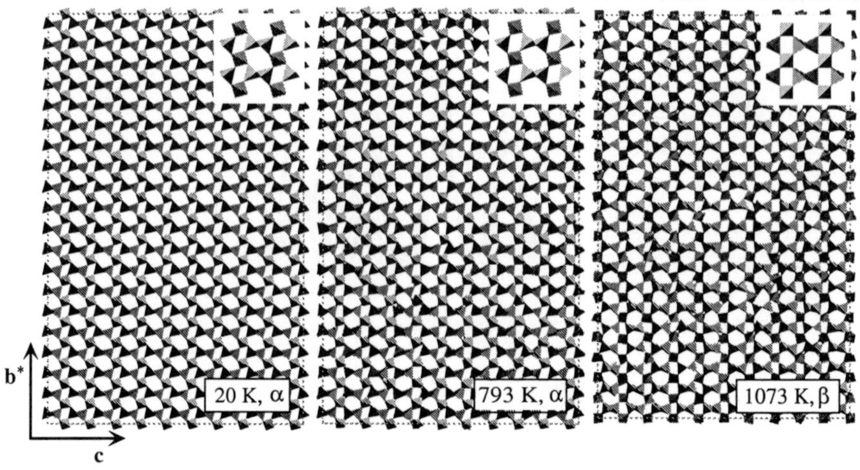

Figure 3. (100) layers of instantaneous RMC atomic configurations of quartz represented by SiO_4 tetrahedra for one temperature above T_c and two below. The inserts show the "average" structures obtained from the same configurations. In this projection the small parallelopiped gaps between tetrahedra become orthogonal in the β-phase, giving a clear representation of the symmetry change at T_c.

It is an artifact of using the average atomic positions to measure the bond distance of this disorder system. The cartoon in figure 2 illustrates the effect, as the temperature increases the oxygen atoms precess and the distance calculated is the distance from the average ⟨Si⟩ atom position to the average ⟨O⟩ position which is smaller that the actual average distance between the atoms ⟨Si–O⟩. The average distance between atoms ⟨Si–O⟩ extracted from RMC shows the true slight expansion of the Si–O bond with increasing temperature.

Figure 3 shows three of the configurations produced for quartz at 20 K, 793 K and 1073 K. The 20 K configuration is remarkably ordered considering the configuration has undergone over a million atom moves. The amount of disorder at 20 K would be minimal, as indeed is shown by the RMC configuration, but since RMC is a random method it could at least in theory introduce disorder where there should be none. This shows that the data and constraints used are sufficient to ensure a highly ordered structure if required. This might imply that the system is in fact over constrained and the structure obtained is simply that provided by the Rietveld analysis. This idea can be quickly dismissed by looking at the 793 K configuration. Here there is considerable disorder within the structure, produced by rotations of the tetrahedra. It should be stressed that RMC is not a dynamic modelling technique in that the configurations obtained are a static snapshots of the final structure and the dynamics of how that structure evolves with time is not information that is obtainable from RMC. However, the final snapshot can be examined in detail and conclusions about the displacements, rotations and distortions needed to produce that structure can be drawn.

The 793 K configuration is still of the α-phase of quartz, so it is clear that the thermal disorder begins to be evident before the actual α–β phase transition occurs. From the 1073K configuration, a high level of disorder in the β-phase of quartz can be seen. Although this is a static snapshot it is clear that at these temperatures the disorder will not be static, but rather consist of the continuous motion of the tetrahedra that make up the structure. This is completely consistent with the RUM model developed by Dove et al.[14–16] The small picture inset in each of configurations of figure 3 are the average structure extracted from the overall RMC configuration. These are the same as the Rietveld structures and show that

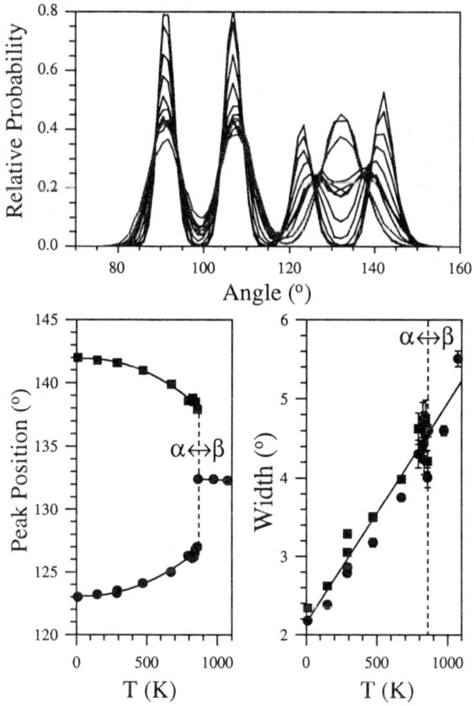

Figure 4. The upper plot shows the nearest-neighbor Si–Si–Si angle distribution in quartz. The lower two plots show the temperature-dependence of the positions and widths of the two peaks that coalesce in the β-phase. In the angle distribution, the temperatures are the same as those in figure 1, with the lowest temperature corresponding to the peaks with higher maximum values and the highest temperature corresponding to the peaks with the lowest maximum values.

the long range average structure has been maintained during the RMC refinement.

From the RMC configurations it is possible to extract three body terms which are not available without a structural model. Figure 4 shows the nearest-neighbour Si–Si–Si angle distribution with temperature. Although this is a short range correlation the two peaks at higher angles clearly show an effect of the α–β phase transition. The effect is that in the α-phase there are four distinct peaks and as T_c is approached the two peaks above 120° shift closer together so in the β-phase they form a single broad peak. The position of these peaks are plotted in the graph in the bottom left corner of figure 4. They clearly show a behaviour consistent with the phase transition, however, the width of the peaks (plotted in the graph in the right hand corner of figure 4) show a linear increase with temperature which is not affected by the phase transition.

4.3. Discussion

It is clear that through the use of neutron total scattering and RMC modelling a unique visualization of the α–β phase transition in quartz from both local and long range length scales has been achieved. Some aspects of the structure are insensitive to the phase transition and considerable dynamic disorder is evident at temperatures well below T_c. The picture obtained is of a dynamic structure of increasing orientational disorder of the SiO_4 tetrahedra with increasing temperature. When T_c is reached enough thermal energy is avail-

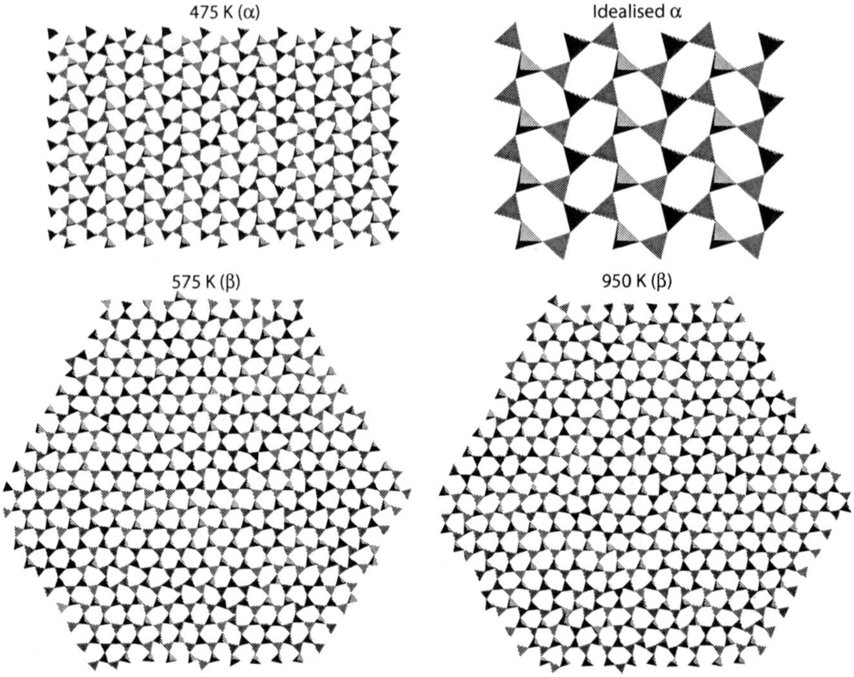

Figure 5. Atomic configurations of cristobalite at the three of the temperatures obtained from the RMC refinement, with the atoms represented as shaded SiO_4 tetrahedra. The views are down the [111] axes of the cubic β-phase, and down the corresponding [201] axis of the tetragonal phase. The figure also shows the corresponding projection of the ideal tetragonal phase with no thermal motion.

able to excite new phonon modes which results in the phase transition to the β-phase. RMC can only produce a static snapshot, but by comparing configurations at various temperatures, this is the conclusion that is drawn. It is clear the new RMC implementation does indeed produce models that are consistent with both the long and the short range order, and new information can be obtained using these models.

5. CRISTOBALITE

Following the success of the quartz study, it was clear that the method works, so it was decided that investigating the α- and β-phases of cristobalite would be a good follow-on study.[17] The aim was again to prove the method works and provide useful information, but also there were interesting questions about the cristobalite structure that remained to be answered. First, is the β-phase made up of domains of the α-phase, or does it indeed have a distinct structure? Secondly, the average β-cristobalite structure has a linear Si–O–Si bond if taken literally, but this is a very high energy configuration of these atoms and so what is the true form of this bond? Finally, since RMC produces three dimensional configurations, the single crystal diffuse scattering can be calculated from these configurations, and does it match the single crystal diffuse scattering measured using TEM?

The data were again collected on the LAD diffractometer at ISIS. Five data sets were collected, one in the α-phase at 475 K and four in the β-phase at 575, 700, 825 and 950 K.

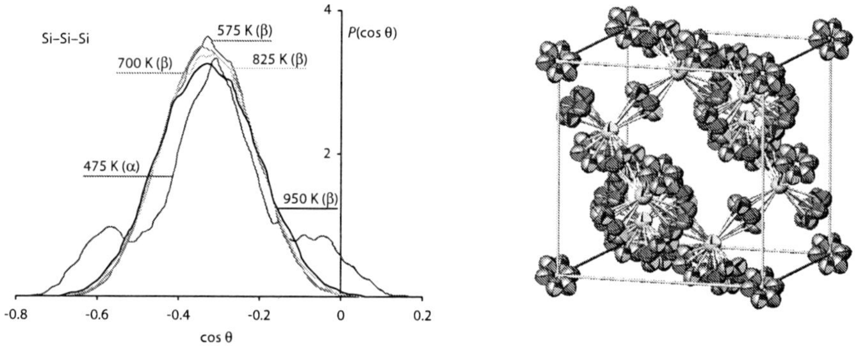

Figure 6. Si–Si–Si angle distribution functions for cristobalite at the five temperatures. The distribution functions for all temperatures in the β-phase are very similar, but significantly different from the distribution functions of the α-phase. The data are given as functions of cos θ, where θ is the particular bond angle in each case. The crystal structure is of β-cristobalite refined using the six-site model for the oxygen atoms. The thermal displacement parameters are represented by ellipsoids. The thermal displacement ellipsoids of the oxygen atoms in a small ring overlap.

The data were corrected in the standard way to produce $i(Q)$'s and $G(r)$'s were obtained using MCGR. RMC was then run starting with a configuration $10 \times 10 \times 10$ unit cells of the average structure found by Rietveld refinement of the data.

5.1. Results of the Cristobalite study

Figure 5 shows sections from three of the five RMC configurations produced. The ideal tetragonal α-cristobalite structure is also shown in figure 5, and the RMC configuration at 473 K matches this structure very well. There is some slight thermal disorder in the structure as would be expected, but there is little deviation from the ideal structure. At 575 K cristobalite is in the cubic β-phase as can be seen from the configuration shown. The ideal cubic structure should consist of perfect hexagonal rings formed by the tetrahedra in this projection. However, the thermal disorder results in the tetrahedra rotating, and a structure consisting of distorted hexagonal rings results. As the temperature is increased the true β-structure does not change much, although the amount of rotation of the tetrahedra increases slightly and the distortion of the hexagonal ring increases as can be seen from the 950 K configuration in figure 5. The β-cristobalite structures do not appear to contain any obvious domains and the certainly do not appear to be made up of various orientations of the α-cristobalite structure.

As with the quartz data, angle correlations can be obtained easily from the RMC configurations. Figure 6 shown the angle distribution for the Si–Si–Si angles. The Si–Si–Si angle distribution once again appears to be sensitive to the phase transition, where there are three resolvable peaks in the α-phase but only one broad peak in the β-phase. There are no dramatic changes with increasing temperature in the β-phase, just a slight broadening of the angle distribution with increasing temperature. The fact that there is a clear difference between the angle distribution in the α-and β-phases of cristobalite seems to dismiss the idea that the β-phase is made up of domains of the α-phase, since if this were the case

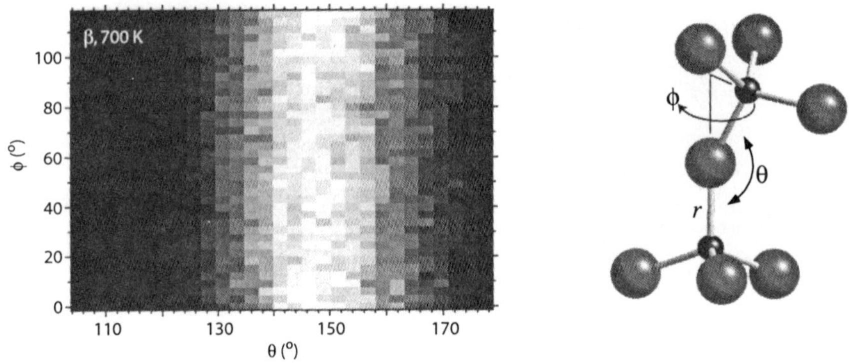

Figure 7. Maps of the distributions of the θ and φ angles define in the cartoon . The higher values of the distributions are indicated by the lighter shading of the map.

the angle distribution would have the same the peak structure throughout the temperature range.

The average structure of β-cristobalite obtained from a Rietveld refinement contains linear Si–O–Si bonds, as discussed earlier. It is generally accepted that this is not physical since this would result in a very high energy atom arrangement. To accommodate this a split-site Rietveld refinement is often carried out. The resulting structure in illustrated in figure 6, the single oxygen between the Si atoms is replaced with six oxygen sites each $\frac{1}{6}^{\text{th}}$ occupied. Figure 7 shows a plot of φ against θ obtained from the 700K RMC configuration. The cartoon illustrates how φ and θ are defined. θ is the Si–O–Si angle between two tetrahedra and φ relates the orientation of two tetrahedra, effectively it describes the precession of the oxygen atoms around the average 'linear' bond between the two Si atoms. Considering the θ scale it is clear the distribution of Si–O–Si is not centered at 180° but at about 150°. This agrees with the picture of a dynamically disordered structure where the linear Si–O–Si bond is energetically unfavourable so the tetrahedra rotate to avoid such a bond. This supports the split-site refinement of the Rietveld structure. However, if we now look at the φ scale of figure 7 we can see the distribution with φ is smooth (apart from some statistical noise) with no preferred angles. This shows the six site model is a simplification of the true system which consists of a continuous distribution of oxygen atoms. This is in fact supported by the split site refinement since the thermal ellipsoids overlap indicated a continuous distribution.

The configurations produced by RMC fit the experimental data well and seem to produce data that is consistent with a dynamically disordered structure, so the possibility of calculating the three dimensional diffuse scattering and comparing it to existing data measured from a single crystal of cristobalite presented itself. We calculate

$$S(\mathbf{Q}) = \sum_{i,j} \bar{b}_i \bar{b}_j \left\langle \exp(i\mathbf{Q} \cdot [\mathbf{r}_i - \mathbf{r}_j]) \right\rangle \tag{18}$$

where $\mathbf{r}_i$ is the instantaneous position of atom, i, and b_i is the corresponding neutron scattering length. Note that we have a grid of values of $\mathbf{Q}$ as given by the size of the RMC simulation sample.

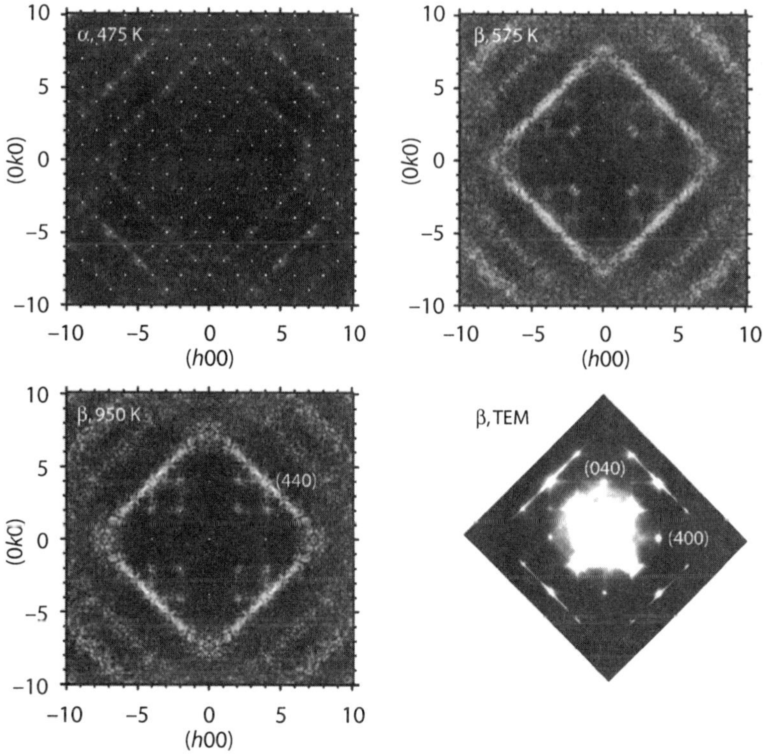

Figure 8. Maps of the three-dimensional diffuse scattering from cristobalite, showing the a*–b* plane in reciprocal space. The plot for α-cristobalite is indexed as β-cristobalite in order to highlight the common aspects. The single-pixel white spots are the Bragg peaks. The intensity of the diffuse scattering, as indicated by the shading (light for higher intensity, dark for low intensity) is the same for all plots. The plot in the bottom-right corner shows experimental TEM measurements for β-cristobalite.[18]

The configurations are small when compared to a real single crystal, just 10 unit cells in each direction, so the statistics are quite low. To improve the results we averaged over symmetry to reduce the stastical noise, and used a Gaussian smoothing over neighboring grid points. Taking these factors in account, there is nevertheless a very good agreement between the general features in the TEM measurements of diffuse scattering from β-cristobalite and the calculated data in figure 8. This gives us confidence that the RMC refinements are indeed reproducing the three dimensional properties of the real structure. It might be suspected that the random nature of RMC will result in diffuse scattering even when it is not supposed to be present. To check for this the diffuse scattering was calculated from the α-cristobalite configuration. α-cristobalite will have some diffuse scattering but it will be much lower in intensity, and this is indeed what RMC produces, as can be seen in figure 8.

5.2. Conclusion

The true structure of β-cristobalite is dynamically disordered with tetrahedra moving to avoid formation of linear Si–O–Si bonds. The experimental data presented now rules out the interpretation of β-cristobalite being based on domains of the α-phase. It is clear that,

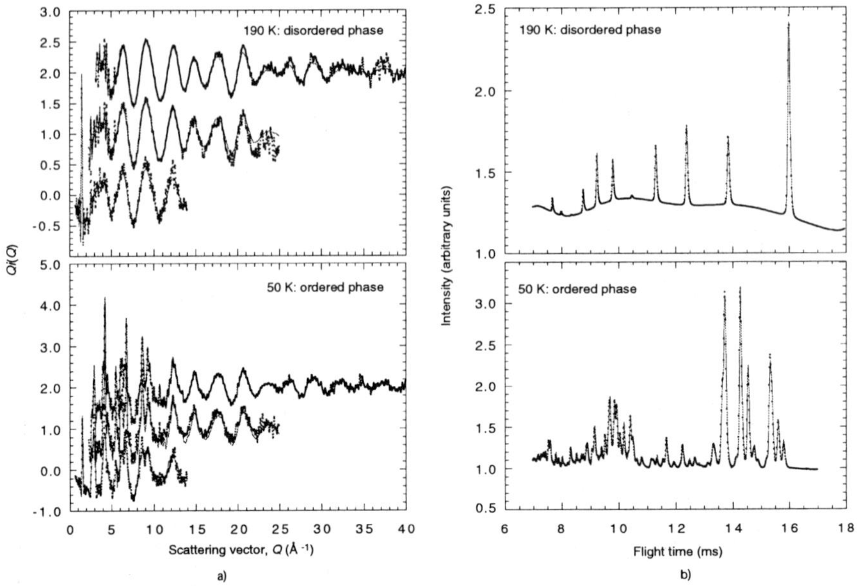

Figure 9. a) $Qi(Q)$ functions for the two phases of SF_6, measured using three different banks of detectors (data points), and compared with the MCGRtof calculations taking account of instrument resolution (continuous curves). b) Bragg diffraction obtained on one bank of detectors (points) for the two phases of SF_6, and compared with RMCprofile calculations (continuous curve).

at least to a limited extent, diffuse scattering can be calculated from RMC configurations and produce data that is consistent with single crystal diffuse scattering measurements.

6. SULPHUR HEXAFLUORIDE

From the studies on silica we have demonstrated that total scattering and RMC provide a very powerful tool for structural analysis. When GEM at ISIS was brought online it was clear that to utilize this powerful machine fully the RMC code and our methodology would have to be updated. To this end we developed the latest versions of the RMC code RMCprofile and MCGRtof as discussed earlier. The first system we investigated using GEM and the new software was the molecular crystal SF_6. This work has not been discussed elsewhere so it will be described here in more detail than the previous two studies.

The crystal structure of SF_6 between 90–230 K is body-centred cubic, with one molecule per lattice point.[19] This material has been studied in detail by Dove and Pawley[20,21] using molecular dynamics simulation. It was found that, although the S–F bonds lie along the ⟨100⟩ direction in the average structure, in the real structure there is considerable orientational disorder. In the average structure, the shortest distance between the closest F atoms of next nearest neighbouring molecules, along the unit cell edge, would be too short. This results in the molecules constantly pushing each other out of the way, resulting in a large amount of orientational disorder caused by the tumbling motion of the molecules. As the temperature is reduced below 90 K this conflict between neighboring molecules is resolved by a phase transition to an ordered structure with monoclinic symmetry.[22,23]

The aim of the RMC study was to investigate this system using experimental data and

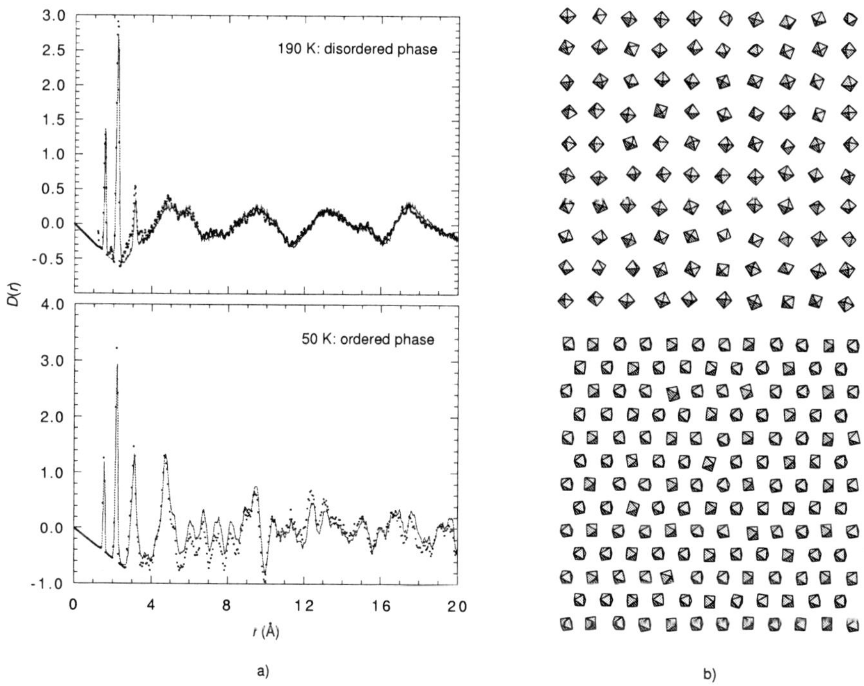

Figure 10. a) Pair distribution functions $D(r)$ for the two phases of SF_6 as obtained using the MCGRtof program (points), and compared with the RMCprofile calculations. b) Layers of molecules from the RMC configurations for the disordered (top)and the ordered(below) phases of SF_6.

then to use the configuration of SF_6 to confirm or disprove the MD simulations. To this end SF_6 was measured in the disordered bcc phase at 190 K and in the order monoclinic phase at 50K. A secondary aim was to demonstrate that the new MCGRtof and RMCprofile work well with GEM data produce useful configurations.

6.1. Results of the SF_6 study

Figure 9 show the MCGRtof fits to the $Qi(Q)$ data and the RMCprofile fit to the normal diffraction patterns for the two phases of SF_6 measured. Both these fits are obtained taking into account the instrument resolution of GEM. The $D(r)$ data and RMCprofile fits are shown in figure 10 along with example layers of molecules from the resultant RMC configurations. The $Qi(Q)$ and $D(r)$ fits are reasonable but they could be improved with longer runs and more careful data correction. The fits to the Bragg diffraction are very good, which indicates the average structure is being maintained. However, the configuration pictures in figure 10b show considerable orientational disorder of the molecules in the high temperature disordered phase. The low temperature phase appears much more ordered, with only a few molecules displaced from their ideal average positions, which is consistent with the molecular dynamics results.

The key to the orientation disorder in the high temperature phase is the intermolecular F...F pair distribution, particularly the shortest F...F contacts along the $\langle 100 \rangle$ directions at the unit cell edge. The S–F distance obtained from the total $T(r)$ at 190 K is 1.565 Å and the unit cell parameter is 5.89 Å. These combine to give a shortest F...F distance of 2.76Å if

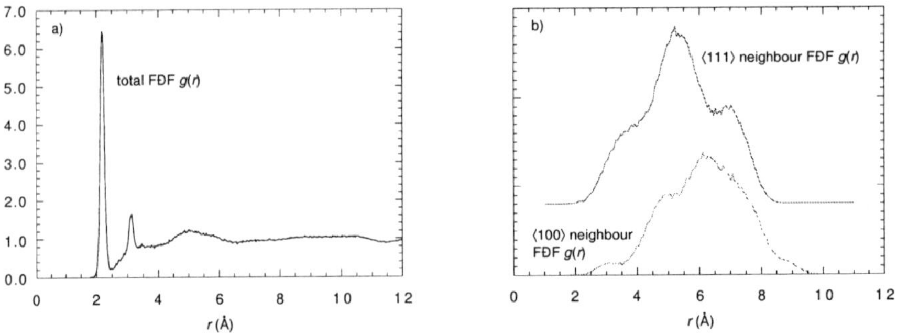

Figure 11. a)F...F partial distribution function of the disorder phase of SF_6 at 190K obtained from the RMC simulation and containing both intra-molecular and inter-molecular distances. b) Corresponding inter-molecular distribution functions for $\langle 111 \rangle$ and $\langle 100 \rangle$ neighboring molecules.

the molecules are in their ideal ordered positions. The molecular dynamics simulations[20,21] indicate that this contact distance would lead to the electronic distribution of the two F atoms overlapping too closely. All that is available directly from the data is the total pair distribution function $D(r)$, as in figure 10 a, which contains the correlation between all pairs of atoms in the system. This does not provide much information on the F...F correlations, since for just two molecules there will be 36 distances involved and information on any particular distance will be lost amongst the other correlations. However, from the RMC configurations it is possible to extract more detailed information.

Figure 11a shows the F...F partial distribution function extracted from the 190K configuration. The sharper peaks in this function arise from intra-molecular correlations, whereas the inter-molecular correlations form the broader more complicated distribution beyond the first peak. This function is still too complicated to provide useful information, but with the RMC configuration we can go further. From the configurations the inter-molecular distribution function can be extracted separately from the intra-molecular function as shown in Figure 11b. Here the inter-molecular function has been further divided into two separate components for the nearest neighbor along $\langle 111 \rangle$ and the next nearest neighbor along $\langle 100 \rangle$. The latter function is the most interesting since it contains the closest approach F...F distance along $\langle 100 \rangle$. The small peak below 4 Å corresponds to the distribution of these distance in the configuration, it is clear that position of this peak has shifted from the 2.76 Å distance discussed above to 3 Å.

6.2. Conclusions

The RMC configuration for SF_6 certainly seem to confirm the molecular dynamics prediction and adds experimental evidence to the conclusions drawn. The high temperature form of SF_6 consists of orientationally disordered molecules that act to increase the closest approach distance of nearest F atoms. The important point to note from this study is that useful information that was not directly accessible from experimental data has been obtained though the combination of total scattering and the RMC method.

7. DISCUSSION

The three example studies in this chapter have shown the reader that the combination of total scattering and the RMC method is a very powerful tool for structural studies of materials. The RMC method is fairly simple to implement but the difficulty comes in ensuring the system has enough data based constraints to ensure that physically meaningful atomic configurations are obtained. Using the methodology described here ensures this, and three dimensional configurations that are consistent with both the long and short range information contained within the data are obtained. The new versions of the RMC code, MCGRtof and RMCprofile, now take into account the instrument resolution function for time-of-flight machines and RMCprofile can fit the Bragg diffraction direct from tof data. The original RMC code is available from the web and the new code used here is available from the authors of this chapter. Readers are encouraged to obtain the code and try the method for themselves. For those studying a system which shows diffuse scattering among the Bragg peaks, or an average structure with large or strongly anisotropic temperature factors, then total scattering and RMC could produce very interesting results.

REFERENCES

1. R. L. McGreevy and L. Pusztai, *Molecular Simulations* **1**, 359–367 (1988).
2. D. A. Keen, *J. Appl. Cryst.*, **34**, 172–177 (2001).
3. M. A. Howe, R. L. McGreevy and W. S. Howells, *J. Phys.: Cond. Matter* **1**, 3433–3451 (1989)
4. R. L. McGreevy, *Nucl. Inst. Meth. A*, **354**, 1–16 (1995)
5. D. A. Keen, *Phase Transitions* **61**, 109–124 (1997).
6. D. A. Keen, in: Thorpe M F, Billinge S J L (ed.) *Local structure from diffraction* (Plenum, New York,1998), pp.101–119
7. M. G. Tucker,M. T. Dove and D. A. Keen, *Journal of Applied Crystallography* (submitted) (2000).
8. M. G. Tucker,M. T. Dove and D. A. Keen, *Journal of Applied Crystallography* (in press) (2001).
9. G. S. Pawley, *J. Appl. Crys.* **14**, 357–361 (1981).
10. L. Pusztai and R. L. McGreevy, *Physica B* **234–6**, 357-358 (1997).
11. M. G. Tucker,M. T. Dove and D. A. Keen, *Journal of Applied Crystallography* (submitted) (2002).
12. M. G. Tucker, M. T. Dove and D. A. Keen, *Journal of Physics: Condensed Matter* **12**, L723–L730 (2000).
13. M. G. Tucker,D. A. Keen and M. T. Dove, *Mineralogical Magazine* **65**, 489-507 (2001).
14. M. T. Dove, V. Heine and K. D. Hammonds, *Mineralogical Magazine*, **59**, 629–639 (1995)
15. K. D. Hammonds, M. T. Dove, A. P. Giddy, V. Heine and B. Winkler, *American Mineralogist*, **81**, 1057–1079 (1996)
16. M. T. Dove, K. O. Trachenko, M. G. Tucker and D. A. Keen, *Reviews in Mineralogy and Geochemistry*, **39**, 1–33 (2000)
17. M. G. Tucker, M. .D. Squires, M. T. Dove and D. A. Keen, Tucker, *Journal of Physics: Condensed Matter* **13**, 403–423 (2001).
18. G. L. Hua, T. .R. Welberry, R. L. Withers and J. G. Thompson, *Journal of Applied Crystallography* **21**, 458–465 (1988).
19. G. Dolling, B. M. Powell and V. F, Sears, *Molecular Physics* **37**, 1859–1883 (1979).
20. M. T. Dove and G. S. Pawley, *Journal of Physics C: Solid State Physics* **16**, 5969–5983 (1983).
21. M. T. Dove and G. S. Pawley, *Journal of Physics C: Solid State Physics* **17**, 6581–6599 (1984).
22. M. .T. Dove, B. M. Powell, G. S. Pawley and L. S. Bartell, *Molecular Physics* **65**, 353–358 (1988).
23. B. M. Powell, M. T. Dove, G. S. Pawley and L. Bartell, *Molecular Physics* **62**, 1127–1141 (1987).

NOTES ON THE ANALYSIS OF DATA FOR PAIR DISTRIBUTION FUNCTIONS

M.F. Thorpe, V.A. Levashov, M. Lei and S.J.L. Billinge*

1. INTRODUCTION

In these notes, we collect together some results that we have found useful in the analysis of data for the Pair Distribution Function (PDF). This work is rather general, and we illustrate the results with examples from molecules and crystalline solids. These notes are arranged into five main sections, following some definitions in the next section. In section 3, we analyze a single PDF peak and show how it can be decomposed into zero, one, two and multiphonon contributions. These results are illustrated using the nearest neighbor PDF peak in a Ni crystal. In section 4, we examine the widely used approximation that the PDF lineshape is Gaussian and find that this is indeed a very good approximation under most circumstances. We give explicit expressions that can be used for the deviations from a Gaussian lineshape in terms of the elements of the displacement–displacement correlation matrix. These results are illustrated for benzene and fullerene molecules, where we show that even in these cases, which are highly anisotropic, the deviation from a Gaussian lineshape is small. In the section 5, we show how the intermolecular correlations can sometimes be taken into account by using a continuum approximation. In section 6, we show how the experimental resolution function can be taken directly into account for a special case. In the discussion section, we show how these results can be brought together to give both the intra– and inter– fullerene PDF in a crystal of fullerite.

2. BACKGROUND

The multiphonon contributions to the PDF are contained within the expression for the coherent scattering cross-section $I(q)$. This cross section[1] can be written as

$$I(\mathbf{q}) = \frac{1}{N}\sum_{i,j} f_i f_j e^{i\mathbf{q}.(\mathbf{r}_i - \mathbf{r}_j)} \left\langle e^{i\mathbf{q}.(\mathbf{u}_i - \mathbf{u}_j)} \right\rangle, \tag{1}$$

* M.F. Thorpe, V.A. Levashov, M. Lei and S.J.L. Billinge, Physics & Astronomy Department, Michigan State University, East Lansing, MI 48824.

where the angular brackets denote that a thermal average has been taken over the displacements $\mathbf{u}_i$ from the equilibrium or mean atomic positions $\mathbf{r}_i$, so that the instantaneous position of the atom i is given by $\mathbf{r}_i + \mathbf{u}_i$. The atomic form factors $f_i(q)$ are as appropriate for x-ray scattering, where they are the Fourier transform of the atomic charge density, or for neutron scattering where they are the q–independent neutron scattering lengths. It is useful to define a reduced scattered intensity $S(\mathbf{q})$

$$S(\mathbf{q}) = \left\{ I(\mathbf{q}) - \left[\frac{1}{N} \sum_i f_i^2 - \left(\frac{1}{N} \sum_i f_i \right)^2 \right] \right\} \Big/ \left(\frac{1}{N} \sum_i f_i \right)^2 \tag{2}$$

that has the property that after the powder average is taken, $S(q) \to 1$ as $q \to \infty$.

In these notes we will focus on only a single atomic species in a crystal with N sites, as in for example a face centered crystal of Ni, and molecules such as fullerene with only C atoms. The generalization to many atoms is quite straightforward, and all our results can easily be extended to a crystal containing many atomic species. We do discuss the benzene molecule, C_6H_6, in section 4. The Debye-Waller factor $\exp(-2W)$ is extracted explicitly in the front of the expression (1) and then re-inserted, so that a multi-phonon expansion can be made

$$I(\mathbf{q}) = \frac{1}{N} e^{-2W} \sum_{i,j} f_i f_j e^{i\mathbf{q}.(\mathbf{r}_i - \mathbf{r}_j)} \left\{ \left\langle e^{i\mathbf{q}.(\mathbf{u}_i - \mathbf{u}_j)} \right\rangle e^{2W} \right\}. \tag{3}$$

We expand the term inside the braces in (3) to get the Bragg peak, the one-phonon peak etc, after an appropriate spherical powder average is done. To be clear on the notation that is used throughout this paper, we write the spherically averaged $I(q) = \overline{I(\mathbf{q})}$, and $S(q) = \overline{S(\mathbf{q})}$ where the bar denotes the spherical powder average, taken by averaging over the directions of the scattering vector $\mathbf{q}$. The density $\rho(r)$ is then obtained from the transform

$$G(r) = 4\pi r[\rho(r) - \rho_0] = \frac{2}{\pi} \int_0^\infty q[S(q) - 1] \sin qr\, dq \tag{4}$$

where $G(r)$ is the reduced pair distribution function (RPDF). The form (4) is appropriate for an infinite system where the term in ρ_0 is removed from the left and the term in 1 from the right. For a finite system, like a molecule, this is not necessary as shown in Eq. (52). The density $\rho(r)$ can be written in terms of the individual peaks in the PDF as

$$\rho(r) = \frac{1}{4\pi r^2} \sum_{i \neq j} P_{ij}(r) \tag{5}$$

where $P_{ij}(r)$ is the PDF for a pair of atoms i and j, and the density $\rho(r)$ is related to the number $N(r)$ of atoms in the spherical annulus between r and $r + dr$ via

$$N(R + dr) - N(R) = 4\pi r^2 \rho(r) dr. \tag{6}$$

3. MULTIPHONON CONTRIBUTIONS

It is actually more convenient to consider this derivation of the lineshape associated with a single peak $P_{ij}(r)$ in the PDF using a different starting point. The PDF peak can be calculated exactly within the harmonic approximation by the one-dimensional Fourier transform[2]

$$P_{ij}(r) = \frac{1}{2\pi}\int dq e^{-iq\left(r-r_{ij}^{0}\right)} C_{ij}(q), \tag{7}$$

where we define $C_{ij}(q)$ slightly differently than in our previous papers[2-4] [the phase factor is taken out here] as

$$C_{ij}(q) = \left\langle e^{iq\mathbf{u}_{ij}\cdot\hat{\mathbf{r}}_{ij}} \right\rangle = e^{-(1/2)q^2\left\langle\left(\mathbf{u}_{ij}\cdot\hat{\mathbf{r}}_{ij}\right)^2\right\rangle} = e^{-q^2\sigma^2} e^{-(1/2)\lambda q^2\left(\sigma_{ij}^2-2\sigma^2\right)} = e^{-2W} e^{-(1/2)\lambda q^2\left(\sigma_{ij}^2-2\sigma^2\right)}. \tag{8}$$

We use q as the conjugate variable to r. Here $\sigma_{ij}^2 = <(\mathbf{u}_{ij}\cdot\hat{\mathbf{r}}_{ij})^2>$ is the factor that determines the width of the PDF peak and the vector $\mathbf{u}_{ij} = \mathbf{u}_i - \mathbf{u}_j$, with $\hat{\mathbf{r}}_{ij}$ being a unit vector from $\mathbf{r}_{ij} = \mathbf{r}_i - \mathbf{r}_j$. The angular brackets denote a thermal average. It is useful to introduce the phonon expansion parameter λ that we use to pick out the various multiphonon contributions to the lineshape. This occurs as an expansion of the exponential $\exp\left[i\mathbf{q}.\left(\mathbf{r}_i + \mathbf{u}_i\right)\right]$ in (1) in powers of the displacements $\mathbf{u}_i$, which is equivalently an expansion in q. Because two terms are required, one to create a phonon and the other to destroy it, the coefficient of the term involving q^{2n} or equivalently λ^n in (8) gives the contribution from the n–phonon term to the PDF lineshape. The observed scattered intensity is obtained when $\lambda = 1$. In the limit that i and j are well separated, we have $\sigma_{ij}^2 \to 2\sigma^2$, which leads to the usual Debye-Waller factor for the peaks at large r in the PDF. The Debye-Waller factor is defined by

$$2W = <\left(\mathbf{q}.\mathbf{u}_i\right)^2> = q^2 <u_x^2> = \frac{1}{3}q^2 <\mathbf{u}^2> \tag{9}$$

for the case of cubic symmetry as in face centered cubic Ni for example. The expression in (7) becomes

$$\frac{1}{\sqrt{2\pi\sigma_{ij}^2}} e^{-(r_{ij}-r)^2/\left(2\sigma_{ij}^2\right)} = \frac{1}{\sqrt{2\pi\left(2\sigma^2+\lambda\left(\sigma_{ij}^2-2\sigma^2\right)\right)}} e^{-(r_{ij}-r)^2/2\left(2\sigma^2+\lambda\left(\sigma_{ij}^2-2\sigma^2\right)\right)} \tag{10}$$

where we can now make an expansion in the no phonon (elastic or Bragg), one–phonon, two–phonon processes etc. by expanding the cross section in powers of λ. This is conveniently done by using dimensionless units and putting

$$x^2 = \left(r-r_{ij}\right)^2 / 2\sigma^2 \tag{11}$$

and

$$\mu = \lambda\left(2\sigma^2 - \sigma_{ij}^2\right) / 2\sigma^2 \tag{12}$$

so that the peak profile becomes

$$\frac{1}{\sqrt{4\pi\sigma^2(1-\mu)}} e^{-x^2/2(1-\mu)} \tag{13}$$

where we suppress the subscripts i, j for clarity in the derivation that follows. The expression (13) can now be expanded in powers of μ to give

$$\begin{aligned} &\frac{1}{\sqrt{4\pi\sigma^2(1-\mu)}} e^{-x^2/2(1-\mu)} \\ &= \frac{1}{\sqrt{4\pi\sigma^2}} e^{-x^2/2} \left[\begin{aligned} &1 + \mu\left(1-x^2\right)/2 + \mu^2\left(3 - 6x^2 + x^4\right)/8 \\ &\quad + \mu^3\left(15 - 45x^2 + 15x^4 - x^6\right)/48 \\ &\quad\quad + \mu^4\left(105 - 420x^2 + 210x^4 - 28x^6 + x^8\right)/384 + \ldots \end{aligned} \right] \end{aligned} \tag{14}$$

where the first term is the Gaussian from the Bragg scattering (with its associated Debye–Waller factor), the term linear in μ is the one–phonon contribution, the term in μ^2 is the two–phonon contribution etc. This result can be conveniently written in terms of Hermite polynomials[5]. Using the orthogonality relations from Gradhsteyn and Ryzhik[5] – 7.373 and 7.374:

$$\frac{e^{-\frac{x^2}{1+\alpha}}}{\sqrt{1+\alpha}} = e^{-x^2} \sum_{n=0}^{\infty} \frac{\alpha^n}{2^{2n} n!} H_{2n}(x) \tag{15}$$

which can be re-written as

$$\frac{e^{-\frac{x^2/2}{1-\mu}}}{\sqrt{1-\mu}} = e^{-x^2/2} \sum_{n=0}^{\infty} \frac{(-\mu)^n}{2^{2n} n!} H_{2n}\left(x/\sqrt{2}\right) \tag{16}$$

so that (14) becomes

$$\frac{1}{\sqrt{4\pi\sigma^2(1-\mu)}} e^{-x^2/2(1-\mu)} = \frac{1}{\sqrt{4\pi\sigma^2}} e^{-x^2/2} \sum_{n=0}^{\infty} \frac{(-\mu)^n}{2^{2n} n!} H_{2n}\left(x/\sqrt{2}\right) \tag{17}$$

This formula contains all the multiphonon contributions – a plot is shown in Fig. 1 using a parameter $\mu = \left(2\sigma^2 - \sigma_{ij}^2\right)/2\sigma^2 = 0.38$, which is the appropriate value for the first peak in Ni.[4] We have used a model with only nearest neighbor central forces that gives a good fit to the experimental results in Ni.[4] For peaks at larger distances, the value of μ would be smaller, approaching zero at large distances as $\sigma_{ij}^2 \rightarrow 2\sigma^2$. Thus the narrowing of the PDF lines at small distances is a direct consequence of the diffuse scattering, which is in turn a result of the structured diffuse scattering and multi–phonon processes. Note that at $x = 0$, all the multiphonon contributions are positive, so that the multiphonon processes coherently sharpen the peak, but the effect also dies off rapidly as both r and n increase. At $x = 0$, the expression (14) becomes

$$\frac{1}{\sqrt{4\pi\sigma^2(1-\mu)}} = \frac{1}{\sqrt{4\pi\sigma^2}}\left[1 + \mu/2 + 3\mu^2/8 + 15\mu^3/48 + 105\mu^4/384 + \ldots..\right]. \quad (18)$$

The integrated weight under each peak in the PDF is unity and so the normalization is not affected by where the series (14) is terminated. The second moment is $< x^2 >= 1 - \mu$, so that the second moment with only the elastic Bragg contribution is 1, and with one phonon terms also included is reduced to $1 - \mu$, and remains at this value as more multiphonon terms are added to (14) up to the exact final answer. This occurs as the multiphonon contributions all have small negative parts that compensate.

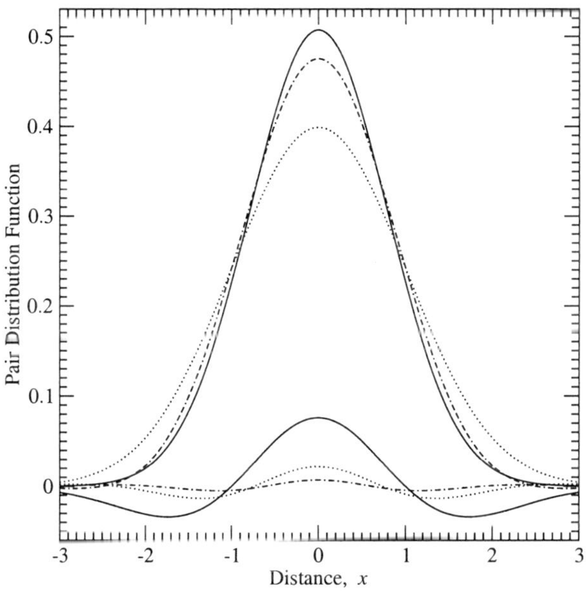

Figure 1. This figure shows the calculated peak in the PDF for the nearest neighbors in Ni at 300K in the upper part as a thick black line[4], with the Bragg contribution dotted, and the sum of the Bragg and one-phonon contribution shown as dot-dash. The separate multiphonon contributions are shown in the lower part of the figure, with the one-phonon contribution as a solid line, the two-phonon contribution as a dotted line, and the three-phonon contribution as a dot-dashed line. The value of $\mu = 0.38$ which comes from $\sigma_{ij} = 0.076$Å and $\sqrt{2}\sigma =$ 0.097Å

Therefore the second moment is not so meaningful as a measure of width, except in the limits of zero or all phonon processes included. In general the multiphonon contributions have positive and negative parts at different values of x, and the area under all the partial phonon contributions, shown in the lower part of Figure 1, is zero.

Note that the peak is only Gaussian if either *only* the Bragg contribution is counted or if *all* the multiphonon contributions are counted. Otherwise the shape is more complex. In order to compare widths, we use the half width at half the maximum in the plots in Fig. 1 of the single peak versus x.

Table 1. The half width and at half the maximum height for the PDF from Fig. 1, with $\mu = 0.38$, for the nearest neighbor peak of Ni at 300K. Note that the width with all the multiphonon terms included is reduced by a factor of $\sqrt{1-\mu}$ over the width when no phonons are included.

Number of terms in series	Half-width at half the maximum height
1 (elastic Bragg term only)	1.178
2 (includes 1-phonon terms)	1.014
3 (includes 2-phonon terms)	0.959
4 (includes 3-phonon terms)	0.938
5 (includes 4-phonon terms)	0.931
…………………………………..	……..
All (includes multi-phonon terms)	0.927

The various multiphonon contributions to the cross section can be obtained from (1), using the expansion for the diffuse scattering $S_D(q)$ as in Warren's book[1] – page 162, where

$$S_D(q) = S_1(q) + S_2(q) + S_3(q) + \dots\dots\dots \tag{19}$$

This decomposition in q space can now be expressed in terms of the density via

$$S(q) = S_B(q) + S_D(q) = 1 + \int_0^\infty dr 4\pi r^2 \frac{\sin(qr)}{qr}\left[\rho(r) - \rho_0\right] \tag{20}$$

where $\rho_0 = \sqrt{2}/a^3$, is the uniform density and a is the nearest neighbor distance in the face centered cubic lattice. Here the density $\rho(r)$ is given by

$$\rho(r) = \frac{1}{4\pi r^2}\sum_{j\neq i}\frac{w_{ij}}{\sqrt{2\pi\sigma_{ij}^2}}e^{-\frac{\left(r-r_{ij}^0\right)^2}{2\sigma_{ij}^2}} \tag{21}$$

We have included a factor $4\pi r^2$ in (21) as required by the definition. If we also define a density with the same Gaussians everywhere, as would be obtained by just using the Bragg scattering, so that all the PDF widths are equal to those at large distances, we then have

$$\rho_0(r) = \frac{1}{4\pi r^2}\sum_{j\neq i}\frac{w_{ij}}{\sqrt{4\pi\sigma^2}}e^{-\frac{\left(r-r_{ij}^0\right)^2}{4\sigma^2}}\ . \tag{22}$$

The Bragg scattering is given by

$$S_B(q) = e^{-2W} + \int_0^\infty dr \frac{\sin(qr)}{qr}\left[\sum_{j\neq i} \frac{w_{ij}}{\sqrt{4\pi\sigma^2}} e^{-\frac{\left(r-r_{ij}^0\right)^2}{4\sigma^2}}\right] \tag{23}$$

so that we can rewrite the total diffuse intensity as

$$S_D(q) = 1 - e^{-2W} + \int_0^\infty dr \frac{\sin(qr)}{qr}\left[\sum_{j\neq i} \frac{w_{ij}}{\sqrt{2\pi\sigma_{ij}^2}} e^{-\frac{\left(r-r_{ij}^0\right)^2}{2\sigma_{ij}^2}} - \sum_{j\neq i} \frac{w_{ij}}{\sqrt{4\pi\sigma^2}} e^{-\frac{\left(r-r_{ij}^0\right)^2}{4\sigma^2}}\right]. \tag{24}$$

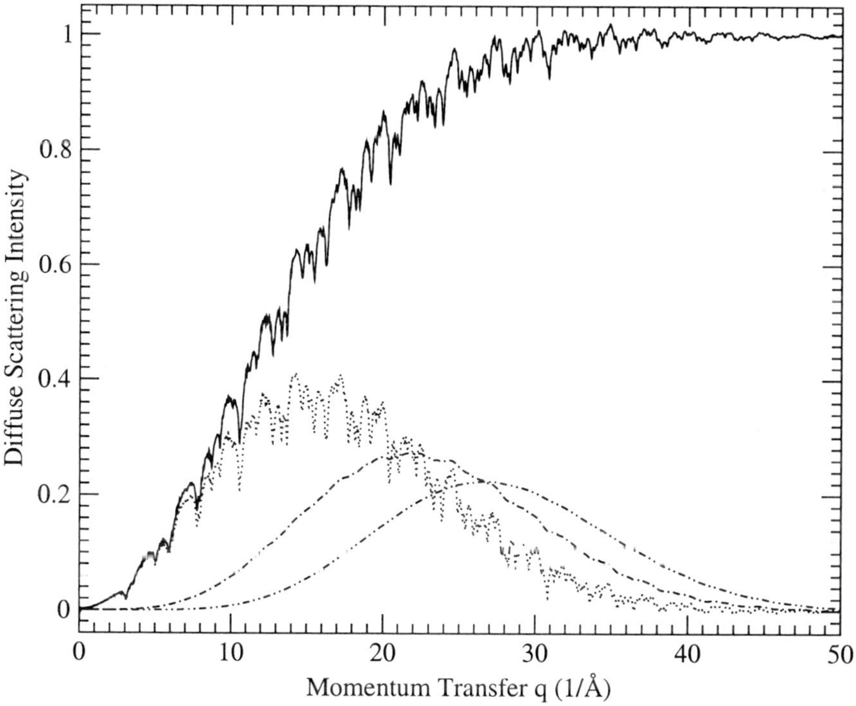

Figure 2. This figure shows the contributions to the multiphonon terms $S_n(q)$ [calculated[4] for Ni at 100K] Shown here are the one-phonon contribution $S_1(q)$ (dotted), the two-phonon contribution $S_2(q)$ (dot-dashed), and the three-phonon contribution $S_3(q)$(double dot dashed) individually. The total diffuse scattering $S_D(q)$ is shown as a solid-line. The oscillatory contributions come from the integrals (26) – (28), although the smooth self terms from (25) dominate for each $S_n(q)$.

The division into the various multiphonon parts requires some care. As well as expanding the integrand in powers of μ as earlier, it is also necessary to expand the first term as

$$1 - e^{-2W} = e^{-2W}\left(e^{2W} - 1\right) = e^{-2W}\left(2W + (2W)^2/2! + (2W)^3/3! + \ldots\right) \tag{25}$$

and so we may expand Eq. (24) as

$$S_1(q) = e^{-2W}(2W) + \int_0^\infty dr \frac{\sin qr}{2qr} \sum_{j \neq i} \frac{w_{ij}}{\sqrt{4\pi\sigma^2}} e^{-x_{ij}^2/2} \mu_{ij}\left(1 - x_{ij}^2\right) \tag{26}$$

$$S_2(q) = \frac{e^{-2W}(2W)^2}{2!} + \int_0^\infty dr \frac{\sin qr}{8qr} \sum_{j \neq i} \frac{w_{ij}}{\sqrt{4\pi\sigma^2}} e^{-x_{ij}^2/2} \mu_{ij}^2\left(3 - 6x_{ij}^2 + x_{ij}^4\right) \tag{27}$$

$$S_3(q) = \frac{e^{-2W}(2W)^3}{3!} + \int_0^\infty dr \frac{\sin qr}{48qr} \sum_{j \neq i} \frac{w_{ij}}{\sqrt{4\pi\sigma^2}} e^{-x_{ij}^2/2} \mu_{ij}^3\left(15 - 45x_{ij}^2 + 15x_{ij}^4 - x_{ij}^6\right) \tag{28}$$

where the first terms are the self-terms from Eq. (25).

In Fig. 2, we show the various contributions to the multiphonon scattering, together with the total diffuse scattering. The broad peak in the multiphonon contribution to the scattered intensity is given by the term involving $(2W)^n$ in Eqs. (25). Differentiation to find the maximum, which occurs at $W = 2/n$, leads to a maximum at $q \sim 1/\sqrt{n}$ as can be seen in Fig. 2.

4. NON-GAUSSIAN LINESHAPE

It is generally assumed that the shape of the individual peaks in the PDF is Gaussian and therefore characterized by the weight, center and width. We have used this approximation extensively and successfully in previous work[2-4, 6] and it is discussed in some detail in an appendix[2]. The Gaussian approximation consists of only taking into account the motion of the two atoms along the vector that connects them, and ignoring the perpendicular motion. In this section we examine this assumption in detail and show that it is good in almost all cases and that deviations from the Gaussian lineshape are small, even in very anisotropic materials. The effect of anisotropy has been recently been discussed by Dimitrov, Röder and Bishop.[7]

We consider a molecule or crystal where the equilibrium position of an atom i is $\mathbf{r}_i = (x_i, y_i, z_i)$ with respect to the other atom j of the pair so that $r_{ij} = \left|\mathbf{r}_i - \mathbf{r}_j\right|$. Atoms vibrate around their equilibrium positions. The probability that the distance between atoms i and j will be found to be $\mathbf{r} = (x, y, z)$ is given by

$$P_{ij}(x,y,z) = \frac{1}{\sqrt{(2\pi)^3 \sigma_x^2 \sigma_y^2 \sigma_z^2}} \exp\left(-\frac{\left(x - x_{ij}\right)^2}{2\sigma_x^2} - \frac{\left(y - y_{ij}\right)^2}{2\sigma_y^2} - \frac{\left(z - z_{ij}\right)^2}{2\sigma_z^2} \right), \tag{29}$$

where the Cartesian coordinates $\mathbf{r}_{ij} = (x_{ij}, y_{ij}, z_{ij})$, as well as the widths $\sigma_{ij} = (\sigma_x, \sigma_y, \sigma_z)$, are given in the frame of the principal axes, defined as those axes where the matrix of displacements

$$\left\langle \left(u_i^\alpha - u_j^\alpha\right)\left(u_i^\beta - u_j^\beta\right) \right\rangle \tag{30}$$

is diagonal. Also note that the contribution to the PDF $P_{ij}(r) = 4\pi r^2 \rho_{ij}(r)$ involves the density $\rho_{ij}(r)$. We are particularly interested in the PDF of a highly anisotropic $(\sigma_x \neq \sigma_y \neq \sigma_z)$ materials in the form of powders, where deviations from Gaussian behavior are likely to be most pronounced. Earlier in the first section when PDF was calculated for crystalline Ni, we used the fact that $\sigma_x = \sigma_y = \sigma_z = \sigma$ and used a Gaussian lineshape $P_{ij}^g(r)$

$$P_{ij}^g(r) = \frac{1}{\sqrt{2\pi\sigma^2}} \exp\left(-\frac{\left(r - r_{ij}\right)^2}{2\sigma^2} \right). \tag{31}$$

Nowadays when experimental techniques for PDF measurement have improved significantly, it is important to examine the assumptions that lead to a Gaussian lineshape like (31). We will do this in order to see the difference between the actual PDF and the Gaussian PDF $P_{ij}^g(r)$.

In order to find the PDF of anisotropic powder materials one should perform the angular average:

$$P_{ij}(r) = \int \frac{r^2 d\Omega}{\sqrt{(2\pi)^3 \sigma_x^2 \sigma_y^2 \sigma_z^2}} \exp\left(-\frac{\left(x - x_{ij}\right)^2}{2\sigma_x^2} - \frac{\left(y - y_{ij}\right)^2}{2\sigma_y^2} - \frac{\left(z - z_{ij}\right)^2}{2\sigma_z^2} \right) \tag{32}$$

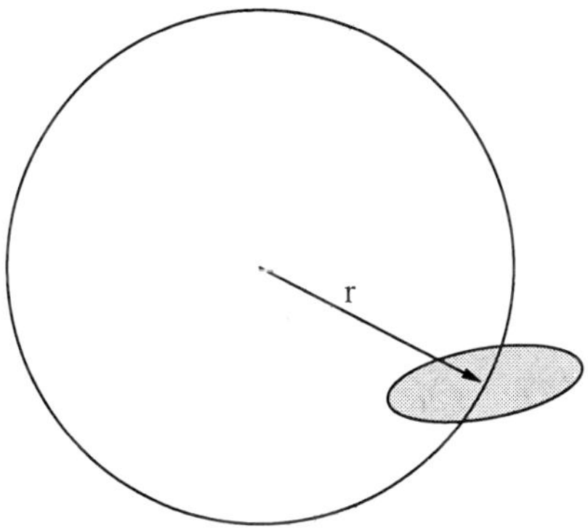

Figure 3. The shaded area represents a cut through the three dimensional ellipse that is the integrand in (32). The spherical average is done so that a sphere of radius *r* cuts through the ellipse.

It is easy to show that in the isotropic case when $(\sigma_x = \sigma_y = \sigma_z = \sigma)$, the average (32) can be done explicitly and this leads to the result[7]

$$
\begin{aligned}
P_{ij}(r) &= \frac{1}{\sqrt{2\pi\sigma^2}}\left(\frac{r}{r_{ij}}\right)\left[\exp\left(-\frac{\left(r-r_{ij}\right)^2}{2\sigma^2}\right)-\exp\left(-\frac{\left(r+r_{ij}\right)^2}{2\sigma^2}\right)\right] \\
&\cong \frac{1}{\sqrt{2\pi\sigma^2}}\left(\frac{r}{r_{ij}}\right)\exp\left(-\frac{\left(r-r_{ij}\right)^2}{2\sigma^2}\right)
\end{aligned}
\tag{33}
$$

where we neglect the second term in square brackets as being exponentially small for positive r. This result (33) can be rewritten as:

$$
P_{ij}(r) = \frac{1}{\sqrt{2\pi\sigma^2}}\exp\left(-\frac{\left(r-r_{ij}\right)^2}{2\sigma^2}\right)\left[1+\left(\frac{r-r_{ij}}{r_{ij}}\right)\right]. \tag{34}
$$

Thus one can see that if peaks are very narrow then the difference between $P_{ij}^{g}(r)$ and $P_{ij}(r)$ as given in Eq. (34) is small, due entirely to the second term in the square brackets in (34) which causes a small upward shift in the line by $\Delta r_{ij} \approx \sigma^2 / r_{ij}$. This is generally quite small [≈ 0.006 Å for the nearest neighbor peak[4] in Ni at T=300K], and gets even smaller for more distant peaks.

In the anisotropic case ($\sigma_x \neq \sigma_y \neq \sigma_z$) the difference between the actual peak shape and its Gaussian approximation could be larger. We first perform the numerical integration over a sphere for some test parameters and the result is shown in Fig. 4.

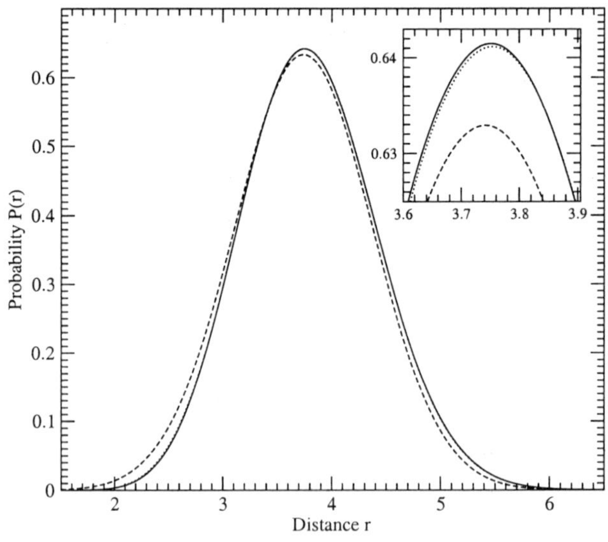

Figure 4. This figure shows the exact result for the PDF lineshape obtained from the integral (32) for the parameters $\mathbf{r}_{ij} = (3, 2, 1)$ and $\sigma = (0.75, 0.25, 0.50)$ as a solid line, where these are written in the coordinate system that makes the displacement-displacement correlation matrix (30) diagonal. The dashed line is the Gaussian approximation (31) and the dotted line is the analytical approximation (44). The insert is a blow-up to show the peak region more clearly.

It is time consuming and inconvenient to do the numerical integral (32) for each individual line in the PDF. We have developed an analytical approximation that can be used more easily once the displacement–displacement correlation matrix (30) has been calculated from the dynamical matrix for the molecule or crystal. We use a coordinate system in which the correlation matrix (30) is diagonal and we use that to write both the separation vector, between the pair of atoms $\mathbf{r}_{ij}$, and the Cartesian components of the widths $\sigma_x, \sigma_y, \sigma_z$. We outline the approach and give the result. In calculating the PDF as given in (32), we first take the back Fourier Transform [see (7)] to obtain

$$C_{ij}(q) = \int P_{ij}(r) e^{iq(r-r_{ij})} dr \tag{35}$$

where we use q as the conjugate variable to r. We expand $r - r_{ij}$ in the exponential of (35) by using

$$r - r_{ij} = r_{ij}\left[\left(1 + \frac{2B}{r_{ij}} + \frac{A^2}{r_{ij}}\right)^{\frac{1}{2}} - 1\right] = B + \frac{(A^2 - B^2)}{2r_{ij}} + \frac{(B^3 - A^2 B)}{2r_{ij}^2} + O\left(\frac{1}{r_{ij}^3}\right) \tag{36}$$

where

$$\begin{aligned} A^2 &= (x - x_{ij})^2 + (y - y_{ij})^2 + (z - z_{ij})^2 \\ B &= \frac{x_{ij}}{r_{ij}}(x - x_{ij}) + \frac{y_{ij}}{r_{ij}}(y - y_{ij}) + \frac{z_{ij}}{r_{ij}}(z - z_{ij}) \end{aligned} \tag{37}$$

so that (35) becomes

$$C_{ij}(q) = \overline{e^{iqB}\left[1 + iq\frac{(A^2 - B^2)}{2r_{ij}} + iq\frac{(B^3 - BA^3)}{2r_{ij}^2} - q^2\frac{(A^2 - B^2)^2}{8r_{ij}^2} + O\left(\frac{1}{r_{ij}^3}\right)\right]}, \tag{38}$$

where the bar over the top of the expression denotes a spherical average that is defined for any variable X as

$$\overline{X} = \int X P_{ij}(r) dr. \tag{39}$$

The variables A and B in (37) are most conveniently written in terms of Cartesian variables. It is best to use the form (32) for $P_{ij}(r)$ to exploit the Cartesian axes, so that the x, y, z integrals can be separated. The various moments needed in (38) can be found from the generating function

$$\overline{e^{iqB} A^{2n} B^m} = \lim_{\lambda \to 0}\left\{\frac{\partial^{n+m}}{\partial\lambda^n \partial(iq)^m}\left(\overline{e^{iqB} e^{\lambda A^2}}\right)\right\}. \tag{40}$$

where

$$\overline{e^{iqB}e^{\lambda A^2}} = \frac{\exp\left\{-\frac{q^2}{2}\left[\left(\frac{x_{ij}}{r_{ij}}\right)^2 \frac{\sigma_x^2}{1-2\lambda\sigma_x^2} + \left(\frac{y_{ij}}{r_{ij}}\right)^2 \frac{\sigma_y^2}{1-2\lambda\sigma_y^2} + \left(\frac{z_{ij}}{r_{ij}}\right)^2 \frac{\sigma_z^2}{1-2\lambda\sigma_z^2}\right]\right\}}{\sqrt{(1-2\lambda\sigma_x^2)(1-2\lambda\sigma_y^2)(1-2\lambda\sigma_z^2)}}. \quad (41)$$

For example, the second term in (38) involving $A^2 - B^2$, after the spherical averaging is complete becomes

$$\overline{e^{iqB}\left(A^2 - B^2\right)} = \exp\left(-\frac{q^2\overline{B^2}}{2}\right)\left[\left(\overline{A^2} - \overline{B^2}\right) - \frac{q^2}{2}\left(\left(\overline{A^2B^2} - \overline{B^4}\right) - \left(\overline{A^2} - \overline{B^2}\right)\overline{B^2}\right)\right] \quad (42)$$

Results such as (42) can be thought of as a generalization of the Debye-Waller result.[8] Similar results are obtained for the other terms in (38). We have kept terms in the expansion in $1/r_{ij}$ up to $(1/r_{ij})^2$ in (38), and the back Fourier transform of (38) is performed using

$$\frac{1}{2\pi}\int e^{-\frac{q^2\Sigma^2}{2}} e^{-iq(r-r_{ij})}(iq\Sigma)^n\,dq = \frac{e^{-\frac{(r-r_{ij})^2}{2\Sigma^2}}}{\sqrt{2^{n+1}\pi\Sigma^2}} H_n\left(\frac{r-r_{ij}}{\sqrt{2}\Sigma}\right) \quad (43)$$

where $\Sigma^2 = \overline{B^2}$, and the first few Hermite polynomials are given in (50). We can write the final result for the PDF lineshape $P_{ij}(r)$ as

$$P_{ij}(r) = \frac{1}{\sqrt{2\pi\Sigma^2}}\exp\left(-\frac{(r-r_{ij})^2}{2\Sigma^2}\right)\left[1 + \frac{A_1(r)}{r_{ij}} + \frac{A_2(r)}{r_{ij}^2} + O\left(\frac{1}{r_{ij}^3}\right)\right] \quad (44)$$

where

$$A_1(r) = \left[\frac{\overline{A^2} - \overline{B^2}}{2\sqrt{2}\Sigma}\right]H_1\left(\frac{(r-r_{ij})}{\sqrt{2}\Sigma}\right) + \frac{1}{8\sqrt{2}}\left[\frac{\overline{A^2B^2} - \overline{B^4}}{\Sigma^3} - \frac{\overline{A^2} - \overline{B^2}}{\Sigma}\right]H_3\left(\frac{(r-r_{ij})}{\sqrt{2}\Sigma}\right) \quad (45)$$

and

$$A_2(r)=\left[\frac{\overline{A^4}-6\overline{A^2B^2}+5\overline{B^4}}{16\Sigma^2}\right]H_2\left(\frac{(r-r_{ij})}{\sqrt{2}\Sigma}\right)$$
$$+\left[\frac{3\overline{A^4B^2}-10\overline{A^2B^4}+7\overline{B^6}}{192\Sigma^4}-\frac{\overline{A^4}-6\overline{A^2B^2}+5\overline{B^4}}{384\Sigma^2}\right]H_4\left(\frac{(r-r_{ij})}{\sqrt{2}\Sigma}\right) \tag{46}$$
$$+\left[\frac{5\overline{A^4B^4}-14\overline{A^2B^6}+9\overline{B^8}}{7680\Sigma^6}-\frac{3\overline{A^4B^2}-10\overline{A^2B^4}+7\overline{B^6}}{768\Sigma^4}+\frac{\overline{A^4}-6\overline{A^2B^2}+5\overline{B^4}}{512\Sigma^2}\right]H_6\left(\frac{(r-r_{ij})}{\sqrt{2}\Sigma}\right)$$

Explicit averages that are needed to evaluate the non-Gaussian lineshape, can be obtained in terms of $(\sigma_x,\sigma_y,\sigma_z)$ and (x_{ij},y_{ij},z_{ij}) which are calculated from a model and using the principal axes as explained earlier. Using the definitions of A and B in (37) together with (39) and (32), we have:

$$\begin{aligned}\overline{A^2}&=\sigma_x^2+\sigma_y^2+\sigma_z^2\\ \overline{A^4}&=2\left(\sigma_x^4+\sigma_y^4+\sigma_z^4\right)+\left(\sigma_x^2+\sigma_y^2+\sigma_z^2\right)^2\\ \overline{A^6}&=6\left(\sigma_x^6+\sigma_y^6+\sigma_z^6\right)+9\left(\sigma_x^4+\sigma_y^4+\sigma_z^4\right)\left(\sigma_x^2+\sigma_y^2+\sigma_z^2\right)+6\sigma_x^2\sigma_y^2\sigma_z^2\end{aligned} \tag{47}$$

$$\begin{aligned}\overline{B^2}&=\Sigma^2=\left(\frac{x_{ij}}{r_{ij}}\right)^2\sigma_x^2+\left(\frac{y_{ij}}{r_{ij}}\right)^2\sigma_x^2+\left(\frac{z_{ij}}{r_{ij}}\right)^2\sigma_x^2\\ \overline{B^4}&=3\Sigma^4,\quad \overline{B^6}=15\Sigma^6,\quad \overline{B^8}=105\Sigma^8\end{aligned} \tag{48}$$

$$\begin{aligned}\overline{A^2B^2}&=\overline{A^2}\;\overline{B^2}+2\left(\frac{x_{ij}^2\sigma_x^4+y_{ij}^2\sigma_y^4+z_{ij}^2\sigma_z^4}{r_{ij}^2}\right)\\ \overline{A^4B^2}&=6\left(\frac{x_{ij}^2\sigma_x^6+y_{ij}^2\sigma_y^6+z_{ij}^2\sigma_z^6}{r_{ij}^2}\right)+3\left(\overline{A^2}\right)\left(\overline{A^2B^2}\right)-\frac{9}{2}\left(\overline{A^2}\right)^2\left(\overline{B^2}\right)+\frac{3}{2}\left(\overline{A^4}\right)\left(\overline{B^2}\right)+2\sigma_x^2\sigma_y^2\sigma_z^2\\ \overline{A^2B^4}&=12\left(\frac{x_{ij}^4\sigma_x^6+y_{ij}^4\sigma_y^6+z_{ij}^4\sigma_z^6}{r_{ij}^4}\right)-6\overline{A^2}\left(\frac{x_{ij}^4\sigma_x^4+y_{ij}^4\sigma_y^4+z_{ij}^4\sigma_z^4}{r_{ij}^4}\right)\\ &\quad 9\left(\overline{A^2}\right)\left(\overline{B^2}\right)^2-6\left(1-\frac{x_{ij}^4+y_{ij}^4+z_{ij}^4}{r_{ij}^4}\right)\sigma_x^2\sigma_y^2\sigma_z^2\\ \overline{A^2B^6}&=15\left(\overline{A^2B^2}\right)\left(\overline{B^4}\right)-2\left(\overline{A^2}\right)\left(\overline{B^6}\right)\\ \overline{A^4B^4}&=-\left(\overline{A^4}\right)\left(\overline{B^4}\right)+6\left(\overline{A^4B^2}\right)\left(\overline{B^2}\right)+6\left(\overline{A^2B^2}\right)^2+2\left(\overline{A^2}\right)^2\left(\overline{B^4}\right)-12\left(\overline{A^2}\right)\left(\overline{B^2}\right)\left(\overline{A^2B^2}\right)\end{aligned} \tag{49}$$

The first few Hermite polynomials, and the only ones needed to this level, are given by

$$\begin{aligned} H_1(x) &= 2x \\ H_2(x) &= 4x^2 - 2 \\ H_3(x) &= 8x^3 - 12x \\ H_4(x) &= 16x^4 - 48x^2 + 12 \\ H_6(x) &= 64x^6 - 480x^4 + 720x^2 - 120 \end{aligned} \tag{50}$$

This result (44) has corrections to the Gaussian result involving the coefficients $A_1(r)$ and $A_2(r)$. The leading term, obtained when $A_1(r)$ and $A_2(r)$ are both zero, is the Gaussian lineshape with a width Σ given in (48) and in a previous paper[2]. The corrections are most important for the nearest neighbor peak when the terms in $1/r_{ij}$ in (44) are largest. However, as can be seen from Fig. 4 and Fig. 5 for benzene, the non–Gaussian effects are never great. Our conclusion then is that the Gaussian approximation can be used with some confidence, except for very accurate measurements on low r peaks for highly anisotropic materials. Note that for isotropic materials, if we compare the result (34) with the expression (44), we have $A_1(r) = (r - r_{ij})$ and $A_2(r) = 0$, which provides a useful limiting case of the results in this section.

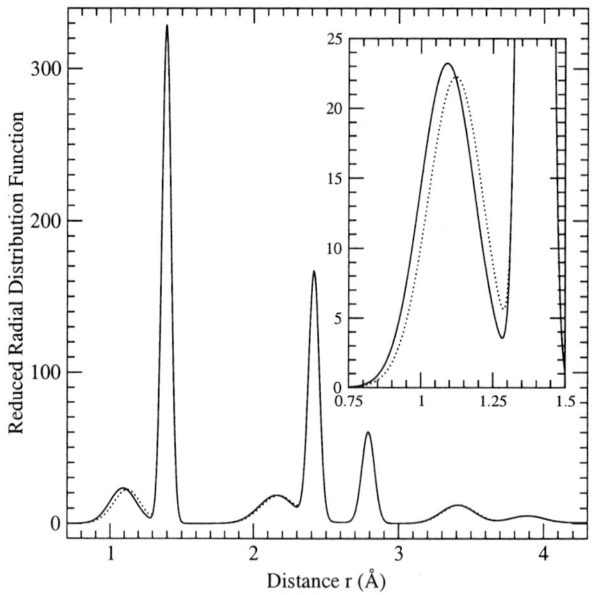

Figure 5. This figure shows the calculated RRDF of a benzene molecule C_6H_6 at 300K. The solid line is the Gaussian approximation (31) and the dotted line is the improved approximation using (44). We used the AM1 method within the Gaussian98 program[9] to calculate atomic distances, eigenfrequencies and eigenvectors of the molecular vibrations. Using these eigenvalues and eigenvectors and Eq. (76), the various σ were calculated, and then used as input in the calculation of the reduced radial distribution function shown here. The first peak at around 1.1Å is the nearest neighbor C–H peak and the second around 1.4Å is the nearest neighbor C–C peak.

5. INTERMOLECULAR CORRELATIONS

In this section we derive RPDF of two identical spherical molecules, on the assumption that the RPDF of one isolated molecule is known. We will use a classical continuum approach that can be useful under some circumstances, as we will show in section 7. The RPDF of one molecule with spherical symmetry can be obtained by a Fourier transform from the diffraction pattern of the single molecule. There is usually a gap in real space between the RPDF of one single molecule and the RPDF of pairs of molecules in a solid. One example is the crystal lattice of fullerene molecules, at sufficiently high temperature that all the buckyballs are spherical rotators, so the charge density can be thought of as a spherical shell, for purposes of calculating the RPDF between pairs of buckyballs. The RPDF of one single spherical distribution is easy to calculate. However the PDF between two separated spherical distributions is a little harder to derive. We do that in this section.

Let $S(q)$ be the diffracted intensity from the powder and $\rho(r)$ be the density. When the density distribution is finite, it relates to the diffracted intensity by the Fourier transformation:

$$\rho(r) = \frac{1}{(2\pi)^3}\int S(q)e^{-i\mathbf{q}\cdot\mathbf{r}}d^3\mathbf{q} = \frac{1}{2\pi^2 r}\int_0^\infty S(q)\sin(qr)qdq \tag{51}$$

which is a modified form of (4), where the effects of the infinite sample that are manifest through the term in ρ_0 from the left hand side, and the term in 1 from the right hand side, are not necessary. In this section, we will focus on the RPDF instead of the density function via

$$G(r) = 4\pi r\rho(r) = \frac{2}{\pi}\int_0^\infty S(q)\sin(qr)qdq \tag{52}$$

as modified from (4). Let us suppose that two identical molecules are separated by a distance D, and consider the case when these two molecules do not overlap with each other. We have concentrated on molecules with spherical symmetry only in this paper to avoid the complexity involved in treating the relative orientation between the two molecules. The diffracted intensity of the two molecules with spherical symmetry is

$$S(q) = \frac{1}{2}S_1(q)\overline{\left|\sum_{j=1}^{2} e^{-i\mathbf{q}.\mathbf{r}_j}\right|^2} = S_1(q) + S_1(q)\frac{\sin(qD)}{qD} \tag{53}$$

where $S_1(q)$ is the diffracted intensity from one single molecule with spherical symmetry. The bar represents the spherical average appropriate for a powder. Substituting Eq. (53) into Eq. (52) we get the RPDF of the two molecules as:

$$G(r) = G_1(r) + G_2(r) \tag{54}$$

with

$$G_1(r) = \frac{2}{\pi}\int_0^\infty S_1(q)\sin(qr)qdq$$

$$G_2(r) = \frac{2}{\pi D}\int_0^{\infty} S_1(q)\sin(qD)\sin(qr)dq \tag{55}$$

The $G_1(r)$ term is the self-term that describes the RPDF of a single molecule. We are interested in the $G_2(r)$ term that is the intermolecular RPDF. The $G_2(r)$ can be expressed as an integral involving the self term:

$$G_2(r) = \frac{1}{2D}\int_0^{r} \mathrm{sign}(D-r')G_1(|D-r'|)dr' + \frac{1}{2D}\int_0^{r} G_1(D+r')dr'. \tag{56}$$

Eq. (56) shows the relation involving the RPDF between two identical molecules and the RPDF of a single molecule. Thus if we can work out the RPDF of a single molecule, which is relatively easy, we can work out the RPDF between two identical spherical molecules, and hence the RPDF of molecules in an infinite super lattice as a superposition of such pair terms.

Note that the second term in Eq. (56) is zero, when the two molecules don't overlap as is generally the case. The RPDF of one single molecule, $G_1(r)$, is non-zero only in the range $r \in [0, 2R]$ where R is the limit of the mass distribution of the molecule. Since we assume the two spherical molecules do not overlap with each other, the distance D is always greater than $2R$. Hence the second term must be zero. The formation of the first term clearly shows that the RPDF between two molecules is symmetric about the distance D. Dropping the second term from Eq. (56) we have:

$$G_2(r) = \frac{1}{2D}\int_0^{r} \mathrm{sign}(D-r')G_1(|D-r'|)dr' \tag{57}$$

As an example we calculate the RPDF of a single spherical shell,

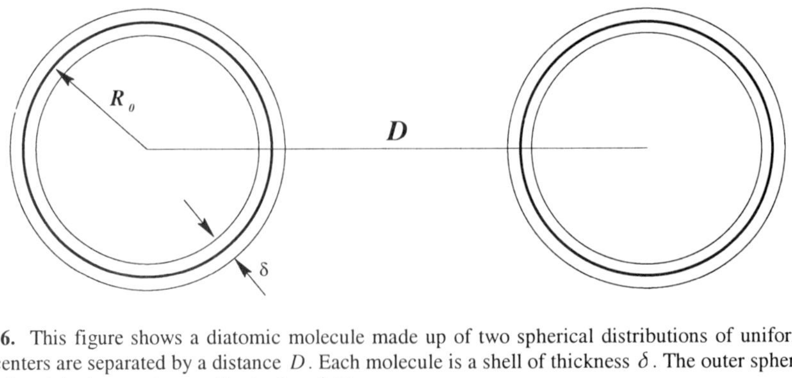

Figure 6. This figure shows a diatomic molecule made up of two spherical distributions of uniform density, whose centers are separated by a distance D. Each molecule is a shell of thickness δ. The outer sphere radius is $R_0 + \delta/2$ and the inner sphere radius is $R_0 - \delta/2$.

$$G_1(r) = \frac{3M}{2\left(12R_0^2 + \delta^2\right)^2 \delta} \left\{ \begin{array}{c} r\left[16R_0^3 + 12R_0\delta(\delta - r) + 36R_0^2(2\delta - r) + 3(\delta - r)^2(2\delta + r)\right] \\ + 2(\delta - r)^2\left[r(2\delta + r) - 12R_0^2\right]\operatorname{sign}(\delta - r) \\ + r(4R_0 - 2\delta + r)\left[2R_0 - \delta - r\right]^2 \operatorname{sign}(2R_0 - \delta - r) \\ - 2(2R_0 - r)^2\left[r(4R_0 + r) - 3\delta^2\right]\operatorname{sign}(2R_0 - r) \end{array} \right\} \tag{58}$$

where R_0 is the radius of the center of the shell, δ is the thickness of the shell, and M is the total mass of the shell, as shown in Fig. 6. Thus the outer radius of the shell is $R_0 + \delta / 2$, and the inner radius is $R_0 - \delta / 2$

By using the Eq. (58) we can work out the RPDF between the two shells in a closed form. However the closed form is rather long. In this paper we give the results for two limiting cases: the infinitely thin shell limit and the homogeneous solid sphere. When the thickness of the shell δ is infinitely thin, the RPDF between the two infinitely thin shells is of radius R_0, with the centers separated by D is:

$$G_{thin\ shell}(r) = \frac{M}{4DR_0^2}\left[2R_0 - |r - D|\right] \tag{59}$$

In the solid sphere limit, the RPDF between two solid spheres of radii R_s, with the centers separated by D is:

$$G_{sphere}(r) = \frac{3M}{160DR_s^6}\left[\left(2R_s - |D - r|\right)^3\left(\left(3R_s + |D - r|\right)^2 - 5R_s^2\right)\right]. \tag{60}$$

Both the RPDF of a pair of infinitely thin shells and the RPDF of a pair of solid spheres are shown in Fig. 7.

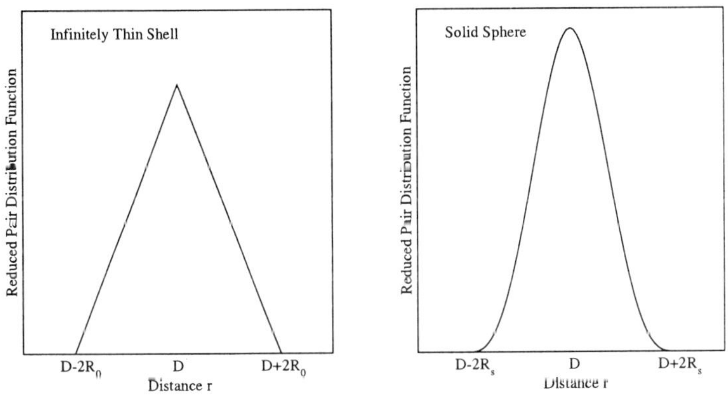

Figure 7. This figure shows the intermolecular reduced radial distribution function $G(r)$ for a pair of spherical molecules with mass M as shown in Fig. 6, whose centers are separated by D. The left panel is for a pair of infinitely thin shells with radii R_0 and the right panel for a pair of solid spheres with radii R_s

6. RESOLUTION FUNCTION

In comparing experimental PDF's and theoretical predictions using models, one is often ultimately limited by the instrumental resolution function, with good experimental data and when a reliable model is available. Surprisingly little attention has been given to this in the past, and it is often hard to obtain a resolution function for a given instrument. In this section we explore a particular resolution function that we have found to be tractable mathematically and useful experimentally.

Suppose in an experiment, the actual intensity $S(q)$ measured is convoluted with an instrumental resolution function to give the measured intensity $S_c(q)$

$$C(q,q') = \frac{q'}{q}\frac{1}{\sqrt{2\pi\sigma_q^2}}e^{-\frac{(q-q')^2}{2\sigma_q^2}} \tag{61}$$

which is a modified Gaussian that is normalized to unity when integrated over q'. The factor q'/q is included to simplify the analysis later. The effect of the extra factor q'/q is to increase the convolution function for $q' > q$ and to decrease the convolution function for $q' < q$, but this effect is small as $C(q,q')$ is strongly peaked around $q' \approx q$. We choose the width σ_q to be

$$\sigma_q^2 = \sigma_0^2 + \delta^2 q^2 \tag{62}$$

so that the width increases as q increases. We adopt this form primarily as it is easy to work with mathematically, as well as being useful in some real experimental situations.[10] The observed convoluted intensity $S_c(q)$ then becomes

$$S_c(q) = \int S(q')C(q,q')dq'. \tag{63}$$

This expression can be rewritten as

$$q[S_c(q)-1] = \int q'[S(q')-1]\frac{q}{q'}C(q,q')dq' \tag{64}$$

where the factor q/q' in (64), removes the factor q'/q introduced in (61).

We are interested in the effect of this convolution in real space, and so write

$$\begin{aligned} q[S(q)-1] &= \int G(r)e^{iqr}dr \\ q[S_c(q)-1] &= \int G_c(r)e^{iqr}dr \end{aligned} \tag{65}$$

The exponentials in (65) can be replaced by sines if both $G(r)$ and $q[S(q)-1]$ are anti-symmetrized so that (4) is obtained. This is always possible as both $G(r)$ and $q[S(q)-1]$ have no physical meaning for negative arguments r and q respectively. The companion equation to (64) in real space is

$$G_c(r) = \frac{1}{2\pi}\int G(r')C(r,r')dr' \tag{66}$$

where

$$\begin{aligned} C(r,r') &= \frac{1}{2\pi}\int e^{-iqr}e^{iq'r'}F(q,q')dqdq' \\ &= \frac{1}{2\pi}\int e^{iq(r'-r)}e^{-r^2\sigma_q^2/2}dq \end{aligned} \tag{67}$$

Using the particular choice for σ_q given in (62), we find that

$$C(r,r') = \frac{1}{\sqrt{2\pi r^2\delta^2}}e^{-\frac{r^2\sigma_0^2}{2}}e^{-\frac{(r-r')^2}{2r^2\delta^2}}. \tag{68}$$

Thus the result in real space is that the observed RPDF $G_c(r)$ is obtained from the actual RDPF $G(r)$ via (66) and (68).

The function $C(r,r')$ splits up into a multiplicative part involving σ_0, which is a result that is often used, and corresponds to assuming that the width σ_q is independent of q. The additional term involving δ in (62) leads to a broadening of the peaks in the RDDF; a broadening that increases as the separation r increases. This can be seen most easily if we use a single Gaussian peak with width σ_{ij} whose lineshape before convolution is

$$\frac{1}{\sqrt{2\pi\sigma_{ij}^2}}e^{-\frac{(r-r_{ij})^2}{2\sigma_{ij}^2}} \tag{69}$$

and which after convolution with (68) becomes

$$\frac{1}{\sqrt{2\pi(\sigma_{ij}^2 + r^2\delta^2)}}e^{-\frac{r^2\sigma_0^2}{2}}e^{-\frac{(r-r_{ij})^2}{2(\sigma_{ij}^2+r^2\delta^2)}}. \tag{70}$$

Thus the total *weight* in the peak is reduced by the term involving σ_0. This gets more pronounced as r increases. The *width* of the peak is increased by the δ term from σ_{ij} to

$$\sqrt{\sigma_{ij}^2 + r^2\delta^2}\ , \tag{71}$$

which also becomes greater at larger r. Interestingly, the larger width at larger q in (62) also results in a larger peak width at large r in the PDF. This is a little counter intuitive, as there is usually an inverse relation with Fourier transforms. The form (71), for the width in the Gaussian PDF peak, has been shown empirically to provide an improved fit in some cases by Toby and Egami[10]. In this section, we have provided a justification for the origin of this term in real space, by finding the partner in q space (61) and (62) that leads to this result.

This result can be used to obtain better fits to experimental PDF peaks, when using a model, particularly at larger distances r as we show in the next section. It is also possible to generalize (61) to involve a series of similar terms with varying widths and weights via

$$C(q,q') = \sum_l \frac{w_l}{\sqrt{2\pi\sigma_{lq}^2}} e^{-\frac{(q-q')^2}{2\sigma_{lq}^2}} \tag{72}$$

where the weights w_l are normalized to unity $\sum_l w_l = 1$ and the widths σ_{lq} are given by

$$\sigma_{lq}^2 = \sigma_{l0}^2 + \delta_l^2 q^2 \tag{73}$$

so that the convolution (68) generalizes to

$$C(r,r') = \sum_l \frac{w_l}{\sqrt{2\pi r^2 \delta_l^2}} e^{-\frac{r^2\sigma_{l0}^2}{2}} e^{-\frac{(r-r')^2}{2r^2\delta_l^2}} \tag{74}$$

and for the case of the Gaussian peak (69), the result (70) generalizes to

$$\sum_l \frac{w_l}{\sqrt{2\pi(\sigma_{ij}^2 + r^2\delta_l^2)}} e^{-\frac{r^2\sigma_{l0}^2}{2}} e^{-\frac{(r-r_{ij})^2}{2(\sigma_{ij}^2+r^2\delta_l^2)}} \tag{75}$$

which is now a sum of Gaussians of varying weights and widths. We have not used the form (75) yet as the single Gaussian proved adequate, as shown for the example in the next section.

The convolution function (68) should be applied to the RPDF $G(r)$ as derived here. The RPDF is strictly Gaussian for the isotropic case $\sigma_x = \sigma_y = \sigma_z$, which gives the result (33) [remember that the PDF $P(r) = rG(r)$]. We have made extensive numerical tests, using the data discussed in the next section and shown in Fig. 9, and we find that there is no discernible difference between doing a convolution involving (68) with either the RPDF $G(r)$, the PDF $P(r)$ or the density $\rho(r) - \rho_0$ with of course appropriate factors of r included before and after the convolution. The same is also true for $\rho(r)$ if the uniform density ρ_0 is added to $\rho(r) - \rho_0$ before the convolution, and then subtracted again afterwards. Thus with a series of Gaussian peaks comprising the PDF as is often assumed, one can proceed to use the broadened and reduced result (70) for each individual peak in the PDF. If the peaks are not Gaussian, as would be the case for example with flexible molecules like ethane at high temperature, then the convolution integral involving (68) must be done numerically. This is also true for other lineshapes that are not Gaussian like those shown in Fig. 7.

7. DISCUSSION

In the previous sections, we have given notes on some thoughts that arise when fitting experimental PDF's using models of known structure and some appropriate dynamical matrix. In this section we bring together a number of these points to obtain a fit to the PDF of a fullerite crystal, at a high enough temperature that the buckyballs can be thought of as free rotators.

In order to see maximum deviations from Gaussian lineshape, the system should exhibit significantly different values of $\sigma_x, \sigma_y, \sigma_z$. We have seen in section 4, that these deviations seem to be small. Another anisotropic molecule is a buckyball where the atomic vibrations in the C_{60} molecule are different in the tangential and radial directions. The chemistry of C_{60} molecules leads to situation when intramolecular interactions are significantly stronger then intermolecular interactions. Thus in the first approximation the intermolecular interaction can be neglected.

First we compute the RRDF of the single buckyball. In order to verify the precision of our calculations, we calculate the RPDF of the fullerite crystal and compare it with high quality RPDF measurements. In order to take into account contributions to the PDF from different molecules, we will adopt the thin shell continuum approach discussed in section 5. We will also make corrections to the calculated PDF due to the finite resolution of the experimental measurements. We find that the corrections to the Gaussian approximation are so small as not to be visible on the scale of Fig. 8 and so are not shown. This result is quite surprising, as some anisotropy is present as shown in Table 2, but is consistent with our general conclusions.

Table 2. This table shows the anisotropy of the widths $(\sigma_x, \sigma_y, \sigma_z)$ associated with various pairs of atoms within a single buckyball listed in order of increasing separation, and given in the principal axes of the displacement–displacement matrix (30). The calculations are for a temperature of 320K using the semi-empirical AM1 method[9] within Gaussian98. The other peaks at larger distances have comparable widths and anisotropies.

Distance [Å]	PDF widths $(\sigma_x, \sigma_y, \sigma_z)$ [Å]
1.38	0.0664, 0.0526, 0.0411
1.46	0.0654, 0.0506, 0.0447
2.37	0.0730, 0.0550, 0.0504
2.47	0.0722, 0.0543, 0.0522
2.85	0.0729, 0.0588, 0.0527

In order to calculate the RRDF of the molecule, we have to calculate average (relative) displacements of the atoms due to their vibrations, and then obtain $(\upsilon_x, \upsilon_y, \upsilon_z)$. The matrix of (relative) average displacements can be calculated if the eigenfrequencies and eigenvectors of molecular vibrations are known

$$<(u_i - u_j)_\alpha (u_i - u_j)_\beta> = \sum_\lambda \frac{\hbar}{\omega_\lambda}\left(n_\lambda + \frac{1}{2}\right)\left(\frac{e^\lambda_{i\alpha} e^\lambda_{i\beta}}{m_i} + \frac{e^\lambda_{j\alpha} e^\lambda_{j\beta}}{m_j} - \frac{e^\lambda_{i\alpha} e^\lambda_{j\beta}}{\sqrt{m_i m_j}} - \frac{e^\lambda_{j\alpha} e^\lambda_{i\beta}}{\sqrt{m_i m_j}}\right) \quad (76)$$

where $<(u_i-u_j)_\alpha(u_i-u_j)_\beta>$ is the matrix of relative atomic displacements. The frequencies of molecular vibrations ω_λ. $e^\lambda_{i\alpha}$ are orthogonal. Normalized Cartesian components of the 174 atomic displacements correspond to the 174 eigenfrequencies ω_λ. The mass of atom i is $m_i$and n_λ is the Bose-Einstein function.

The matrix of atomic displacements can be diagonalized and its eigenvalues are the squares of the widths in the principal axes $\sigma_x, \sigma_y, \sigma_z$. The coordinates of pairs of atoms in the principal axes of the pair can be found. After that the application of the derived formulas for non-Gaussian line shape, as given in section 4 using (44) is straightforward. In order to calculate eigenfrequencies and eigenvectors of a single buckyball, we use two semi–empirical methods[9] AM1 and CNDO in Gaussian98. The results for the PDF were quite similar in both cases and so we give only the results for the buckyball at 320K obtained with the AM1 method.

We see from table 1 that although there is some anisotropy in the widths σ, nevertheless the RPDF shown in Fig. 8 is almost identical for the Gaussian, using the first term in (44), and non-Gaussian lineshapes (44). In the insert in Fig. 8, we show the small differences in more detail.

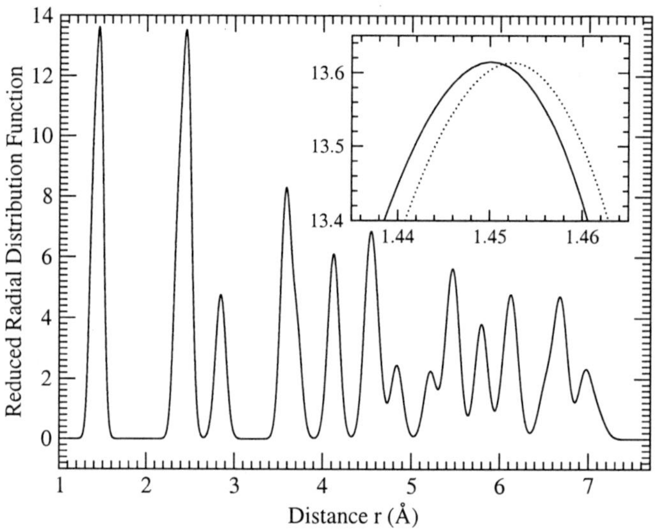

Figure 8. The RRDF [see (4)] of a single buckyball at 320K is shown. The solid line is the Gaussian approximation which is the first term in (44), while the dashed line also has the corrections involving $A_1(r)$ and $A_2(r)$ from (44). It can be seen that the corrections are small as emphasized for the first peak in the insert.

The picture shows that the correction to the widely used Gaussian line shape approximation due to the anisotropy of vibration in case of the fullerene molecule is negligible. We found similar results in section 4 for benzene. The difference between the Gaussian approximation and the more precise approximation can be seen only on the inset.

As well as the contribution to the RPDF from a single buckyball, another contribution comes from the intermolecular correlations. At room temperature, fullerene molecules rotate around their centers. The shape of the molecule is approximately spherical. Thus we use approximation of the thin shell (59) in order to take into account the contribution to the RPDF from contributions involving pairs of buckyballs. These contributions are then summed over all pairs of buckyballs in the fullerite crystal. The total RPDF includes both of these contributions and is shown in Fig. 9.

The C_{60} neutron powder diffraction data were collected from ~ 2 gm. of C_{60} powder on the GLAD diffractometer at the Intense Pulsed Neutron Source (IPNS) at room temperature. The data were corrected for background, absorption, multiple scattering and normalized using the program PDFgetN.[13] There has been extensive diffraction studies on fullerite over the past decade. A good overall reference is Nield and Keen[14], and references therein. An earlier PDF study was done by Li et al[15] and by Soper et al[16] Intermolecular correlations have been studied by Copley et al[17] and as well as high temperature studies[18-20].

In the fullerite crystal, the centers of the fullerene molecules are situated at the sites of a face centered cubic lattice[11] with the distance between the centers of nearest neighbors being 10Å. We first calculate the reduced pair distribution function $G(r)$ without any correction due to the final resolution of experimental measurements. We then convolute our result for the RPDF with the resolution function (68). We calculate the integral (66) numerically with the values of parameters used being $\sigma_0 = 0.020\text{Å}^{-1}$, $\delta = 0.015$, which give the remarkable agreement with the experiment as seen in Fig. 9. It is particularly interesting that the triangular type peaks from the left-hand panel of Fig. 7 can be clearly seen beyond 8Å.

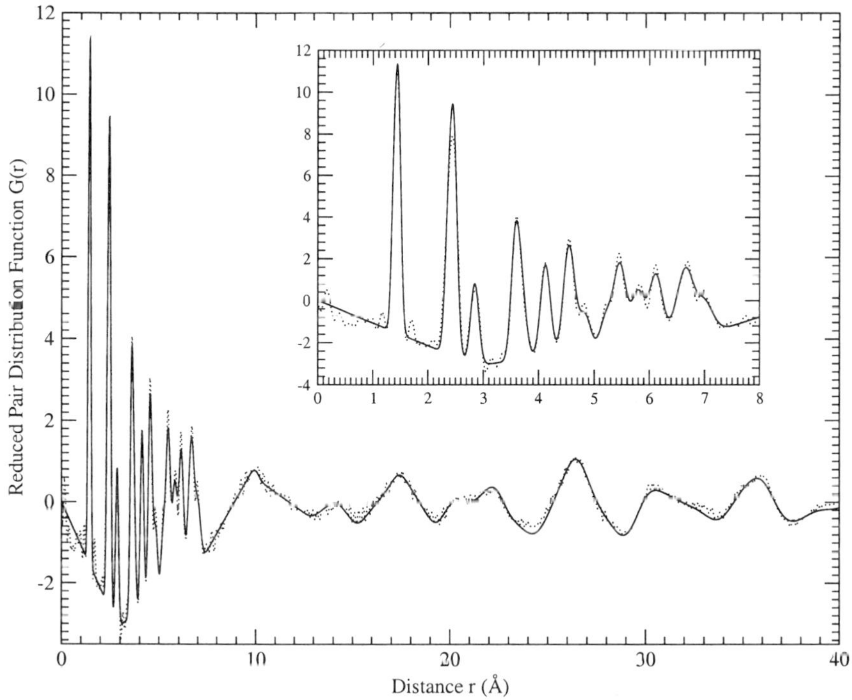

Figure 9. The RPDF of a fullerite crystal 300K is shown. The dashed line is the experimental result of Billinge and Petkov[11] and the solid line is the fit discussed in this section. The single buckyball results are taken from Fig. 8, and the other peaks are from the continuum approximation discussed in section 5, using infinitely thin spherical shells. The instrumental resolution function is also included in the calculated solid line as discussed in the text.

8. CONCLUSIONS

In these notes we have collected together some ideas that we have found useful in analyzing PDF data. With the improvements in experimental data that have taken place in the last few years, notably getting data out to larger wave-vectors q, many corrections that could previously be neglected are now required. When the experimental data is transformed into real space data, it is up to the modelers to take account of these effects. In this paper, we have assumed that the vibrations of the molecule or solid is described by a harmonic potential. At elevated temperatures, or where the molecule or solid contains flexible units, this approach would have to be modified.

9. ACKNOWLEDGEMENTS

We should like to thank Valeri Petkov for useful discussions and the DOE for financial support under contract number # DE-FG02-97ER45651. SJLB would like to thank Dr. A.P. Ramirez and Dr. R.C. Haddon for supplying the C_{60} sample. The IPNS facility is operated by the USDOE Basic Energy Sciences under IPNS contract # W-31-109-Eng-38.

REFERENCES

1. B. E. Warren, *X-Ray Diffraction*, (Dover Publications, Inc, New York, 1990).
2. Jean Chung and M. F. Thorpe, Phys. Rev. **B 55**, 1545 (1997).
3. Jean S. Chung and M. F. Thorpe, Phys. Rev. **B59**, 4807 (1999).
4. M.F. Thorpe, J.S. Chung, S.J.L. Billinge and F. Mohiuddin-Jacobs in *Local Structure from Diffraction,* Ed. By S. J. L. Billinge and M. F. Thorpe (Plenum Press, New York, 1998) page 157.
5. I. S. Gradshteyn and I.M. Ryzhik, *Table of Integrals Series and Products* (Fourth Edition, Academic Press, New York, 1965)
6. V. Petkov, I-K. Jeong, J.S. Chung, M. F. Thorpe, S. Kycia and S.J.L. Billinge, Phys. Rev. Letts. **83**, 4089 (1999).
7. D.A. Dimitrov, H. Röder and A.R. Bishop, Phys. Rev. **B 64**, 14303, (2001).
8. A.A. Maradudin, E.W. Montroll, G.H. Weiss and I.P Ipatova, Solid State Physics, Supplement 3, *Theory of Lattice Dynamics in the Harmonic Approximation* (Academic Press, New York, 1971).
9. M. J. Frisch et al., *Gaussian 98*, Revision A.7, Gaussian, Inc., Pittsburgh PA, 1998; M. J. S. Dewar, E. G. Zoebisch and E. F. Healy, J. Amer.Chem. Soc. **107**, 3902 (1985).
10. B.H. Toby and T. Egami, Acta Cryst. **A 48**, 336 (1992).
11. S.J.L. Billinge and V. Petkov, unpublished.
12. *The Fullerenes*, edited by Harold R. Kroto, John E. Fischer and David E. Cox, (Pergamon Press, Oxford, (1993)
13 P. F. Peterson, M. Gutmann, Th. Proffen and S. J. L. Billinge, J. Appl. Crystallogr. **33**, 1192 (2000)
14. V. M. Nield and D. A. Keen, in *Diffuse Neutron Scattering from Crystalline Materials*, p 251 (Oxford Science publications, Oxford, 2001).
15. J.-C. Li, D. Ramage, J. S. Lannin and J. Conceicao, Phys. Rev. **B 44**, 13167 (1991).
16. A. Soper, W. I. F. David, D. S. Sivia, T. J. S. Dennis, J. P. Hare, K. Prassides, J. Phys: Condensed Matter, **4**, 6087 (1992).
17. J. R. D. Copley, D. A. Neumann, R. L. Cappelletti, W. A. Kamitakahara, E. Prince, N. Coustel, J. P. McCauley, N. C. Maliszewskyj, J. E. Fischer, A. B. Smith, K. M. Creegan and D. M. Cox, Physica **B 180**, 706 (1992); J. R. D. Copley, D. A. Neumann, R. L. Cappelletti and W. A. Kamitakahara, J. Phys. Chem. Solids, **53**, 1353 (1992).
18. F. Leclercq, P. Damay, M. Foukani, P. Chieux, M. C. Bellissent-Funel, A. Rassat and C. Fabre, Phys. Rev. **B 48**, 2748 (1993).
19. P. Damay and F. Leclercq, Phys. Rev. **B 49**, 7790 (1994).
20. L. Pintschovius, Phase Transitions, **67**, 295 (1998).

LOCAL STRUCTURE: THE REALM OF THE CHEMIST?

Simon J. Hibble and Alex C. Hannon*

1. INTRODUCTION

1.1. The Chemists' View Of Structure

Chemists have always thought about local structure, and indeed this is often all that interests them. For example, given the crystal structure[1] of $NaAg(CN)_2$, the chemist will extract the $[Ag(CN)_2]^-$ complex ion and produce a structural diagram of the ion, Fig. 1. Structure at the local level (i.e. the connectivity of atoms, interatomic distances, and bond angles) is what interests the average chemist. This is what a chemist means by "chemical structure". Chemists also believe they have, or they seek to gain, an understanding of structure at this level. When they examine the structure of the $[Ag(CN)_2]^-$ ion they see the linear co-ordination around silver. This is what the chemist expects for Ag^I, a d^{10} ion, from their knowledge of the chemistry of silver and of the other elements, copper and gold, in the same group.

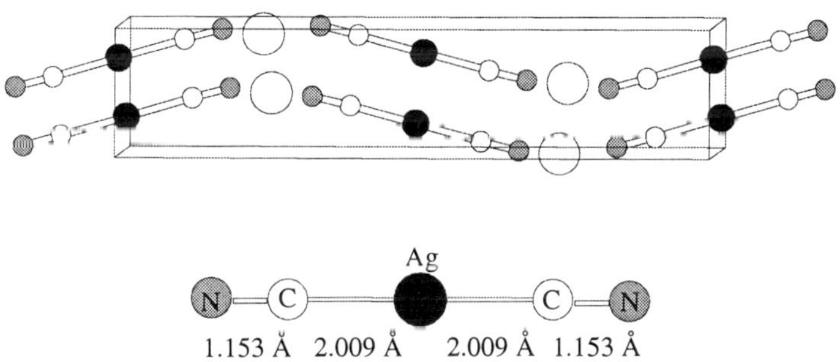

Figure 1. The structure of $NaAg(CN)_2$ and the $[Ag(CN)_2]^-$ ion (large open circle Na, others as in ion). Light lines indicate the unit cell.

* Simon J. Hibble, Department of Chemistry, University of Reading, Whiteknights, Reading, RG6 6AD, UK. Alex C. Hannon, ISIS Facility, Rutherford Appleton Laboratory, Chilton, Didcot, Oxon OX11 0QX, UK.

Even solid state chemists, who deal with extended solids, break down structures into units and describe structures in terms of units joined by shared atoms. For example, they view platinum sulfide[2], PtS, as a network formed from square planar PtS_4 units linked together by sharing sulfur atoms in an ordered fashion, Fig. 2. They focus on the square planar coordination around Pt^{II}, which has the outer electronic configuration of d^8. They understand that the geometry adopted is a result of the greater importance of the ligand field stabilisation energy relative to electron-electron repulsion energies for this third row transition metal.

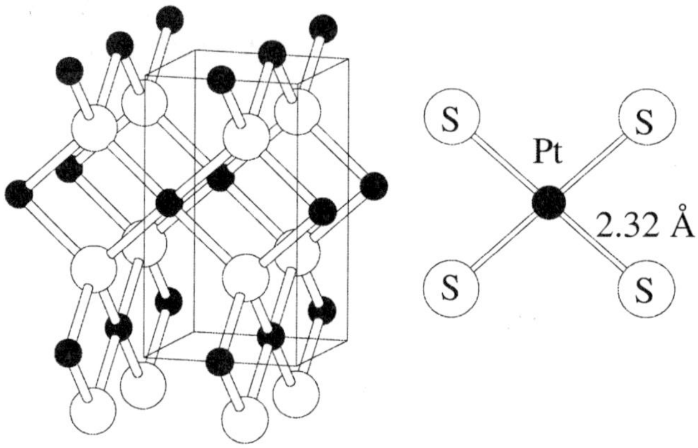

Figure 2. The PtS structure, with square planar PtS_4 groups joined to form a network in an ordered fashion. Light lines indicate the unit cell.

It is perhaps the way in which chemists describe and view structures that differentiates them from physicists. The chemist starts with the molecular, whereas the physicist will have begun his scientific life considering single atoms and collections of atoms without ever drawing bonds between them.

1.2. Average Structure And The Chemist

The average chemist never thinks about the average structure of a material, the meaning of this term, or its implications. Exceptional chemists know that the structure of a material obtained from diffraction studies is an average over many units. However, even they normally assume that the units are identical and that the average structure, given by a set of average atomic positions obtained from the interpretation of Bragg diffraction experiments, is an accurate representation of the local structure. There may be the odd problem, e.g. disorder of a large cation, such as the tetraphenylphosphonium ion $(C_6H_5)_4P^+$, but these can often be ignored since chemists are not interested in this part of the structure. They are content that the problem "disappears", absorbed into high thermal parameters, or is dealt with by disorder over many sites, so long as it does not affect the part of the structure that concerns them.

It is, however, becoming clear, even to chemists, that there are many materials of interest for which the average structure determined from Bragg scattering gives a misleading or incorrect picture of the local structure. In this work, we will discuss the determination of the local structure in disordered crystalline materials and how relating the local and average structures gives a more complete structural description of materials.

Having first disparaged chemists, we will demonstrate in this article, using examples from our own work, that not only does the chemist have much to gain from the study of disordered materials, but also much to contribute. Chemists have an insight into local structure which can be extremely useful in our aim to progress beyond the average atomic structure in the description of materials. Not only does the chemist's insight help us to describe and understand local structures, but it is also an essential tool in determining the structure of disordered materials. We will show how this approach can be extended to the study of amorphous materials.

The approach we adopt in determining the structure of disordered crystalline materials is summarised in Fig. 3. Our goal is to determine the true structure of a material. If we knew this, we would know the local structure in the material, the goal of the chemist. The experiments we carry out either yield a crystal structure (from Bragg scattering), which contains information on the average positions of atoms, or correlation functions (from total diffraction or extended X-ray absorption fine structure (EXAFS) studies), which contain information on the internuclear distances in the material. By combining the information from these two types of experiment with chemical knowledge, we aim to determine the true structure of materials. In reality, we can of course only move nearer to this ideal. However, we will show that by combining these different types of information, and using them to constrain our final model, we can make significant advances towards that goal.

2. BEYOND THE AVERAGE STRUCTURE

2.1. Disordered Crystalline Materials

The average structure of a disordered crystalline material determined from Bragg scattering data conceals the true local structure within the material[3]. The description of

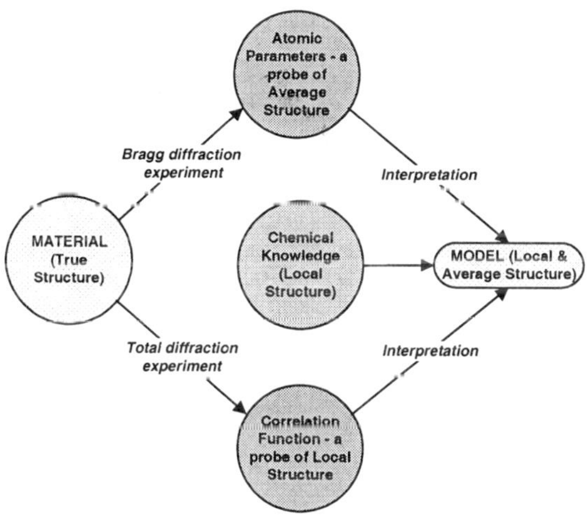

Figure 3. A flowchart showing how we produce a model that describes both the average and local structure of disordered crystalline materials.

the structure is deficient. Although the space group, unit cell parameters, and atomic parameters describe the average structure, they give an incorrect picture at the local level. Obvious problems arise when sites are fractionally occupied; on a local scale they are either occupied or vacant, and they are only occupied by one type of atom at a time. Although some thermal disorder must always be present, even in a perfectly crystalline material, abnormally high atomic displacement factors are a sign of problems in the description of the structure. An anomalously large thermal factor from an analysis of the Bragg diffraction is an indication that at a local level the structure is different from that predicted by the average structure. In this way, high atomic displacement factors absorb deficiencies in the model and mask the true magnitude of the inadequacy of this form of structure description. Studies at low temperature, in which true thermal disorder is minimised, provide useful information on the magnitude of the problems remaining in describing the structure, and offer insights into which parts of the model are most problematical. They also provide high quality total diffraction data for use in a more complete description of the structure. For this reason, we normally make low temperature (~10K) measurements to solve structural problems, with additional measurements at higher temperatures if we are interested in how the structure changes with temperature.

2.2. The Use Of Total Diffraction Data And The Total Correlation Function, *T(r)*

Total diffraction data, which includes the Bragg scattering, can be used to yield the total correlation function, $T(r)$, for a material: A total diffraction experiment measures the differential cross section[4, 5]

$$\frac{d\sigma}{d\Omega} = I^S(Q) + i(Q), \tag{1}$$

where $I^S(Q)$ is the self scattering, $i(Q)$ is the distinct scattering, and Q is the magnitude of the scattering vector (momentum transfer) for elastic scattering, given by

$$Q = \frac{4\pi \sin\theta}{\lambda}. \tag{2}$$

The self scattering is calculated within an approximation and then subtracted. The distinct scattering is multiplied by Q to give the interference function, $Qi(Q)$, which may be Fourier transformed to yield the total correlation function

$$T(r) = T^0(r) + \frac{2}{\pi}\int_0^{Q_{max}} Qi(Q)M(Q)\sin(rQ)dQ\ , \tag{3}$$

where

$$T^0(r) = 4\pi r g^0\left(\sum_l c_l \bar{b}_l\right)^2, \tag{4}$$

and $M(Q)$ is a modification function (used to reduce termination ripples due to the finite maximum momentum transfer, Q_{max}, of the experimental data), $g^0(=N/V)$ is the macroscopic number density of scattering units, and $\overline{b}_l$ is the coherent neutron scattering length for element l for which the atomic fraction is c_l ($= N_l/N$). The modification function used in this work is that due to Lorch[6]

$$M(Q) = \frac{\sin(Q\Delta r)}{Q\Delta r}, \tag{5}$$

where $\Delta r = \pi/Q_{max}$.

For an ordered crystalline material, extracting the total correlation function, $T(r)$, from experimental data is a time consuming task which provides no extra information beyond that available from an analysis of the Bragg diffraction. Indeed, $T(r)$ can be calculated easily from the crystal structure[7]. This is done by calculating the partial pair distribution functions, $g_{ll'}(r)$, from the lattice parameters and atomic coordinates of the model according to

$$g_{ll'}(r) = \frac{1}{\sum_{i=1}^{N_l} w_i} \sum_{j=1}^{N_l} \sum_{\substack{j'=1 \\ j \neq j'}}^{N_{l'}} w_j w_{j'} \left\langle \delta\left(\mathbf{r} + \mathbf{R}_j - \mathbf{R}_{j'}\right) \right\rangle, \tag{7}$$

where $\mathbf{R}_j$ and w_j are the position and occupancy for the jth atom respectively. The j,j' sums are over the $N_l,N_{l'}$ atoms of element l,l' and $j \neq j'$ means that j and j' are not allowed to refer to the same atom. $g_{ll'}(r)$ may be interpreted as the number density of atoms of element l' at a distance r ($= |\mathbf{r}|$) from an origin atom of element l, averaged over all possible origin atoms and directions of $\mathbf{r}$. The weighted partial correlation functions

$$t_{ll'}(r) = 4\pi r g_{ll'}(r) \tag{8}$$

are then summed to yield the total correlation function

$$T(r) = \sum_{ll'} c_l \overline{b}_l \overline{b}_{l'} t_{ll'}(r), \tag{9}$$

where the l and l' summations are both over the elements of the sample. The functions $t_{ll'}(r)$ must be broadened to include the effect of thermal motion in the material, and in the harmonic approximation this is achieved using a Gaussian function.

3. AN ORDERED CRYSTALLINE MATERIAL, $Zn_2Mo_3O_8$

Figure 4 shows the interference function, $Qi(Q)$, for $Zn_2Mo_3O_8$. A comparison[8] of the experimentally determined $T(r)_{exp}$, calculated from $Qi(Q)$, with $T(r)_{mod}$, calculated from the crystal structure, is given in Fig. 5. The individual pair correlation functions have each been broadened using mean square atomic displacements derived from the temperature factors given by a Rietveld refinement of the Bragg scattering data. Figure 5 shows the quality of agreement that should be achievable when one has a good

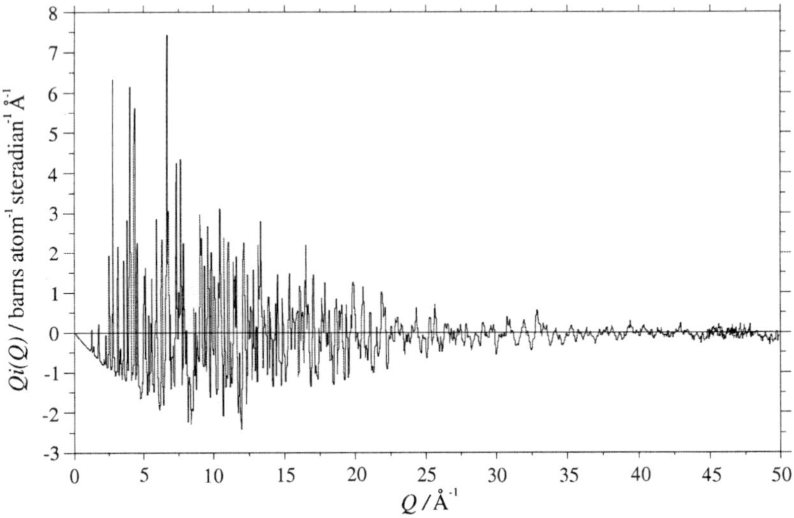

Figure 4. The interference function, $Qi(Q)$, for $Zn_2Mo_3O_8$.

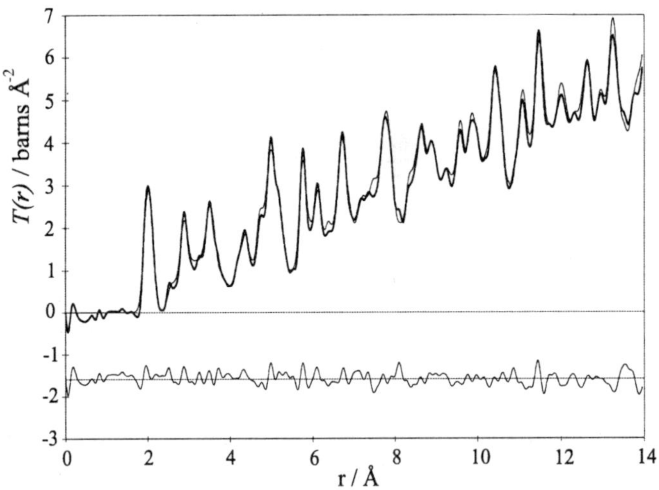

Figure 5. The total correlation function, $T(r)_{exp}$, measured by total neutron diffraction (heavy line), and $T(r)_{mod}$ (light line) calculated from the crystal structure for $Zn_2Mo_3O_8$.

description in the local structure in a material. Such a description is readily available in the case of ordered crystalline materials such as $Zn_2Mo_3O_8$.

The compound zinc molybdate[9], $Zn_2Mo_3O_8$, is of particular interest to chemists because it contains Mo_3O_{13} groups in which the molybdenum atoms are bonded together to form metal-metal bonded ($d_{Mo\text{-}Mo}$ = 2.58 Å) triangles, Fig. 6. It is the prototype for what is now an extensive class of reduced oxides of molybdenum which contain Mo-Mo bonded clusters. The Mo_3O_{13} unit is found in many solids and can also be produced in solution as the $[Mo_3O_4(OH_2)_9]^{4+}$ ion. In $Zn_2Mo_3O_8$, the Mo_3O_{13} groups are linked together to form $Mo_{3/4}O_2$ layers, as shown in Fig. 7(a), and these layers are held together by the zinc ions which are on octahedral and tetrahedral sites between the layers.

4. TOTAL DIFFRACTION STUDIES OF DISORDERED CRYSTALLINE MATERIALS

4.1. Li_2MoO_3

For a disordered crystalline material, the average structure obtained from Bragg diffraction alone does not provide a good description of the local structure, even though it may give a reasonable fit to the Bragg peaks. Figure 8 shows the results of a Rietveld refinement using the model of James and Goodenough[10] and our neutron diffraction data collected on Li_2MoO_3, a disordered crystalline material, at 12K[11]. Although the fit to the

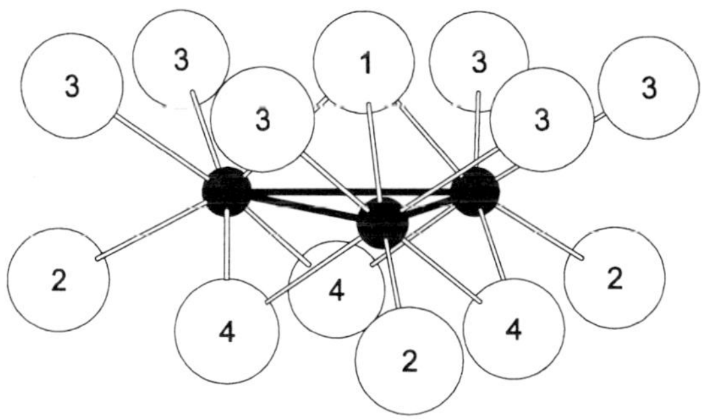

Figure 6. A Mo_3O_{13} cluster (solid circles, Mo; open circles, O) showing the numbering scheme used in Table 1.

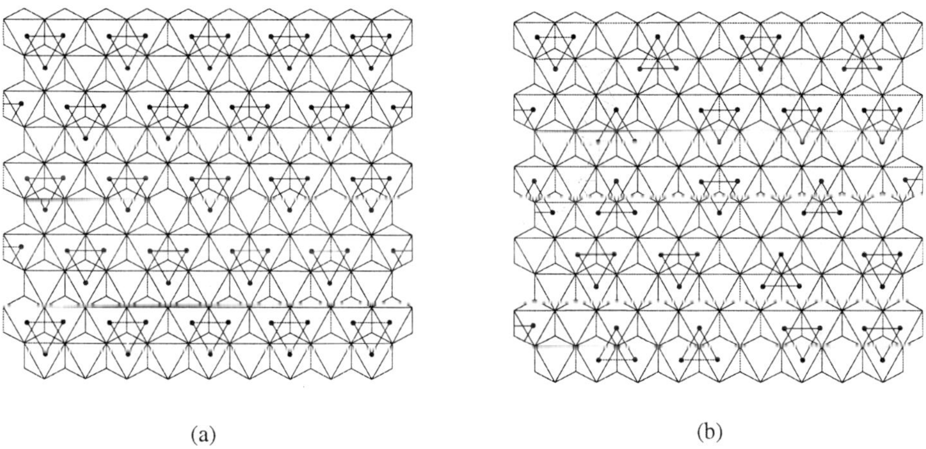

Figure 7. (a) Ordered Mo_3 triangles in the $[Mo_{3/4}O_2]$ layers in $Zn_2Mo_3O_8$; (b) Disordered Mo_3 triangles in the $[Li_{1/3}Mo_{2/3}O_2]$ layers in Li_2MoO_3.

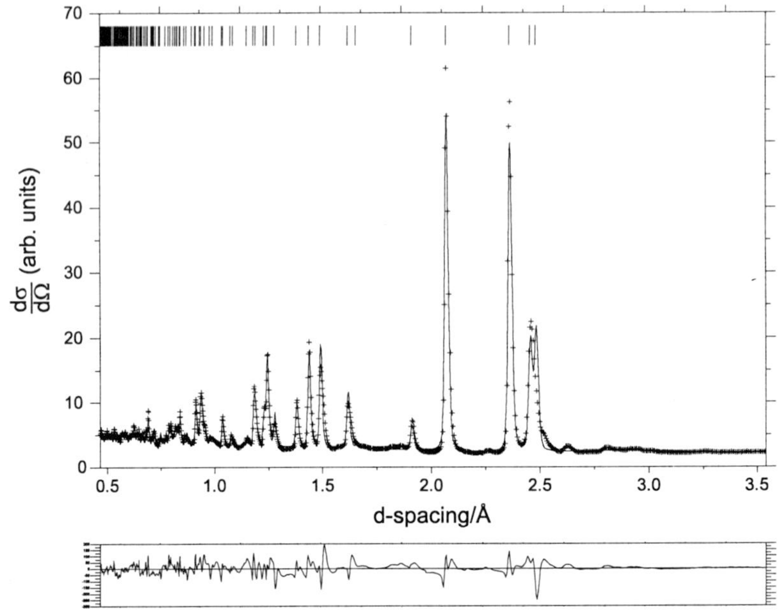

Figure 8. A neutron diffraction Rietveld profile refinement of the average structure of Li_2MoO_3. Observed data are indicated by crosses and calculated data by a solid line. Tick-marks directly above the pattern indicate the positions of reflections. A difference curve appears at the bottom.

powder diffraction profile appears acceptable, a comparison of $T(r)_{exp}$ with $T(r)_{mod}$ for Li_2MoO_3, calculated from the structural parameters from Rietveld refinement, Fig. 9, shows that the description of the local structure by this model is totally inadequate.

Although it is not immediately obvious, the lithium molybdate Li_2MoO_3, is closely related to $Zn_2Mo_3O_8$. Both materials adopt layer structures based on MO_2 sheets and both contain Mo^{IV}. In $Zn_2Mo_3O_8$, the sheets have the composition, $Mo_{3/4}V_{1/4}O_2$, where V is a vacancy and the Mo atoms are bonded to form Mo-Mo bonded triangles in the ordered manner shown in Fig. 7(a). In Li_2MoO_3, the sheets have the composition, $Mo_{2/3}Li_{1/3}O_2$, and the sheets are held together by additional interlayer lithium ions; this is perhaps clearer in the alternative formulation of Li_2MoO_3 as $Li[Mo_{2/3}Li_{1/3}]O_2$.

To the chemist, the poor agreement between $T(r)_{exp}$ with $T(r)_{mod}$ for Li_2MoO_3 is not surprising, since the average structure does not contain the Mo_3 metal-metal bonded units that the chemist "knows" will be present. So confident were James and Goodenough that, without any direct structural evidence, they drew a structure containing disordered Mo_3 triangles, Fig. 7(b), in their paper reporting the structure[10]. It is unfortunate that in their Bragg diffraction study there was no evidence for the formation of Mo_3 triangles, with all Mo-Mo(Li) distances in the $[Li_{1/3}Mo_{2/3}O_2]$ layers equal at $d_{Mo\text{-}Mo(Li)} = 2.88$ Å. Although both their Rietveld refinement and ours gave reasonable agreement with the Bragg scattering, close examination of the structural parameters indicates that there are shortcomings in the normal crystallographic structure description in terms of an average unit cell. For example, the high thermal factor, $B_{Mo} = 1.2$ Å^2, found for molybdenum in our Rietveld refinement of the powder neutron diffraction data collected at 12K, is a typical symptom of such problems.

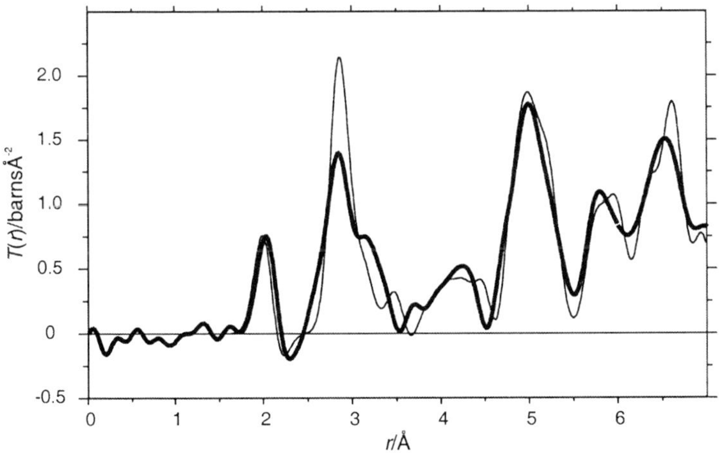

Figure 9. The total correlation function, $T(r)_{exp}$, measured by total neutron diffraction (heavy line), and $T(r)_{mod}$ (light line) calculated from the crystal structure from Rietveld refinement for Li_2MoO_3.

Extended X-ray absorption (EXAFS) studies do yield local information, and Mo K-edge EXAFS[12] provided the first direct evidence that short Mo-Mo bonds, $d_{Mo\text{-}Mo} = 2.58$ Å, and longer non-bonded distances, $d_{Mo\text{-}Mo} = 3.17$ Å, were present in Li_2MoO_3, Fig. 10. Comparison with the Mo K-edge EXAFS results for the ordered crystalline material, $Zn_2Mo_3O_8$, strongly supported the idea that there were indeed Mo_3 triangles present.

Definitive proof of this interpretation comes from producing models including Mo-Mo bonded Mo_3 triangles which can account for both the local structure seen in $T(r)$ and the average structure revealed from Bragg scattering. Figure 11 shows a comparison

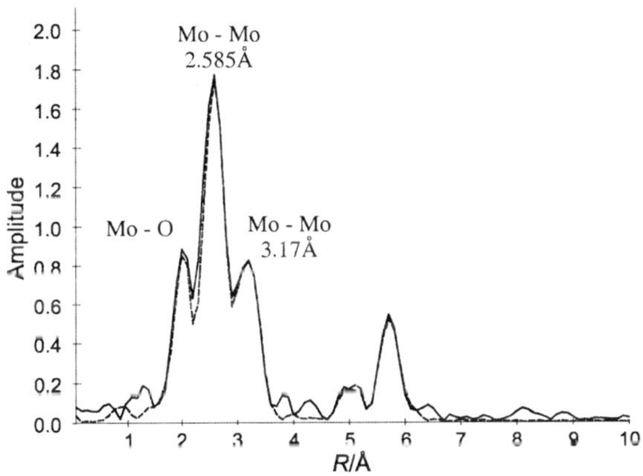

Figure 10. The Fourier transform of the Mo K-edge k^3-weighted EXAFS data for Li_2MoO_3, heavy line experimental, dotted line model.

of $T(r)_{exp}$ with $T(r)_{mod}$ for a model including Mo_3O_{13} Mo-Mo bonded units. This model uses a super-cell of the cell used in the Rietveld refinement, allowing molybdenum and oxygen atoms to be displaced from their average sites to form Mo_3O_{13} clusters. Table 1 shows how the final model, which explains the Bragg scattering, total diffraction and EXAFS results, yields a very different local structure to that obtained using only Bragg scattering, and the distances of interest to the chemist.

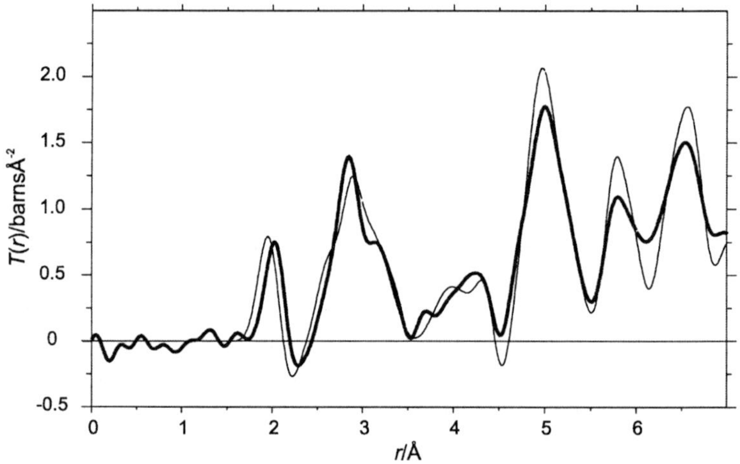

Figure 11. The total correlation function, $T(r)_{exp}$, measured by total neutron diffraction (heavy line), and $T(r)_{mod}$ (light line) calculated from the Li_2MoO_3 model incorporating the local structure.

In this case, it is easy to see how the inherent disorder in the system (the impossibility of placing Mo_3 triangles in the MO_2 layers in such a way as to occupy 2/3 of the M sites in an ordered fashion) effectively makes the Mo_3 triangles invisible to Bragg diffraction studies. By combining information from EXAFS studies with the results from Bragg diffraction studies, and adding chemical knowledge, we produced a model of the local structure in Li_2MoO_3. This model gives much better agreement with $T(r)$ than that calculated from the average structure from Bragg diffraction and we were

Table 1. Selected interatomic distances (Å) in Li_2MoO_3 for the average and final models. The oxygen labels correspond to those in Fig. 6.

Atom pair	Original model fitting only Bragg diffraction	Improved model fitting Bragg diffraction, $T(r)$ and Mo K-edge EXAFS
Mo–Mo	2.878 (×3.8 on average)	2.585 (×2) 3.170 (×1.8 on average)
Mo–O	2.031 (×6)	1.937 (x2) (O4) 1.993 (x1) (O1) 2.052 (×2) (O3) 2.079 (x1) (O2)

able to produce direct structural evidence that Li_2MoO_3 does indeed contain Mo_3O_{13} Mo-Mo bonded units. Nevertheless, there are still shortcomings in our model, and the fit to $T(r)_{exp}$ is not as good as that obtained for the model compound, $Zn_2Mo_3O_8$. Improvements to the model might be made by varying the Li positions to refine the fit to the first peak in $T(r)$. The exact form of this peak is very sensitive to lithium positions, because of the negative scattering length of Li and the overlap of Li-O and Mo-O distances at around 2 Å. It is likely that the lithium ions will relax from the idealised positions in our model. To solve these remaining problems, other statistical approaches, for example Reverse Monte Carlo (RMC) methods[13], using our model as the starting configuration might prove profitable. However, we chose to stop our modelling at this stage, having obtained the chemist's answer to the structure of Li_2MoO_3 - the identification and characterisation of the Mo_3O_{13} structural unit. Note that the first peak in the Fourier transform of the EXAFS signal, which arises only from Mo-O correlations, is well fitted by our model.

4.2. $LaMo_2O_5$

More interesting to the chemist is the use of total diffraction to determine the structure of materials containing new structural units. The structure determination of $LaMo_2O_5$ is such an example. This compound was synthesised by chemists trying to produce new molybdates containing metal-metal bonded clusters[14]. To do this, they had to produce compounds containing molybdenum in oxidation states of 4 or less. They were therefore delighted to obtain what appeared to be single crystals of a new material, $LaMo_2O_5$, with Mo in an average oxidation state of 3.5, and were confident that the material contained Mo-Mo bonded clusters. However, the material was disordered and it was to take over thirteen years to solve the structure. Single crystal X-ray diffraction

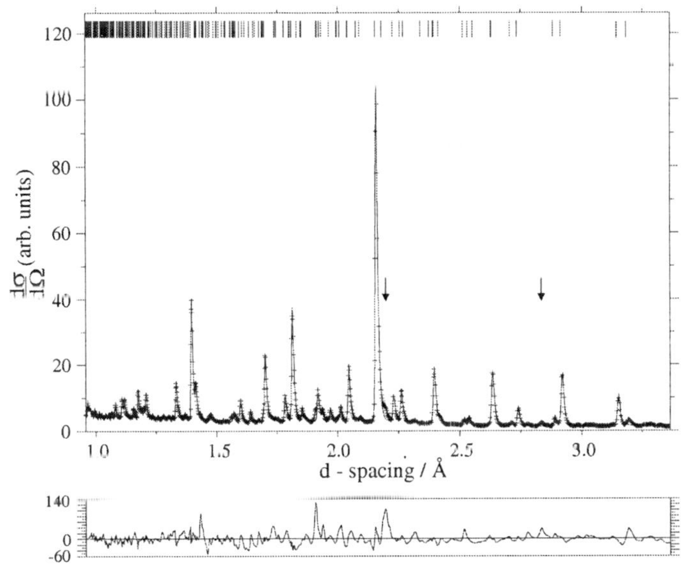

Figure 12. Final fitted profiles (points, observed; line, calculated; lower ((Iobs-Icalc)/esd)) from Rietveld refinement for $LaMo_2O_5$ using the model in space group $P6_3/mmc$. The arrows indicate the positions of two broad Bragg peaks unaccounted for by our models.

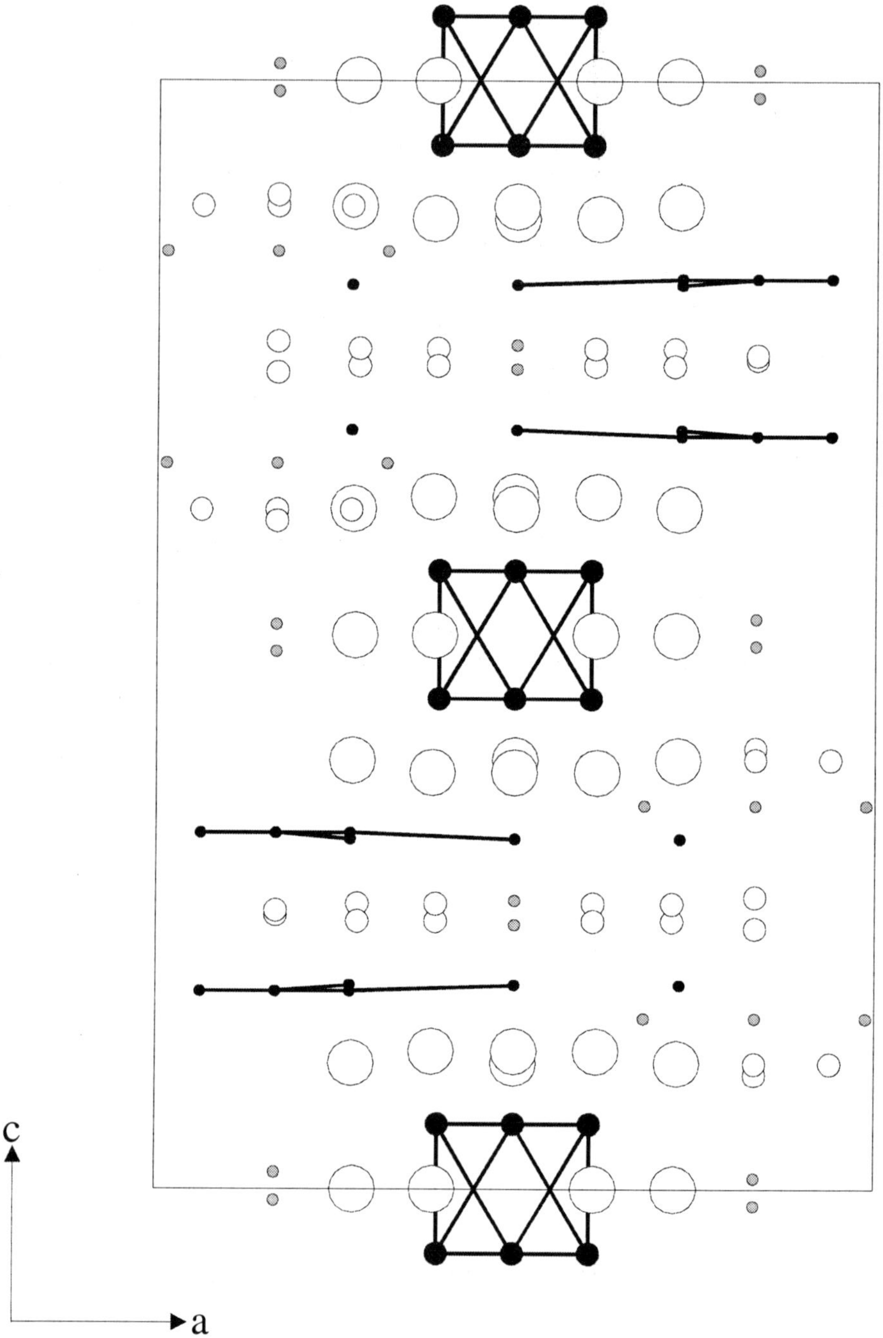

Figure 13. The average unit cell for $LaMo_2O_5$ viewed in the [120] direction. (Space group *P6₃/mmc*. Dark shaded circles, Mo; open circles, O; light shaded circles, La. Small and large circles represent atom sites with occupancies 0.5 and 1 respectively. Only Mo-Mo bonds are shown, light lines indicate the unit cell.)

studies always yielded structural solutions that contained a large number of atoms which were unreasonably close. The likely origin of this problem was thought to be twinning, but powder neutron diffraction work (by others), in which large-scale twinning presents no problem, had not succeeded in solving the problem. We were able to progress by accepting an apparently unpromising structural solution of the type produced from the single crystal X-ray diffraction studies, as the starting point for our structure determination[8, 15]. We produced a model in $P6_3/mmc$ that gave an excellent fit to the Bragg diffraction component of our neutron diffraction pattern, Fig. 12, with a weighted profile R-factor (R_{wp}) of 0.048. However, in this model over half the atomic positions were only 50% occupied, Fig. 13.

Unsurprisingly, this model provides an inadequate description of the local structure, and this is clear when $T(r)_{mod}$ is calculated and compared with $T(r)_{exp}$ calculated from our total diffraction data, Fig. 14. Using chemical knowledge, we were able to disentangle the structure, and we found two substructures which could be described in two of the maximal isomorphic subgroups, $P6_3mc$, and $P\text{-}3m1$, of $P6_3/mmc$ as shown in Figure 15. Calculations of $T(r)_{mod}$ from both of these substructures yielded excellent fits to $T(r)_{exp}$, showing that they describe the local structure of $LaMo_2O_5$ well. The two models have identical short-range order and this can be seen in the comparison of their correlation functions in Fig. 16. This suggests that intergrowths of the two structures will

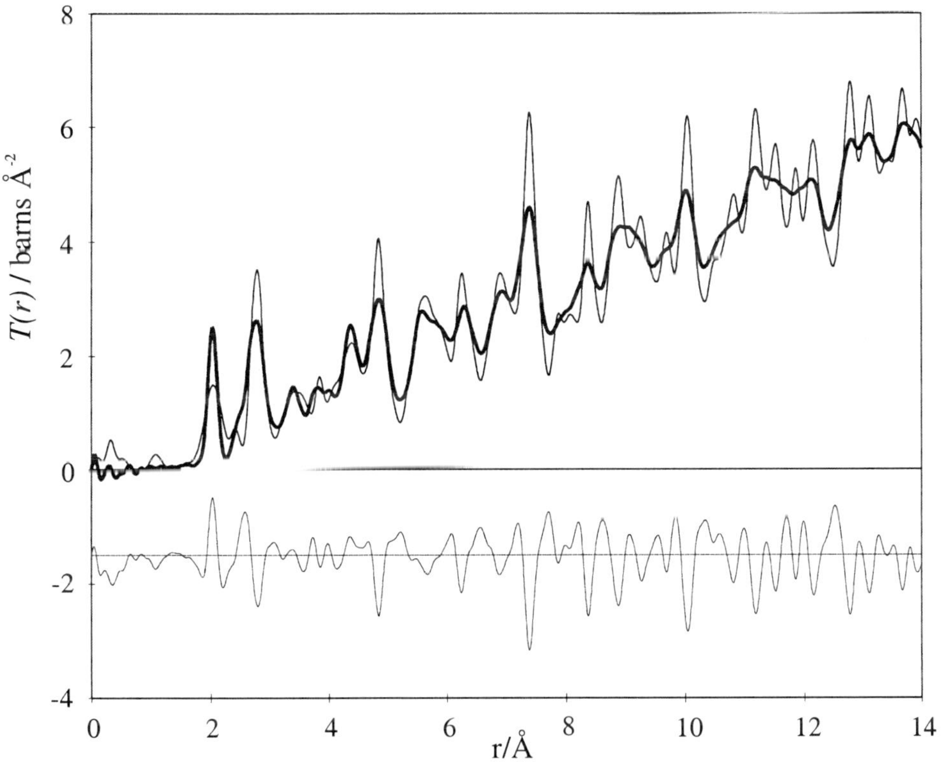

Figure 14. $T(r)_{exp}$, measured by total neutron diffraction (heavy line), and $T(r)_{mod}$ calculated for $LaMo_2O_5$ from the Rietveld refinement in space group $P6_3/mmc$ structure (light line). The difference is shown below.

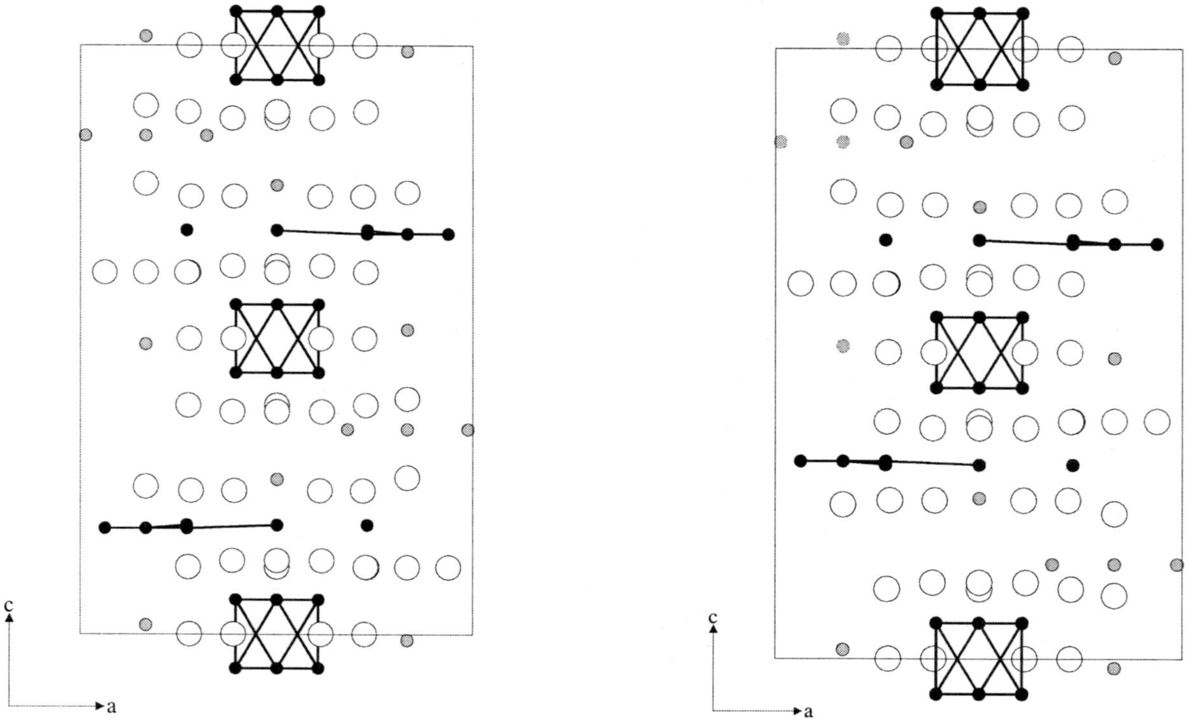

a) b)

Figure 15. Structural models for $LaMo_2O_5$ a) in space group *$P6_3mc$*, and b) in space group *P-3m1*, viewed in the [120] direction. (Dark shaded circles, Mo; open circles, O; light shaded circles, La. Only Mo-Mo bonds are shown. Light lines indicate the unit cell.)

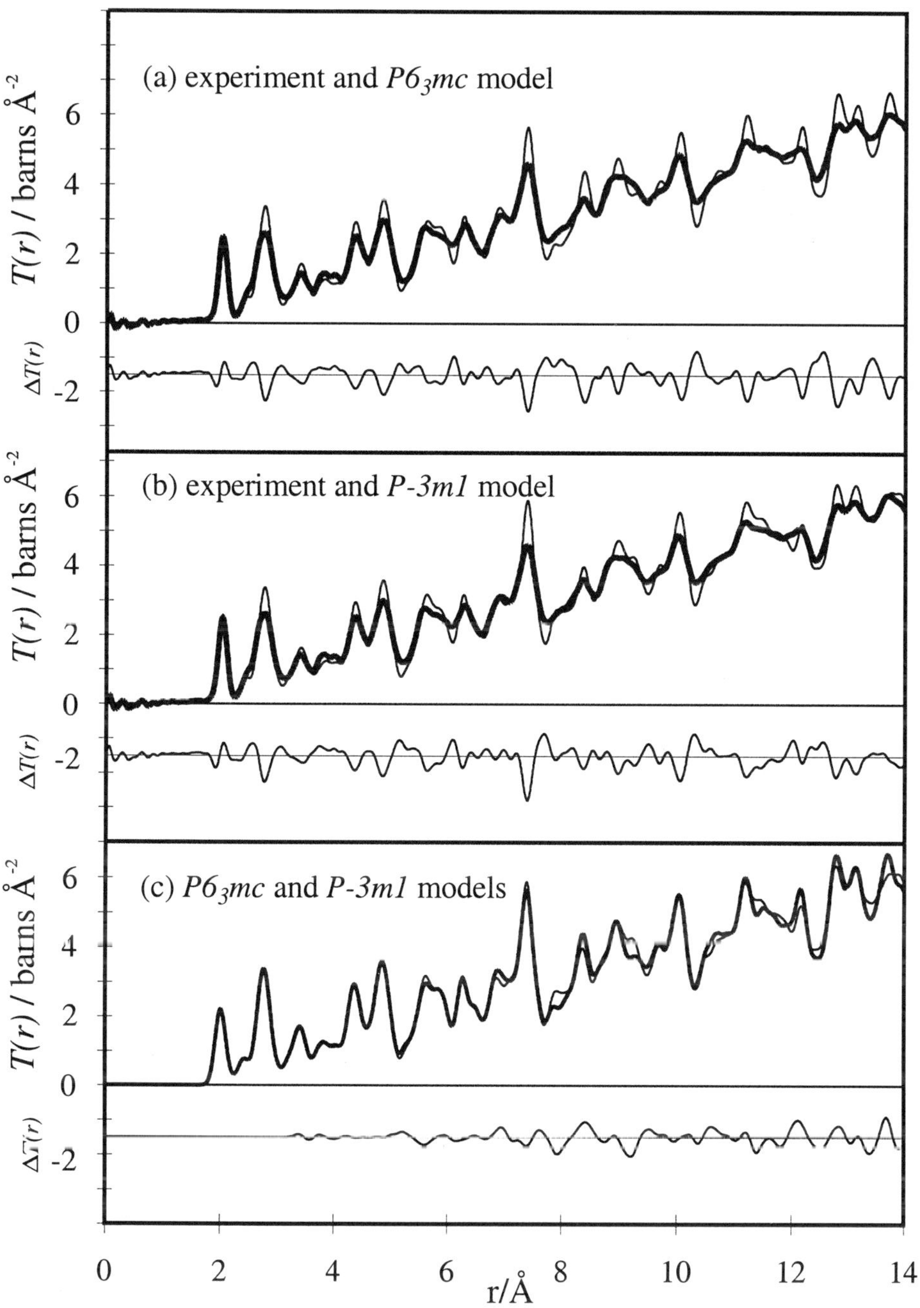

Figure 16. a) $T(r)_{exp}$ for $LaMo_2O_5$ (heavy line), $T(r)_{mod}$ calculated for $P6_3mc$ structure (light line) and difference (below); b) $T(r)_{exp}$ for $LaMo_2O_5$ (heavy line), $T(r)_{mod}$ calculated for *P-3m1* structure (light line) and difference (below); and c) $T(r)_{mod}$ calculated for $P6_3mc$ structure (heavy line) and $T(r)_{mod}$ calculated for *P-3m1* structure (light line) and difference (below).

occur easily, and it is easy to envisage a single unit or block of $LaMo_2O_5$ with the *$P6_3mc$* structure joining on to a unit with the *P-3m1* structure. Twinning on a microscopic scale explains why both the single crystal studies and previous powder neutron diffraction studies were unable to solve the structure of this disordered crystalline solid satisfactorily.

By combining the results of Bragg diffraction with $T(r)$ from total neutron diffraction and chemical knowledge, we have been able to determine the local structure of $LaMo_2O_5$, to explain the origin of the disorder, and to show how the local structure and average structure are related.

The structure was certainly worth solving, since it contained not one, but two unusual Mo-Mo bonded clusters. The Mo_6O_{18} Mo-Mo bonded octahedral cluster, Fig. 17(a), is known only in one other material[16], $Ca_{16.5}Mo_{13.5}O_{40}$, which is also disordered and for which the structural parameters determined from Bragg diffraction data are unreliable. The structure also contains a new type of Mo-Mo bonded sheet, Figure 17(b).

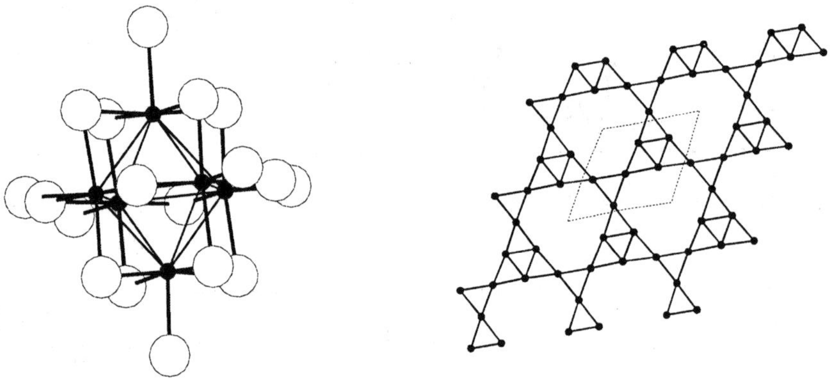

(a) (b)

Figure 17. a) The Mo_6O_{18} Mo-Mo bonded octahedral cluster in $LaMo_2O_5$ (solid circles, Mo; open circles, O), and b) part of the Mo-Mo bonded sheet in $LaMo_2O_5$.

4.3. $LiMoO_2$

Comparison of $T(r)_{mod}$ with $T(r)_{exp}$ provides a sensitive test as to whether the average structure determined by fitting the Bragg scattering is also a good description of the local structure of a material. The two examples above showed that, for disordered crystalline materials, the average structure obtained from Bragg scattering can be totally inadequate in describing the local structure.

An illustration of the power of total diffraction studies in checking whether a structural model is correct is provided by our work[17] on $LiMoO_2$. Figure 18 shows the results of Rietveld profile refinements for two models for $LiMoO_2$. For model A in space group *R-3m* $R_{wp} = 0.1212$, and R_{wp} for model B in space group C2/m 0.0963. Model A is

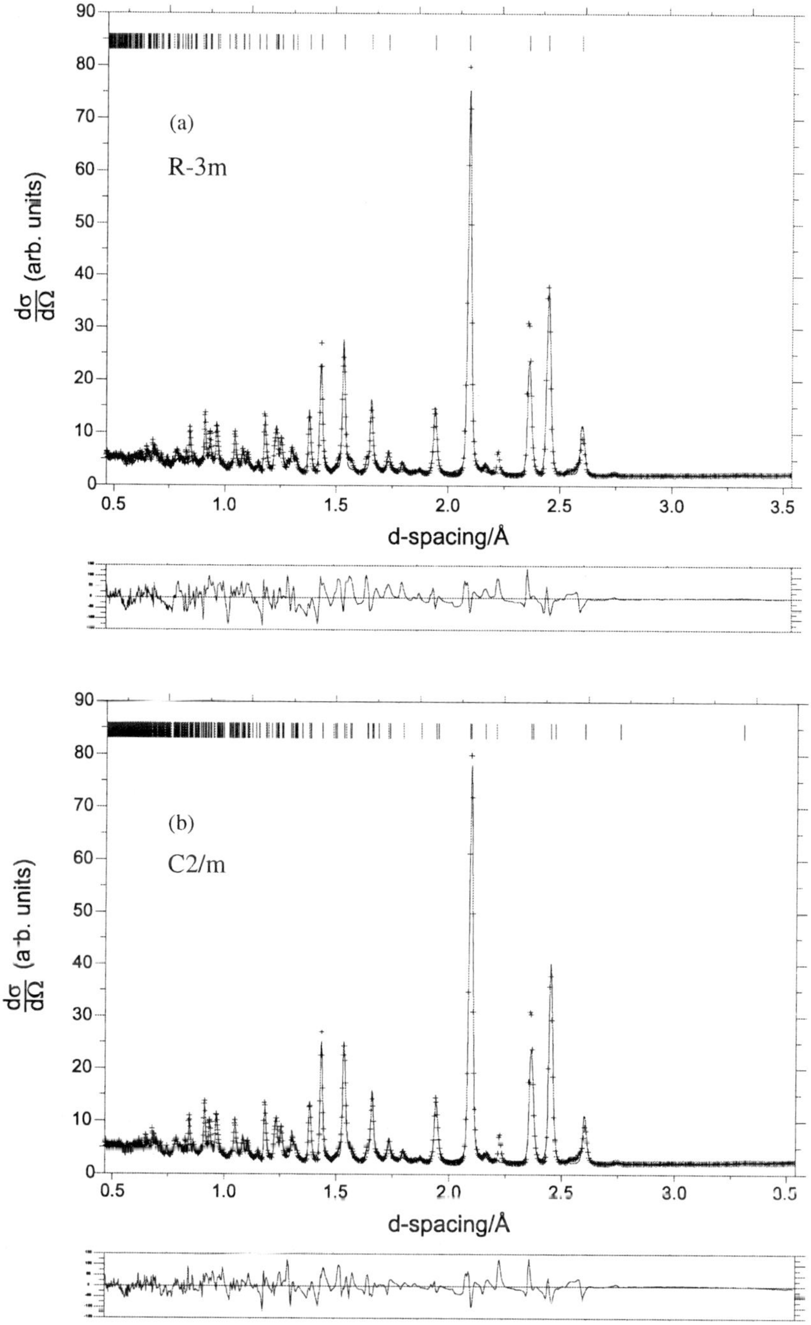

Figure 18. Rietveld neutron diffraction profile refinements of $LiMoO_2$. a) Model A in space group *R-3m*, and b) model B in space group C2/m. Observed data are indicated by crosses and calculated data by a solid line. Marks directly above the pattern indicate the positions of reflections. A difference curve appears at the bottom. An impurity peak due to Mo metal is present at d ~ 2.2 Å.

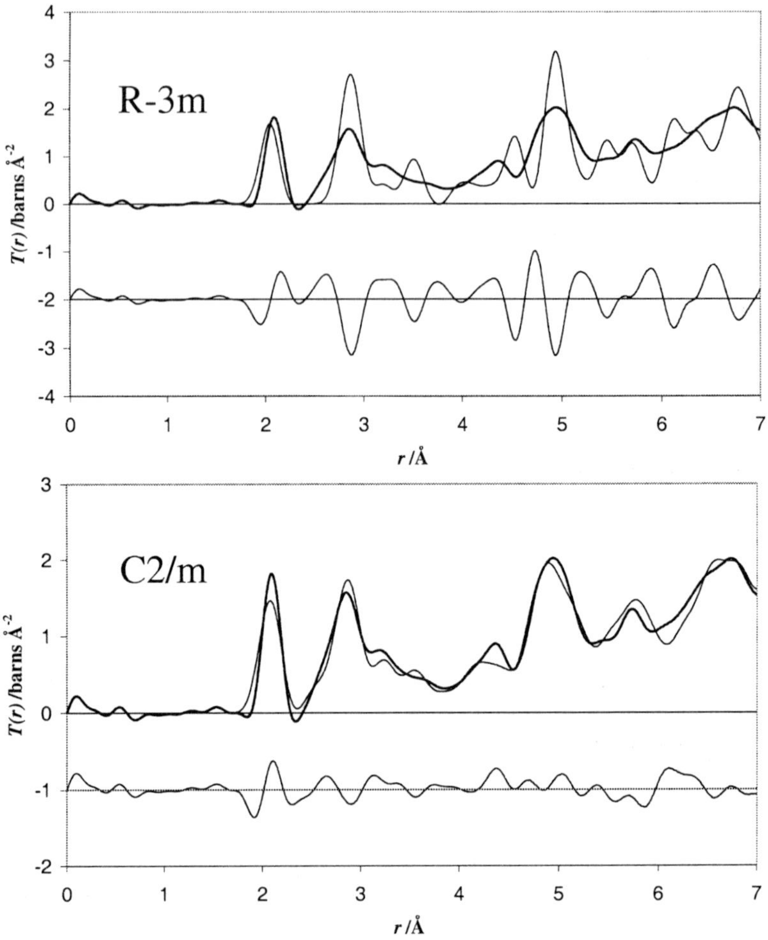

Figure 19. $T(r)_{exp}$ (bold line) and $T(r)_{mod}$ (light line) for $LiMoO_2$. a) Model A in space group *R-3m*, and b) model B in space group C2/m.

incorrect, but the profile refinement appears reasonable; indeed, it looks so plausible that Aleandri and McCarley[18] published this model as the structure of $LiMoO_2$. Model B is the result of our work using total neutron diffraction in which we modelled both *T(r)* and the Bragg scattering. The fit to the Bragg scattering is improved, although not dramatically. However, Fig. 19 shows that, when *T(r)* comparisons are made, model B is far superior to model A. Without any information on the local structure from Mo K-edge EXAFS and *T(r)*, there might never have been any suspicion that model A was wrong, giving a totally inaccurate picture of the local structure in $LiMoO_2$.

Table 2 gives selected interatomic distances for the two models. It is clear that the local structure in the two models is very different. Prior chemical knowledge suggests that the Mo-Mo bonding in model A, with six equal Mo-Mo distances, is unlikely, and indeed Aleandri and McCarley commented that it was rather unusual[18]. Unfortunately, they did not have the benefit of *T(r)* data to check their model.

Table 2. Selected interatomic distances (Å) in $LiMoO_2$ for the two models.

Atom pair	Model A R-3m model	Model B C2/m model
Mo–Mo	2.865 (×6)	2.617 (×2) 2.863 (×2) 3.122 (×2)
Mo–O(1)	2.135 (×6)	2.056 2.078 (×2)
Mo–O(2)		2.094 (×2) 2.102

4.4. AgCN

Reduced molybdates are not the only disordered crystalline materials which can profitably be investigated using a combination of Bragg scattering, total diffraction and chemical intuition. Even apparently simple materials can provide structural challenges. Silver cyanide, AgCN, has a structure consisting of one-dimensional linear chains, Fig. 20. Recently Bowmaker et al.[19] refined the structure using powder neutron diffraction data. We were suspicious of their structural conclusions because the atomic thermal parameters, $B_{Ag} = 7.7$ Å^2 and $B_N = 8.1$ Å^2, are very high, even for a structure determined at 300 K, suggesting the material is highly disordered. We were also alerted to potential problems with the structure determination by their comment that *d*(Ag-C) = 2.15 Å > *d*(Ag-N) = 1.86 Å was surprising.

Figure 21 shows a comparison of $T(r)_{exp}$ for AgCN from our total neutron diffraction data[20] compared to $T(r)_{mod}$ for the crystallographic model of Bowmaker et al. It is immediately apparent that the published structure is in error. Our work shows that *d*(Ag-C) and *d*(Ag-N) are identical, or nearly so, and equal to 2.07 Å. This is vital information to the chemist. Our work also shows that the C N bond length is *d*(C N) = 1.15Å, and not 1.26Å as found in the crystal structure of Bowmaker et al.[19]

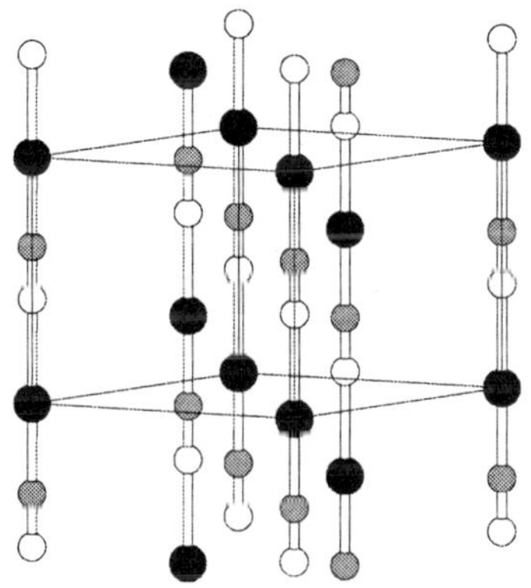

Figure 20. The structure of AgCN (open circle C, solid circle Ag, grey circle N). Light lines indicate the unit cell.

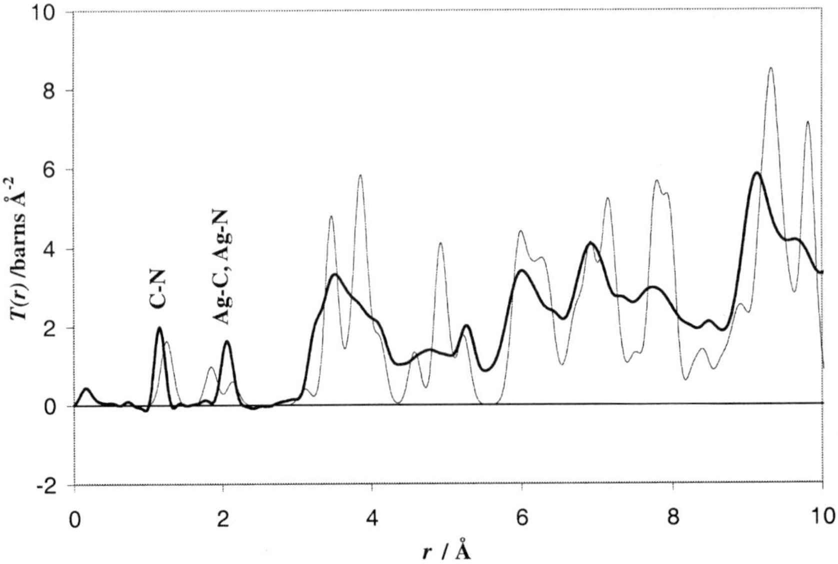

Figure 21. $T(r)_{exp}$ for AgCN (bold line), and $T(r)_{mod}$ (light line) from Bowmaker et al.'s crystal structure[19].

One possible cause of disorder in AgCN is immediately apparent, to the chemist. When the $[NC\text{-}Ag\text{-}CN]^-$ units, see Fig. 1, link together to form infinite chains, -Ag-CN-Ag-CN-Ag-CN-, the CN groups are unlikely to order perfectly in the manner indicated, and head to tail disorder is likely to occur, with some local coordination of the types NC-Ag-CN and CN-Ag-NC.

4.5. Disordered Crystalline Materials: Concluding Remarks

For disordered crystalline materials, the average structure determined from Bragg scattering does not give a full description of the structure. A complete picture requires the determination of the local structure and an understanding of how this relates to the average structure. Information on local structure can be obtained using a number of techniques. EXAFS and total diffraction are particular useful as they provide information on interatomic distances and coordination numbers. We have concentrated on the use of total diffraction measurements, since these can be obtained in the same experiment as the Bragg diffraction. We have used chemical knowledge to construct structural models compatible with the correlation functions and average structure, and we have shown how in a number of cases the local structure and average structure can be reconciled.

When a final structural solution for the structure of a disordered crystalline solid is obtained, one can have a high level of confidence in its correctness. This is because the average structure obtained from Bragg diffraction provides a rather rigid framework in which to work. The problem of working with correlation functions obtained from powder diffraction data or EXAFS is that they contain only one-dimensional data. Using Bragg diffraction data allows us to regain the other two dimensions.

Disordered crystalline materials are just one type of disordered material in which the average structure conceals the true local structure of the material. Amorphous materials are another class of materials in which we would like to determine the local structure. In this case there is no Bragg diffraction yielding three-dimensional information, and diffraction experiments yield only one-dimensional correlation functions.

5. THE LOCAL STRUCTURE OF AMORPHOUS MATERIALS

5.1. Introduction

Chemists have in the past generally left the study of amorphous materials to physicists. This is regrettable, because chemists have much to contribute in terms of experience and expertise. They also produce new amorphous materials which are often neglected by them, because of lack of expertise in structural determination, and neglected by physicists, because of a lack of interest or awareness of these materials. Chemists also have chemical tools that can be used to obtain structural information from amorphous materials.

We have recently been applying a chemist's approach to our structural studies of a number of amorphous transition metal sulfides, including MoS_3[21], $MoS_{4.7}$, WS_3, WS_5 and Re_2S_7[22]. These materials are expected to contain metal-metal bonded units, and there is evidence from previous structural work, particularly from EXAFS studies, for metal-metal bonds in these materials[22, 23]. Ideas as to the type of metal-metal bonded units which may be present can be obtained by analogy with related materials and their chemistry.

5.2. Amorphous Molybdenum Trisulfide, *a*-MoS_3

Molybdenum trisulfide, MoS_3, is not a compound of Mo^{VI} and S^{2-} as might be naïvely imagined, but contains both sulfide, S^{2-}, and disulfide S_2^{2-} ions. The exact number of each is not known, but it appears that the limiting formulations are $Mo^{IV}(S_2^{2-})S^{2-}$ and $Mo^{V}(S_2^{2-})_{1/2}(S^{2-})_2$. Molybdenum in either oxidation state, IV or V, would be expected to form Mo-Mo bonds with $d_{MoMo} \sim 2.8$ Å in a sulfide, and Mo-Mo distances of this length are found in EXAFS studies[23]. For Mo^{IV}, one would expect Mo_3 triangles to form (reminiscent of the Mo_3 units found in the oxides $Zn_2Mo_3O_8$ and Li_2MoO_3 described above) and, for Mo^{V}, one would expect Mo_2 dimers. The apparently simple problem of determining which units are present in MoS_3 has proved difficult to resolve from Mo K-edge EXAFS studies because of the uncertainties in coordination numbers. X-ray and neutron diffraction experiments have yielded $T(r)$ data which seemed to support models based on Mo_2 dimers[24]. However, more recent chemical degradation studies[21] of MoS_3 support the presence of Mo_3 triangles.

We recently carried out quantitative chemical excision reactions on MoS_3, in which we determined that 70% of the molybdenum was present in Mo_3 units[21]. Figure 22 shows the Mo_3S_4 core of the $Mo_3S_4(CN)_9^{5-}$ ion produced in the excision reaction. If the remaining 30% of molybdenum atoms do not participate in Mo-Mo bonding, then our result can be reconciled with previous work, since it yields an *average* Mo-Mo coordination number of 1.4, close to the value found by most workers in EXAFS studies, and this value is consistent with the $T(r)$ studies.

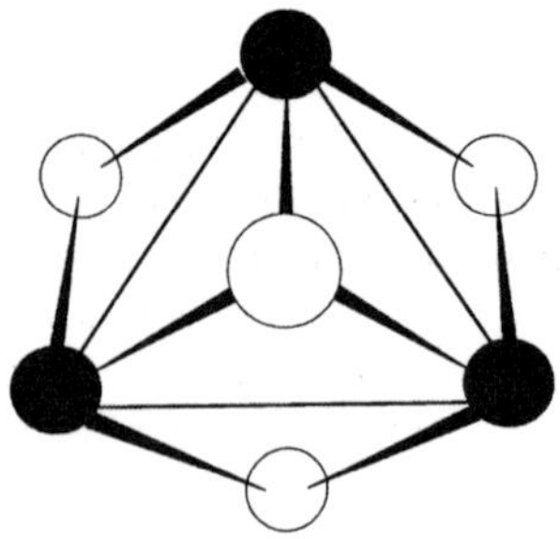

Figure 22. The Mo_3S_4 core.

We concluded that earlier modelling had been flawed by imposing the assumption that all the molybdenum was present in the same form, either as Mo_2 dimers or as Mo_3 triangles. We are now using this additional information, together with McGreevy's RMCA[25] program, to construct structural models. The goal is to construct models compatible with all the experimental information on MoS_3, rather than just the information from diffraction experiments. Figure 23 shows that a model containing 70% of molybdenum atoms in Mo_3 triangles and 30% in monomers is consistent with the neutron diffraction data[26].

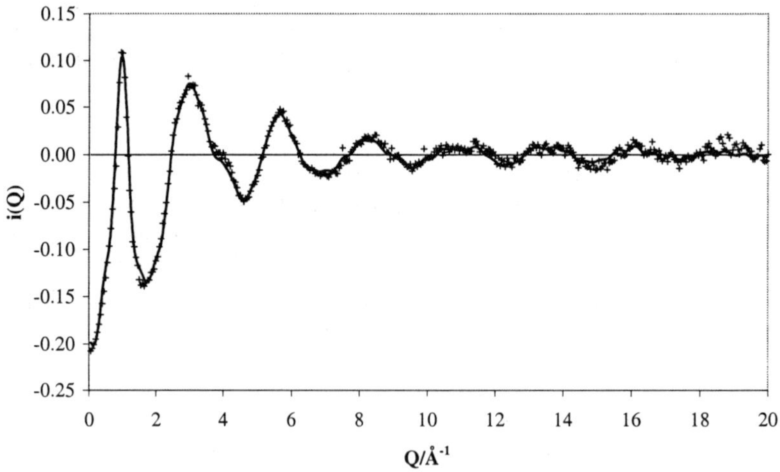

Figure 23. $i(Q)_{exp}$ (crosses) and $i(Q)_{mod}$ (line) for a structural model of a-MoS_3 containing 70% of molybdenum atoms in Mo-Mo bonded triangles and 30% in monomers.

5.3. Amorphous Materials: Concluding Remarks

Determining the structure of amorphous materials is complex. Moving beyond correlation functions to three-dimensional models is fraught with difficulties and uncertainties. Different models may fit the diffraction data, and deciding which is correct needs additional information or the application of additional criteria. In complex systems, the chemist can supply both additional information, from chemical studies, and additional criteria, from chemical knowledge. We have shown how these techniques can be applied to the study of MoS_3.

6. OVERALL CONCLUSIONS

The local structure of materials is of great interest, particularly to chemists. We have shown how total diffraction studies and EXAFS studies can yield accurate local structures for the chemist, and that the chemist is more than a passive observer in this process of structure determination. In the case of disordered crystalline materials, chemists often have a good idea of the type of structural units present, or can use their chemical knowledge to identify the structural units hidden in the average structure determined from Bragg diffraction. Determining both the local and average structures of a material gives a more complete picture of the structure. Understanding the relationship between the local and average structures gives an even more fundamental insight into nature.

ACKNOWLEDGEMENTS

We wish to thank our past and present students and collaborators, including I. D. Fawcett, R. I. Walton, S. P. Cooper, S. M. Cheyne, S. Patat, W. H. McCarroll and D. Hayman, without whom the work described here would not have been possible. We also thank the EPSRC for their support and especially for the provision of neutron and EXAFS beam time.

REFERENCES

1. K. J. Range, S. Kühnel and M. Zabel, *Acta Cryst. C* **45**, 1419-1420 (1989).
2. F. Grønvold, H. Haraldsen and A. Kjekshus, *Acta Chem. Scand.* **14**, 1879-1893 (1960).
3. S. J. Hibble, A. C. Hannon and I. D. Fawcett, *J. Phys.: Condens. Matter* **11**, 9203-9219 (1999).
4. A. C. Hannon, W. S. Howells and A. K. Soper, *IOP Conf. Series* **107**, 193-211 (1990).
5. A. C. Hannon, in: *Encyclopedia of Spectroscopy and Spectrometry*, vol. 2, edited by J. Lindon, G. Tranter, and J. Holmes, (Academic Press, London, 2000) pp. 1493-1503.
6. E. A. Lorch, *J. Phys. C* **2**, 229-237 (1969).
7. A. C. Hannon, Rutherford Appleton Laboratory Report RAL-93-063, 1993.
8. S. J. Hibble, S. P. Cooper, S. Patat and A. C. Hannon, *Acta Cryst. B* **55**, 683-697 (1999).
9. G. B. Ansell and L. Katz, *Acta Cryst.* **21**, 482-485 (1966).
10. A. C. W. P. James and J. B. Goodenough, *J. Solid State Chem.* **76**, 87-96 (1988).
11. S. J. Hibble, I. D. Fawcett and A. C. Hannon, *Acta Cryst. B* **53**, 604-612 (1997).
12. S. J. Hibble and I. D. Fawcett, *Inorg. Chem.* **34**, 500-508 (1995).
13. R. L. McGreevy and L. Pusztai, *Mol. Simulation* **1**, 359-367 (1988).
14. W. H. McCarroll, C. M. Darling and G. J. Jakubicki, *J. Solid State Chem.* **48**, 189-195 (1983).
15. S. J. Hibble, S. P. Cooper, A. C. Hannon, S. Patat and W. H. McCarroll, *Inorg. Chem.* **37**, 6839-6846 (1998).
16. B. Lindblom and R. Strandberg, *Acta Chem. Scand.* **43**, 825-828 (1989).
17. S. J. Hibble, I. D. Fawcett and A. C. Hannon, *Inorg. Chem.* **36**, 1749-1753 (1997).
18. L. E. Aleandri and R. E. McCarley, *Inorg. Chem.* **27**, 1041-1044 (1988).

19. G. A. Bowmaker, B. J. Kennedy and J. C. Reid, *Inorg. Chem.* **37**, 3968-3974 (1998).
20. S. J. Hibble, A. C. Hannon and S. M. Cheyne, unpublished results.
21. S. J. Hibble, M. R. Feaviour and M. J. Almond, *Dalton Trans.* 935-940 (2001).
22. S. J. Hibble and R. I. Walton, *Chem. Comm.* 2135-2136 (1996).
23. S. J. Hibble, D. A. Rice, D. M. Pickup and M. P. Beer, *Inorg. Chem.* **34**, 5109-5113 (1995), and references therein.
24. S. J. Hibble, R. I. Walton, D. M. Pickup and A. C. Hannon, *J. Non-Cryst. Solids* **232-234**, 434-439 (1998), and references therein.
25. R. L. McGreevy, M. A. Howe and J. D. Wicks, NFL Studsvik Internal Report (unpublished) 1993.
26. S. J. Hibble and D. Hayman, unpublished results.

FROM CRYSTALS TO NANOCRYSTALS: SEMICONDUCTORS AND BEYOND

Valeri Petkov* and Simon J.L. Billinge*

1. INTRODUCTION

Knowledge of the atomic-scale structure is an important prerequisite to understand and predict the physical properties of materials. In the case of crystals it is obtained from the positions and the intensities of the Bragg peaks in the diffraction data[1]. However, many materials of technological importance are not perfectly crystalline but contain significant disorder at the atomic scale. The diffraction patterns of such materials show only a few Bragg peaks, if any, and a pronounced diffuse component. This poses a real challenge to the usual techniques for structure characterization. The challenge can be met by employing the so-called atomic pair distribution function (PDF) technique. The atomic PDF gives the number of atoms in a spherical shell of unit thickness at a distance r from a reference atom. It peaks at characteristic distances separating pairs of atoms and thus describes the structure of materials. The *PDF*, $G(r)=4\pi r[\rho(r)-\rho_o]$, is the sine Fourier transform of the so-called total scattering structure factor, $S(Q)$,

$$G(r)=(2/\pi)\int_{Q=o}^{Q_{max}} Q[S(Q)-1]\sin(Qr)dQ, \qquad (1)$$

where $\rho(r)$ and ρ_o are the local and average atomic number densities, respectively, Q is the magnitude of the wave vector and $S(Q)$ is the corrected and properly normalized total powder diffraction pattern of the material[2,3]. As can be seen the atomic PDF is simply another representation of the diffraction data; however, exploring the experimental data in real space is advantageous and helpful for several reasons, especially in the case of materials with significant structural disorder. First, as Eq.(1) implies the *total, not only the Bragg diffracted*, intensities contribute to the PDF. In this way both the average, *long-range* atomic structure, manifested in the well-defined Bragg peaks, and the *local structural imperfections*, manifested in the diffuse component of the diffraction pattern, are projected in the *PDF*. Note that conventional crystallographic studies take only the Bragg peaks into account. Second, the PDF is barely influenced by diffraction optics and experimental factors since these are accounted for in the step of normalizing the raw dif-

* Physics & Astronomy Dept, Michigan State University, East Lansing, MI 48824.

From Semiconductors to Protein, edited by S.J.L. Billinge and M.F. Thorpe
Kluwer Academic/Plenum Publishers, 2002

fraction data and converting it to *S(Q)* data. This renders the PDF a structure-dependent quantity giving directly the relative positions of atoms in materials. Third, by accessing high values of *Q*, experimental PDF's with high real-space resolution can be obtained and, hence, quite fine structural features revealed. Fourth, the PDF is obtained with no assumption of periodicity. Thus, materials exhibiting any degree of structural disorder, ranging from perfect crystals to glasses and liquids, can be studied with the same approach. Indeed the PDF technique has been the approach of choice for characterizing liquids and glasses for a long time[2,3]. However, its application to study crystalline materials with significant disorder has been relatively recent[4,5,6,7]. Last but not the least, when a structural model is available, it is trivial to calculate the corresponding PDF and compare it to the experimentally determined PDF. This enables the convenient testing and refinement of structural models. Here we demonstrate the great potential of the PDF technique by employing it to study the structure of materials showing various degrees of structural disorder: In-Ga-As semiconductor alloys, nanophase $LiMoS_2$, Ca-Al-Si-O glasses, metallic Cu_3Au and organic ferrocene, $Fe(C_5H_5)_2$.

2. EXAMPLES

2.1 Materials showing small deviation from perfect crystallinity: In-Ga-As semiconductor alloys

Ternary semiconductor alloys, in particular $In_xGa_{1-x}As$, have technological significance because they allow important properties, such as band gaps, to be varied continuously between the two end points by varying the composition, *x*. It has been found that $In_xGa_{1-x}As$ alloys crystallize in the cubic zinc-blende structure and their lattice parameters interpolate linearly between those of the end members, GaAs and InAs. Since this behaviour is consistent with the so-called Vegard's law, the atomic arrangement of $In_xGa_{1-x}As$ alloys is often approximated with a virtual zinc-blende type lattice where all bond lengths are taking equal, compositionally averaged, values and the bond angles remain unperturbed from the ideal tetrahedral ones. However, XAFS experiments[8] have shown that Ga-As and In-As bonds indeed remain close to their natural lengths of 2.43 Å and 2.61 Å, respectively. This finding is close to the so-called Pauling limit[9] and implies that the zinc-blende lattice of $In_xGa_{1-x}As$ is distorted locally to accommodate the two distinct Ga-As and In-As bonds lengths present. Knowledge of the fully distorted structure is necessary as input for accurate band structure and phonon dispersion calculations.

Very high real space resolution is required to reveal the individual bond lengths and the associated local structural distortions in $In_xGa_{1-x}As$ alloys. We achieved it by carrying out diffraction experiments using x-rays of energy 60 keV delivered by the A2 24 pole wiggler of the Cornell High Energy Synchrotron Source[10] Experimental reduced structure factors Q[S(Q)-1] are shown in Fig. 1 and the corresponding atomic PDF's G(r) - in Fig. 2. Significant Bragg scattering (well-defined peaks) is immediately evident up to Q ~ 35 $Å^{-1}$ in the diffraction patterns of the end members, InAs and GaAs (Fig. 1). The two samples are perfect crystals. The Bragg peaks disappear at much lower Q values in the diffraction patterns of the alloys: these samples are still long-range ordered but they have significant local positional disorder. The disorder is coming from the presence of two distinct bond lengths, Ga-As and In-As, seen as a split first peak in the experimental PDF's (Fig. 2). The peak remains sharp and the position of it two sub-components hardly changes with *x*. This implies that the immediate atomic ordering in In-Ga-As alloys is quite rigid. The nearest neighbour peak is, however, the only peak which shows such a

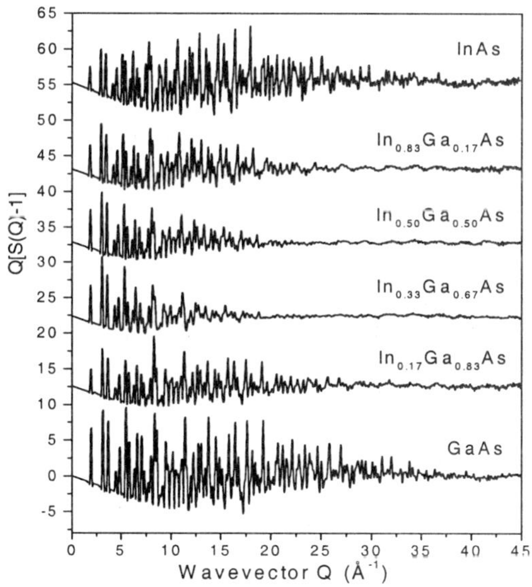

Figure 1. Reduced structure factors Q[S(Q)-1] for $In_xGa_{1-x}As$ at various concentrations.

steady behavior. From the second neighbour peak onwards the significant atomic displacements in the alloy samples result in broad atomic distributions, i.e. broad PDF peaks, without any resolvable splitting. In addition, the position of all higher-r peaks vary

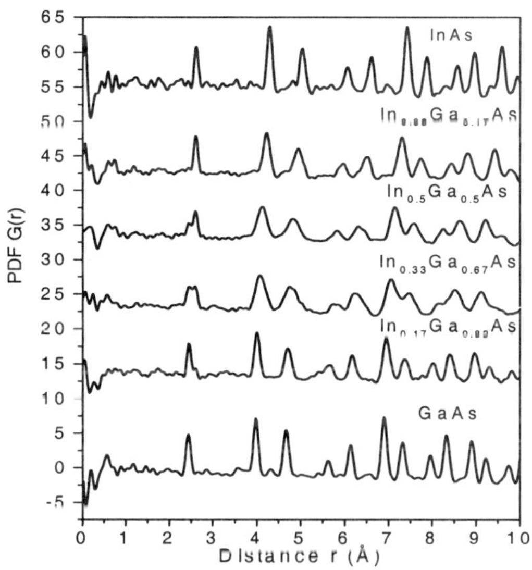

Figure 2. Experimental atomic PDF's, G(r), for $In_xGa_{1-x}As$ at various concentrations.

linearly with x. The bond-length mismatch gives rise to a considerable deformation of the underlying zinc-blende lattice. Thus by collecting data to very high diffraction vectors and obtaining PDF's of high real-space resolution we are able to reveal the presence of two distinct bond lengths and local structural distortions in $In_xGa_{1-x}As$ alloys. To quantify

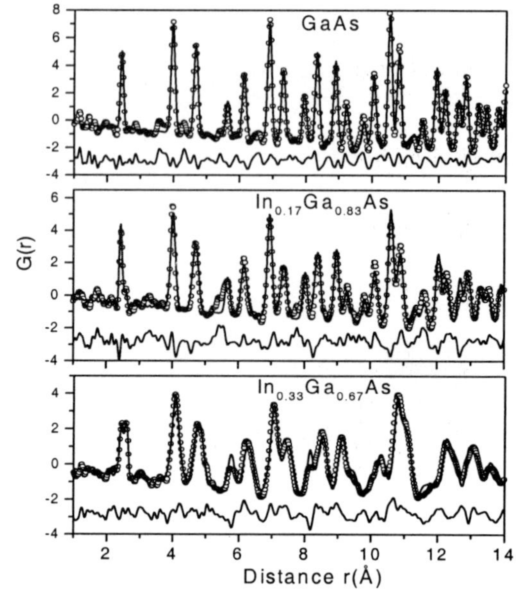

Figure 3. Experimental (symbols) and fitted (solid line) PDF's for $In_xGa_{1-x}As$ alloys. The residual difference between the fit and the data is given in the lower part.

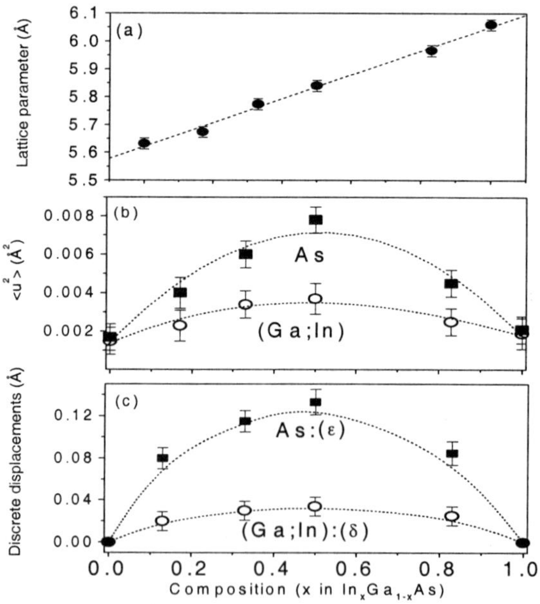

Figure 4. Lattice constants (a), mean-square (b) and discrete (c) displacements of metal (δ) and As (ε) atoms from their positions in an ideal zinc-blende lattice as extracted from 8-atom unit cell refinements guided by the experimental PDFs The results of a linear fit to the lattice constants and a second order polynomial fit to the atomic displacements are also shown (dashed curves).

the structural disorder the experimental PDF's were fit with models based on the 8-atom cubic unit cell of the zinc-blende structure[11]. In accordance with the experiment the overall symmetry of the unit cell was kept cubic during the refinements and so only the length of the cell's edge was varied. The presence of two distinct bond lengths required that metal (In,Ga) and As atoms occupy specific sites in the unit cell. It was achieved by letting the atoms move off their ideal positions. For simplicity all As atoms were constrained to be displaced by the same amount as each other. The metal atoms were likewise constrained to have the same displacement amplitude, but the metal site displacement were independent of the As sublattice displacements. In addition to these discrete atomic displacements, thermal factors (mean-square displacements) of As and metal atoms were refined as well[12]. All refinements were done with the help of the program PDFFIT[13]. Fits are shown in Fig. 3 and results of the refinements are summarized in Fig. 4. As can be seen most of the disorder occurs on the As sublattice ($\varepsilon > \delta$; see Fig. 4) but displacements of metal atoms are non negligible too. This approach to modeling the data uses few parameters yet quantifies the essential features of the local structure. An alternative approach is to use potential based modeling as was successfully done with these materials[10]. In this case large supercells of approximately 10^4 atoms are created and interactions between the atoms specified by a simple potential. The supercell is relaxed to minimize the energy of the system, then the PDF calculated from the relaxed ensemble and compared with the data. Potential parameters can be varied to obtain a good fit to the data. Again, only a few parameters are specified ensuring that the fitting is not over-parameterized. This approach[10] provides complimentary information to that obtained from the simple regression modeling[12] described above.

2.2 Crystalline materials showing significant structural disorder: nanophase $LiMoS_2$

MoS_2 is the key catalyst for the removal of sulfur from crude oil (hydrodesulfurization). Pristine MoS_2 is perfectly crystalline and consists of layers of Mo-S_6 trigonal prisms held together by Van der Waals forces[11]. $LiMoS_2$ has Li intercalated between the MoS_2 layers. It is important as a precursor of stable MoS_2 colloids used to prepare a variety of lamellar nanocomposites[14]. Despite being extensively studied for the last two decades the structure of $LiMoS_2$ has not been determined. The reason is that, on Li intercalation, pristine MoS_2 is dramatically modified resulting in a nanophase product that is too poorly diffracting to allow a traditional structural determination. This leaves unanswered the important question of what exactly happens when MoS_2 gets reduced with Li. Theoretical predictions and XAFS studies suggest the presence of considerable Mo-Mo bonding but no unequivocal experimental evidence has been advanced so far. We tackled the problem by employing the PDF technique. The experiments[15] were done at the beam line X7A of the National Synchrotron Light Source (NSLS), Brookhaven National Laboratory. Experimental structure functions are shown in Fig. 5 and the corresponding PDF's - in Fig. 6.

Sharp Bragg peaks are present in the $S(Q)$ of MoS_2 up to the maximal Q value of 24 Å^{-1} (Fig. 5(b)). The corresponding $G(r)$ also features sharp peaks reflecting the presence of well-defined coordination spheres in this "perfectly" crystalline material (Fig. 6(b)). The well-known 6-atom hexagonal unit cell of MoS_2[11] was fit to the experimental PDF and the structure parameters refined so as to obtain the best agreement between the calculated and experimental data. The fit was done with the program PDFFIT[14] and was constrained to have the symmetry of the $P6_3/mmc$ space group. In comparing with experiment, the model PDF was convoluted with a *Sinc* function to account for the finite Q_{max} of the measurement. The best fit achieved is shown in Fig. 6(b). Refined structure

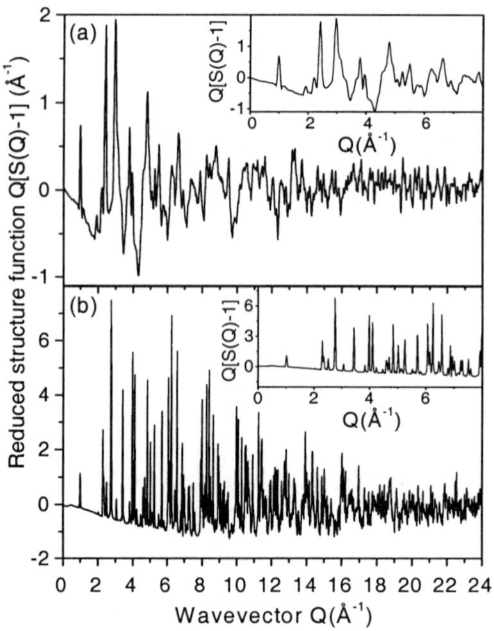

Figure 5. Experimental structure factors of (a) $LiMoS_2$ and (b) MoS_2. Note the different scale between (a) and (b). The data are shown in an expanded scale in the insets.

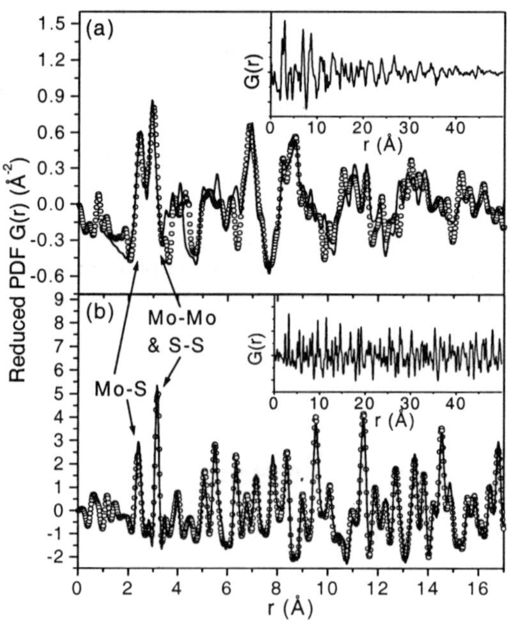

Figure 6. Experimental (dots) and fitted (solid line) PDF's for $LiMoS_2$ (a) and MoS_2 (b). Note the different scale between (a) and (b). The first two peaks in the PDF's are labeled with the corresponding atomic pairs. The experimental data are shown in an expanded scale in the insets.

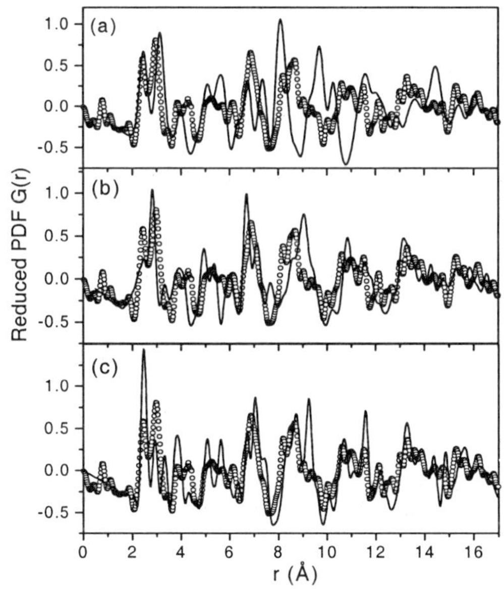

Figure 7. Comparison between the experimental PDF for $LiMoS_2$ (symbols) and model PDF's (full line) for (a) hypothetical hexagonal $LiMoS_2$, (b) hypothetical 1T-$LiMoS_2$ (c) triclinic $LiMoS_2$ as predicted in [17].

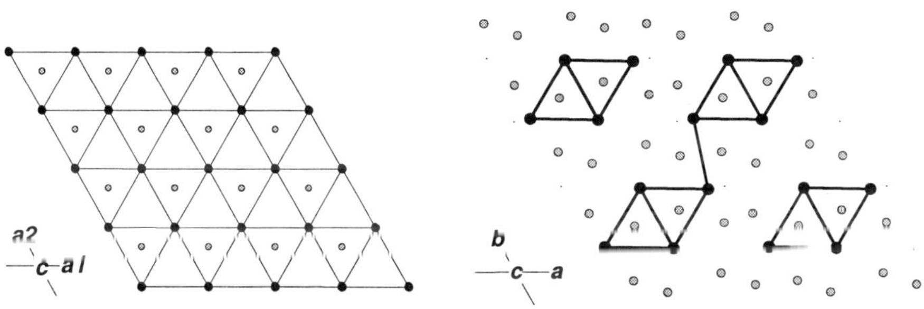

Figure 8. Projection down the *c* axis of the crystal structures of hexagonal MoS_2 (left) and triclinic $LiMoS_2$ (right). The large black circles are Mo atoms and the small gray circles are the S atoms. Li atoms are not shown for the sake of clarity.

parameters of MoS_2 agree well with published results[11] obtained by traditional crystallographic techniques.

In the diffraction pattern of $LiMoS_2$ the Bragg peaks are significantly broadened (see the inset in Fig. 5(a) and already at ~ 8 $Å^{-1}$ merge into a slowly oscillating diffuse component. A diffraction pattern so poor in sharp Bragg peaks lacks the number of "statistically independent reflections" needed to apply the traditional techniques for structure determination. The corresponding PDF is, however, rich in distinct, structure related features and, as we will see, lends itself to structure determination. To determine the parameters of the atomic structure of $LiMoS_2$ we explored several structure models as follows: First, a model based on the hexagonal MoS_2 structure (*S.G. $P6_3/mmc$*) was attempted. It was constructed by inserting 2 Li atoms into the 6-atom unit cell of MoS_2 so that they resided between the Mo-S_2 layers. This model, however, could not reproduce the experimental data as can be seen in Fig. 7(a). A previously proposed[16] structure for $LiMoS_2$ is based on trigonal 1T-MoS_2 (*S.G. P3*) which can be considered as built of layers of distorted Mo-S_6 octahedra. This structure model was also tested. It performed somewhat better but still could not reproduce the details in the experimental PDF as can be seen in Fig. 7(b). Next we attempted a model based on the "exfoliated-restacked" MS_2 structure (M= W, Mo)[18] which features zig-zag metal-metal chains. This was an excellent starting point and as the refinement proceeded the Mo atoms shifted slightly but significantly to new positions that define the chains of Mo atoms shown in Fig. 4. This model fits all the important details in the experimental PDF as can be seen in Fig. 7(a). Refined structure parameters are reported in Petkov *et al*[15].

In pristine MoS_2 Mo atoms from a single MoS_2 layer arrange in a regular hexagonal lattice (see Fig. 8) and are all separated by the same distance of 3.16 Å. In contrast, the Mo atoms in $LiMoS_2$ occupy two distinct positions in the triclinic unit cell[15] giving rise to short (2.9-3.10 Å) and long (3.44-4.07 Å) Mo-Mo distances. As a result, a structural motif emerges made up of chains of diamonds of bonded Mo atoms (Fig. 8). This is in qualitative agreement with the structure proposed in an earlier XAFS study[16] and predicted theoretically[17] though the details are significantly different. Using the refined atomic coordinates[15] we calculated the electronic band-structure. We found that a gap of about 0.2 eV is present in the electronic density of states at the Fermi level. The result suggests that $LiMoS_2$ should exhibit semiconducting properties which is in agreement with experiment[14].

2.3 Completely disordered materials: Ca-Al-Si-O glasses

Silicate glasses in general, and aluminosilicate glasses in particular, are of great industrial and geological importance. Despite this, some very basic issues about their structure are still not well established. It is well known that silica glass is a continuous network of corner-shared Si-O tetrahedra. At the vertex of each tetrahedron, and shared between two tetrahedra, is an oxygen atom known as a bridging oxygen (BO). When Si^{4+} is partially replaced by another network former such as Al^{3+} and, in addition, a modifier cation such as Ca^{2+} is introduced to balance the charge, the glass structure changes. The change involves a gradual substitution of Al-O polyhedra for Si-O ones and an emergance of so-called non-bridging oxygens (NBO) linking one Si -O or Al-O polyhedral unit with a Ca-based one. Since Si/Al-O-Ca linkages are much weaker than Si-O-Si ones the tetrahedral network disrupts. The breaking of the network, i.e. the presence of NBO's, has a large effect on the thermodynamics of the glass and on important properties such as the viscosity of the melt[19]. The exact nature of the modifier ion (Ca) coordination and of the way the NBOs are introduced into the network has been the subject of intensive research yet is still not fully resolved. The coordination of Al, too, has been an issue for quite a time. We addressed these issues by carrying out a PDF study on a series of $Ca_{x/2}Al_xSi_{1-x}O_2$ glasses with x=0,0.25,0.5, 0.67. The experiments were performed at the beam line 1-ID of the Advanced Photon Source, Argonne with the use of x-rays of energy 80.6 keV[20]. Experimental structure factors $Q[S(Q)-1]$ are shown in Fig. 9. As can be seen $Q[S(Q)-1]$ data exhibit prominent oscillations up to the maximum Q value reached. These oscillations may only come from the presence of well-defined coordination polyhedra building the glassy structure. The polyhedra are seen as well defined peaks in the low-r region of the experimental PDF's shown in Fig. 10. The first peak in the PDF for SiO_2 glass is positioned at 1.61(1) Å and is very sharp. By fitting the peak with a Gaussian one finds that in silica glass the Si-O coordination number, Z, is 4.0(1) (see Fig. 11)

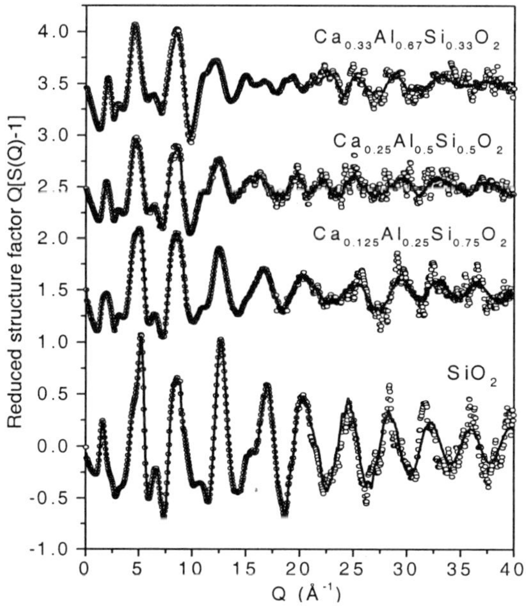

Figure 9. Reduced structure factors Q[(SQ)-1] for calcium aluminosilicate glasses (symbols) together with the optimum smooth line.

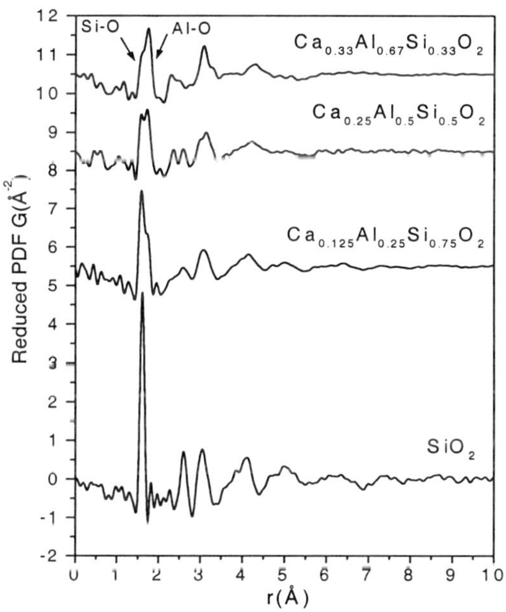

Figure 10. Reduced atomic PDF's for $Ca_{x/2}Al_xSi_{1-x}O_2$ glasses (x=0,0.25,0.5,0.67) obtained by Fourier transforming the smoothed data of Fig. 9.

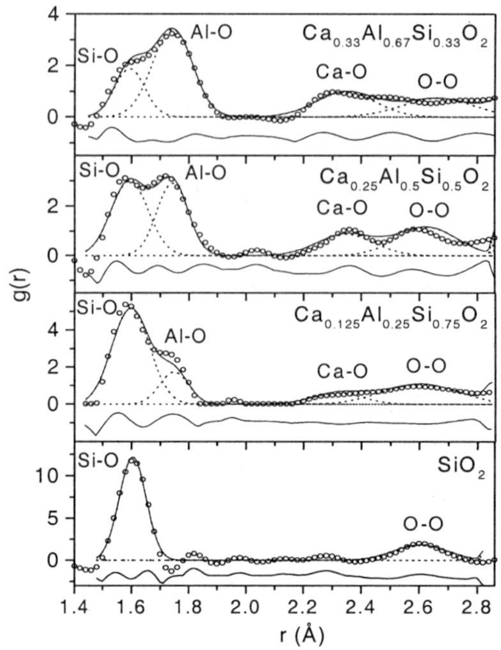

Figure 11. Gaussian fit to the first peaks in the PDFs, $g(r) = \rho(r)/\rho_o$, for $Ca_{x/2}Al_xSi_{1-x}O_2$ (x=0,0.25,0.5, 0.67) glasses. Experimental data: symbols; fitted data: full line; individual Gaussians: broken line; residual difference: full line (bottom). Peaks are labeled with the corresponding atomic pairs.

and the root-mean-square deviation, $(\langle u^2\rangle)^{1/2}$, from the average Si-O distance is 0.040(2) Å. The present results are in rather good agreement with those of Grimley et al.(Si-O = 1.609(4) Å ; Z = 3.85(16) and $\langle u^2\rangle^{1/2}$ =0.047(4) Å) obtained by time-of-flight neutron diffraction experiments[21].

As can be seen in Fig. 10 the first peak in the experimental PDFs for $Ca_{x/2}Al_xSi_{1-x}O_2$ glasses (x=0.25,0.5,0.67) is composed of two components. The first one, peaking at approximately 1.6 Å, becomes lower in intensity while the intensity of the second one, peaking at approximately 1.75 Å, increases with x. They can be unambiguously attributed to Si-O and Al-O coordination units. The corresponding first coordination numbers and distances determined by a Gaussian fit to the data, as shown in Fig. 11, are summarized in Table 1. A careful analysis of the Si-O and Al-O coordination numbers reveals the following characteristics of the atomic ordering in $Ca_{x/2}Al_xSi_{1-x}O_2$ glasses. Both Si and Al atoms are coordinated by four oxygens and thus participate in a continuous tetrahedral network when Al/Ca content is low (x=0, 0.25). By x=0.5 the network begins to break as the slight decrease in the average coordination number of Si indicates. When more than

TABLE 1. First neighbour Si-O, Al-O, Ca-O and O-O distances and average coordination numbers in $Ca_{x/2}Al_xSi_{1-x}O_2$ glasses (x=0,0.25,0.5,0.67) obtained by a Gaussian fit to the corresponding high-resolution PDF's, $g(r) = \rho(r)/\rho_o$.

x	Si-O	Al-O	Ca-O
0	1.61(1) Å / 4.0(1)		
0.25	1.60(1) Å / 3.95(10)	1.75(1) Å / 3.95(10)	2.32(2) Å / 5.3(2)
0.5	1.60(1) Å / 3.85(10)	1.75(1) Å / 4.0(1)	2.36(2) Å / 5.2(2)
0.67	1.60(1) Å / 3.2(1)	1.75(1) Å / 3.95(10)	2.34(2) Å / 5.3(2)

half of the Si ions are replaced by Al (x=0.67) the number of NBO's in the first coordination sphere of Si increases further making the average Si-O coordination number as low as 3.2(1). Al-O tetrahedra, however, remain free of NBOs. The result suggests that the disintegration of the tetrahedral network proceeds mainly by the creation of Si-O-Ca and not Al-O-Ca linkages. The disintegration of the tetrahedral network, however, does not seem to affect the Ca-O first neighbour distribution. As data in Table 1 shows the oxygen coordination of Ca does not change even though the Si/Al ratio and the number of NBO's change significantly. The result implies that Ca has a well-defined and stable oxygen coordination. By demanding and acquiring it Ca atoms would inevitably affect the way the tetrahedral backbone of $Ca_{x/2}Al_xSi_{1-x}O_2$ glasses arranges in space.

2.4 Materials showing no positional but chemical disorder: metallic Cu_3Au

In many materials two or more types of atomic species occupy a prefect lattice. When cooled down below a critical temperature particular atomic species can arrange themselves in an orderly manner on the lattice. The material is said to be in a chemically ordered state. When the material is heated above the critical temperature the atomic species can rearrange in a more or less random manner and the material is said to be in a chemically disordered state. Such a transformation from a chemically ordered to disordered state does not involve distortions or modifications of the underlying lattice but does affect the physical and mechanical properties of materials. Chemically ordered materials are usually stronger and harder, have a lower electrical resistivity, and a higher thermal conductivity than their disordered counterparts. Chemical order-disorder transitions have been studied by traditional diffraction techniques with success for many years[22]. Long-range chemical ordering has been associated with the appearance of superlattice peaks in the diffraction patterns of materials. Short-range chemical ordering has been studied by single crystal diffuse scattering experiments[1,23]. However, single crystal diffuse scattering experiments are quite difficult and time consuming. We carried out a PDF study on Cu_3Au alloy to ascertain the sensitivity of the technique to chemical ordering in materials. At high temperatures the alloy is chemically disordered with Cu and Au atoms occupying the sites of a fcc lattice at random. Below 391° C the alloy undergoes a transition into an ordered state with copper atoms occupying the corners and gold atoms the face centers of the cubic unic cell. Two samples were investigated: ordered and disordered Cu_3Au. The disordered sample was made by quenching Cu_3Au from above the ordering transition temperature. The experiments were carried out at the beam line X7A of the National Synchrotron Light Source, Brookhaven with the use of x-rays of energy 30 keV. Experimental powder diffraction patterns are shown in Fig. 12. Bragg peaks in the diffraction pattern of disordered Cu_3Au can be indexed and fit with a cubic unit cell (S.G. Fm-3m) of edge length a_o =3.7587(2) Å. Ordered Cu_3Au produces superlattice peaks with mixed values of Miller indices (*hkl*). The diffraction pattern can be indexed and fit with a cubic unit cell (S.G. Pm-3m) of edge length a_o= 3.7542(2) Å. The presence of such superlattice peaks clearly confirms the existence of chemical ordering in the sample. Their absence in the other sample shows that it is truly disordered. Experimental PDF's extracted from the powder diffraction patterns are shown in Fig. 13. As can be seen the PDF's have the same number of peaks positioned at the same interatomic distances. This reflects the fact that Cu and Au atoms occupy the same cubic lattices in both ordered and disordered Cu_3Au. The amplitudes of the corresponding PDF's peaks are, however, quite different. The difference is due to the fact that the same coordination spheres comprise different number of Cu and Au atoms in ordered and disordered Cu_3Au. The difference is

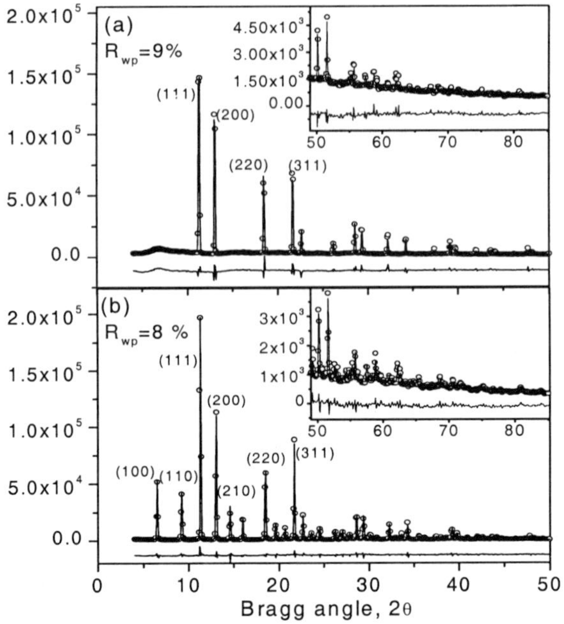

Figure 12. Experimental (circles) and Rietveld fitted (solid line) XRD patterns of (a) disordered and (b) ordered Cu_3Au. The residual difference is given in the lower part. Peaks are labeled with the corresponding Miller indexes. Data are given in an expanded scale in the insets.

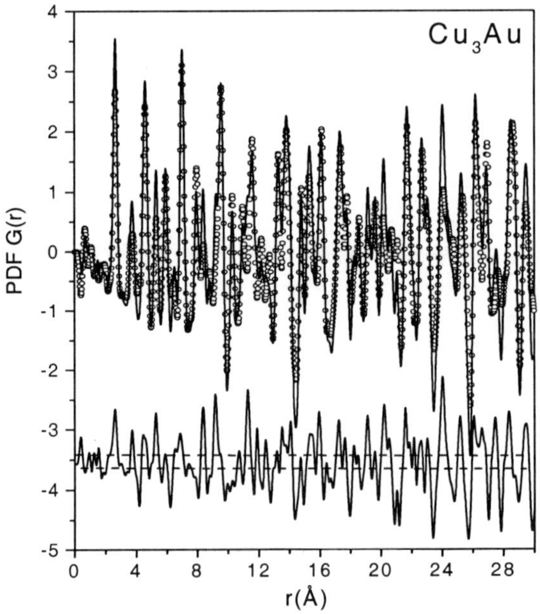

Figure 13. Experimental PDF's of ordered (solid line) and disordered (circles) Cu_3Au. Their difference is shown in the lower part (solid line). The estimated experimental uncertainty is marked by horizontal broken lines.

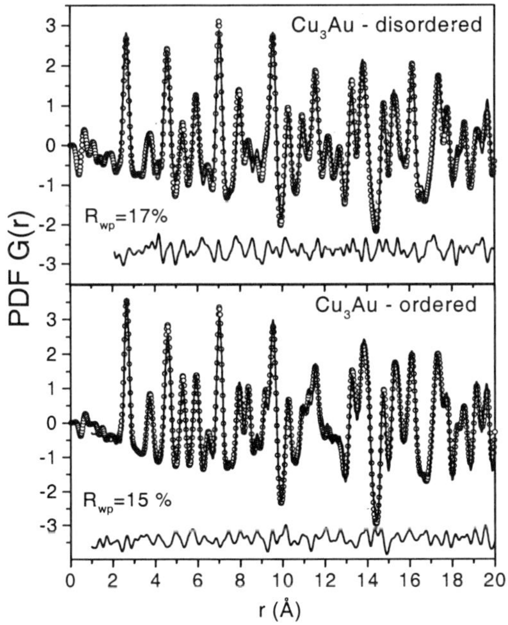

Figure 14. Experimental (circles) and fitted (solid line) PDF's for ordered and disordered Cu_3Au. The residual difference is shown in the lower part.

much bigger than the experimental uncertainty. This makes it possible to detect and quantify chemical ordering in materials, in particular Cu_3Au, by employing the PDF technique. The degree of chemical ordering in the samples was determined by fitting the experimental PDF's with models based on the 4-atom unit cell of Cu_3Au. Fits were done with the help of the program PDFFIT[13]. Results are given in Fig. 14. In full agreement with Rietveld refinements (see Fig. 13) PDF fits showed the presence of complete chemical order in one of the samples and no signatures of chemical ordering in the other[24].

2.5 Disorder in organic materials: ferrocene

Traditionally, the structures of organic and biological materials are determined by single crystal diffraction experiments[1]. However, single crystals of many organic and biological materials may not be grown at all. The PDF technique may be a useful approach to obtain information about both the average structure and local structural imperfections in such cases. The performance of the technique was tested on ferrocene, $Fe(C_5H_5)_2$. This is an organometallic compound used as a burning rate catalyst in fuels and raw material for the production of antibiotics and blood-nourishing agents. At the molecular level ferrocene is a sandwich made of two cyclopentadienyl rings with an iron atom between them. A fragment of the crystal structure[25] of ferrocene, with several molecules shown, is depicted in Fig. 15. The molecule of ferrocene is, however, not very rigid. The two cyclopentadienyl rings are rotated with respect to each other with the angle of rotation changing with temperature[25]. Experiments on ferrocene were carried out at the beam line 1-ID of the Advanced Photon Source, Argonne National Laboratory, with the use of x-rays of energy 80 keV. The experimental structure factor obtained is shown in Fig. 16 and the corresponding atomic PDF - in Fig. 17. A PDF calculated from the crystal

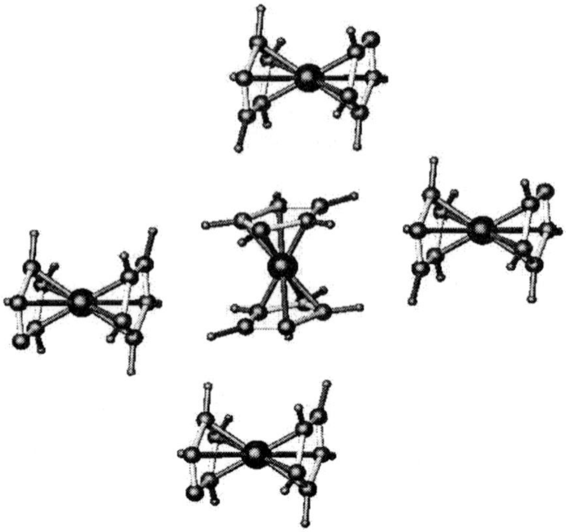

Figure 15. Fragment of the crystal structure of ferrocene. Iron atoms (big circles) are centrally placed between two staggered rings of carbon (medium size circles) and hydrogen atoms (small circles).

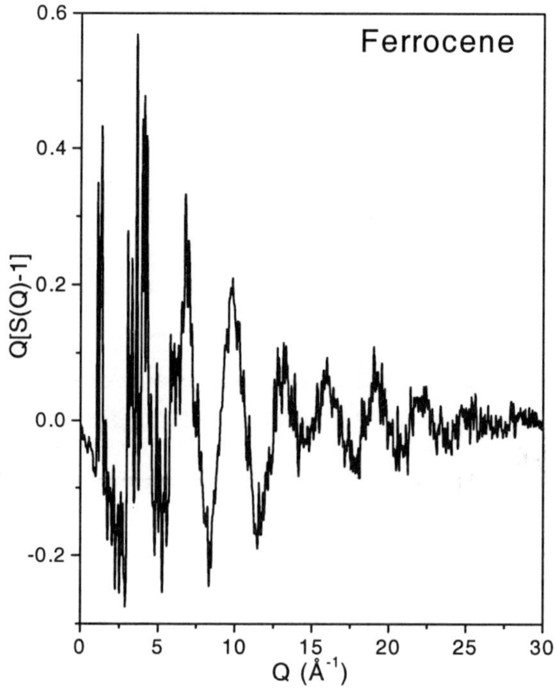

Figure 16. Experimental reduced structure factor Q[S(Q)-1] for ferrocene.

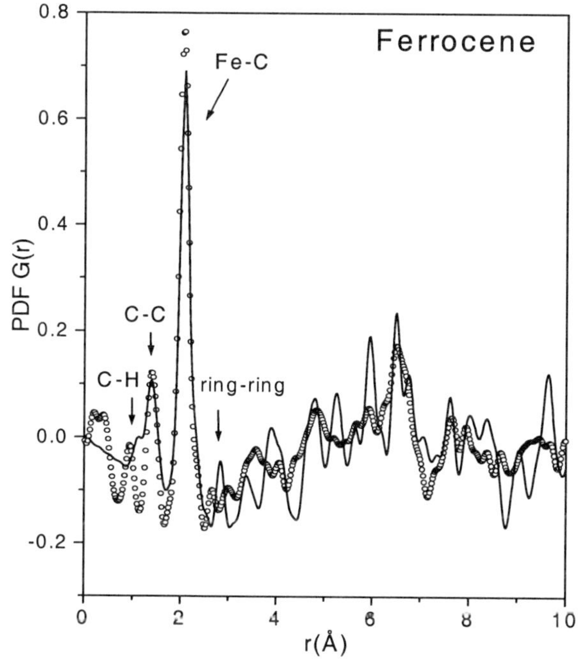

Figure 17. Experimental (symbols) and calculated (solid line) PDF for ferrocene. Peaks are marked with the corresponding intramolecular atomic-pair correlations.

structure of ferrocene with all molecules considered as completely rigid bodies is shown in Fig. 17 as well. As can be seen the experimental PDF clearly reveals the presence of well defined carbon-hydrogen and carbon-carbon bonds within the cyclopentadienyl rings as well as the well defined iron-carbon interatomic distances. From those distances onwards the orientational disorder in ferrocene gives rise to a considerable broadening of the PDF peaks. No such broadening is seen in the calculated PDF assuming rigid ferrocene molecules. Thus atomic PDF's appear to be quite sensitive to both well defined structural fragments and disorder in organic materials.

CONCLUSIONS

The atomic PDF technique can be applied with success to study the structure of complex materials with intrinsic positional and/or chemical disorder. Key to the success is that both Bragg and the diffuse component of diffraction data are collected over a wide range of wave vectors before converted into the corresponding PDF. This makes the PDF a structure-sensitive quantity reflecting both the average and local structure of materials. By employing new sources of radiation such as x-ray synchrotrons, diffraction data extended to at least 40 Å $^{-1}$ can be collected even for materials composed of light, weakly scattering, atomic species and fine structural features differing in as little as 0.1 Å revealed. We hope that the examples shown here will motivate further development of the PDF technique as a routine tool in materials structure studies.

Acknowledgments

Thanks are due to M. G. Kanatzidis, K. K. Rangan, B. Himmel, S.D. Shastri, T. Vogt, S. Kycia, Th. Proffen , I-K. Jeong, P. F. Peterson, M. F. Thorpe and V. Levashov. The work on the semiconductor alloys, ferrocene and Cu_3Au was supported by US-DOE through Grant DE-FG02-97ER45651, and the other work in the Billinge group by NSF Grant CHE 99-03706. APS is supported by DOE under contract No. W-31-109-Eng-38. CHESS is operated by NSF through Grant No. DMR97-13424. NSLS is supported by DOE through DE-AC02-98-CH10886.

REFERENCES

1. M.M. Woolfson, *An introduction to X-ray Crystallography* (Cambridge, University Press 1997); B.E. Warren, *X-ray diffraction* (New York, Dover, 1969).
2. H.P. Klug and L.E. Alexander, *X-ray diffraction procedures for polycrystalline materials* (Wiley, New York 1974).
3. C.N.J Wagner, *J.Non-Cryst. Solids* **31**, 1 (1978).
4. T. Egami, *Mater. Trans.* **31**, 163 (1990).
5. S.J.L.Billinge, R.G. DiFrancesco, G.H. Kwei, J.J. Neumeier and J.D. Thompson, *Phys. Rev. Lett.* **77**, 715 (1996).
6. S.Teslic and T.Egami, *Acta Cryst.* **B 54**, 750 (1998).
7. M.Tucker, M. Dove and D.A. Keen, *J. Phys. Condens. Matter* **12**, L723 (2000).
8. J.C. Mikkelson and J.B. Boyce, *Phys. Rev. Lett.* **49,** 1412 (1982).
9. L. Pauling, *The Nature of the Chemical Bond* (Cornell University Press, New York 1967).
10. V. Petkov, I-K. Jeong, J.S. Chung, M.F. Thorpe, S. Kycia and S.J.L. Billinge, *Phys. Rev. Lett.* **83**, 4089 (1999); I-K. Jeong, F. Mohiuddin-Jacobs, V. Petkov, S.J.L. Billinge and S. Kycia, *Phys. Rev. B.* **63,** 205202 (2001).
11. R.W.G. Wyckoff, *Crystal Structures* ((Wiley, New York 1967).
12. V. Petkov and S.J.L. Billinge, *Physica B* **305**, 83 (2001).
13. Th. Proffen and S.J.L. Billinge, *J. Appl. Cryst.* **32**, 572 (1999).
14. W.M.R Divigalpitiya, R.F. Frindt, S.R. Morrison, *Science* **246**, 369 (1989).
15. V. Petkov, S.J.L. Billinge, P. Larson, S.D. Mahanti, T. Vogt, K.K. Rangan and M.G. Kanatzidis, *Phys. Rev. Lett.* (2000), submitted (cond-mat/0106303).
16. M. A. Py and R. Haering, *Can. J. Phys.* **61**, 76 (1983).
17. X. Rocquefelte, F. Boucher, P. Gressier, G. Ouvrard, P. Blaha and K. Schwartz, *Phys. Rev B.* **62,** 2397 (2000).
18. V. Petkov,S.J.L. Billinge, J. Heising and M.G.Kanatzidis, *J. Am.Chem. Soc.* **122**, 11571 (2000).
19. B.O. Mysen, *Structure and properties of silicate melts* (Elsevier, Amsterdam, 1988).
20. V. Petkov, S,J.L. Billinge, S.D. Shastri and B. Himmel, *Phys. Rev. Lett.* **85**, 3436 (2000).
21. D.I. Grimley, A.C. Wright and R. N. Sinclair, *J. Non-Cryst. Solids* **119**, 49 (1990).
22. J.M. Cowly, *Diffraction physics*, (Elsevier, 1995).
23. M. Bessiere, S. Lefebre and Y. Calvayrac, *Acta Cryst.* **B39**, 145 (1983).
24. Th. Proffen, V. Petkov and S.J.L. Billinge, *in preparation.*
25. P. Seiler and D. Dunitz, *Acta Cryst.* **B35**, 1068 (1979).

STRUCTURAL MODELING OF NANOPOROUS CARBON: A REVIEW OF APPROACHES TO SIMULATING AN APERIODIC AND NON-EQUILIBRIUM SOLID

Michael S. Strano, Jane Rempel, John Halverson, Chris Burket, Jonathan Mathews and Henry C. Foley*

1. INTRODUCTION

Nanoporous materials as a class are the subjects of intense research at present because they offer the potential for substantially new technologies in areas as diverse as electronics and medicine. Carbons derived by pyrolysis of polymeric precursors are an interesting case in point of a nanotechnological application which is already practiced.[1] The separation of air has been done for over a century by brute force, and energetically intensive, method of cryogenic distillation. Zeolites (Li-X) offer an alternative means to producing oxygen via an adsorptive procedure. Nitrogen is preferentially attracted to and held the lithium cations. Nanoporous carbons (NPC) produce enriched nitrogen by a process which is quite different. Experiments show that oxygen is adsorbed substantially faster than nitrogen at pressures from 101 to 10,100 Kpa or more at room temperature.[2] More recently, membranes consisting of fibrous or supported forms of nanoporous carbon have been shown to display a similar preference to the transport of oxygen over nitrogen.[3] While much recent work has been done, the fundamental aspects of why oxygen is more rapidly transported than nitrogen remains elusive. As to whether this difference in rate is caused by a lower enthalpy barrier or an increased entropy of transition for oxygen at the narrowest constrictions of the pore structure,[4] there is still considerable disagreement. On one point there is no disagreement and that is that the narrowest pores, those with dimensions lying between 0.33nm and 0.4nm, are both necessary and sufficient to give rise to this significant separative effect. To understand the more fundamental aspects of this process, one must have a significantly better model

*Henry C. Foley, Department of Chemical Engineering, The Pennsylvania State University, University Park, PA USA 16801

From Semiconductors to Protein, edited by S.J.L. Billinge and M.F. Thorpe
Kluwer Academic/Plenum Publishers, 2002

for the structure of NPC than those which presently exist. This provides a very strong motivation for the pursuit of new and more sophisticated structural simulations.

In contrast to sophisticated synthetic procedures used to prepare zeolites and mesoporous inorganic oxides, such as MCM-48 and related materials, the nanoporous carbons are prepared by the simplest of methodologies at high temperatures via pyrolysis. No templating whatsoever it necessary. Instead the structures "emerge" at temperatures between 400 and 800°C from the apparent chaos of the pyrolysis process.[5] Furthermore although one might expect that the structure of the precursor would be directly responsible for the structure and porosity of the final material, in fact this is not borne out by experiment. Instead it is observed that the same structures are formed by pyrolysis of many different precursors,[6] so long as the polymer is one which undergoes carbonization (solid carbon formation) in parallel with and at rates comparable to gasification.

That a synthetic process in which no external structure-directing force or agent is utilized leads reproducibly to the same nanoporous network, and one in which the pores are so narrowly distributed around a mean value of approximately 0.5 nm, is, to say the least, unexpected. The facts point to a unique high temperature process from which the structure emerges as a result of the kinetics, transport and local chemistry. Testing this hypothesis requires that we have a much more detailed structural understanding of nanoporous carbon. This, then, provides a second and very powerful motivation for the development of efficient and valid structural simulations.

So the detailed structural modeling of nanoporous carbon is motivated both by a desire to better understand the physical processes which take place in its single most important technological application and to identify those features of the physical and chemical processes which lead inexorably to formation of a regular, nanoporous network at high temperature and in the absence of any structure directing force or agent. Deeper knowledge of this kind should prove useful for the design and application of new nanoporous materials in general, not just carbons.

2. A BRIEF REVIEW OF THE ESSENTIAL PROPERTIES OF NPC

The nanoporous carbons are prepared by the pyrolysis of a polymer which forms carbon in fairly high yields. In the work which we have done we have focused upon polyfurfuryl alcohol as the NPC precursor. The chemistry of the decomposition is complex but it follows a set of reactions of the kind shown below with increasing temperature. (**Figure 1**) Added to these is the equilibrium associated with the water gas shift reaction.

The most prominent feature of the NPC materials is the narrow pore size distribution with a mean dimension of approximately 0.5 nm.[7, 8] Both x-ray and neutron diffraction analyses indicate that material lacks long range order. From neutron diffraction it is observed that materials prepared at temperatures between 400 and 800°C show little in the way of structure beyond about 0.7 nm according to the very low intensities of features in $G[r]$ beyond this value of r.[2] Consistent with this are the HRTEM images of these materials which show no structural features other than those which can be described as homogeneous and "worm-like" on a 1 nm length scale.[2] Earlier work had shown that according to nuclear magnetic resonance spectroscopy, the carbon atoms were mostly and

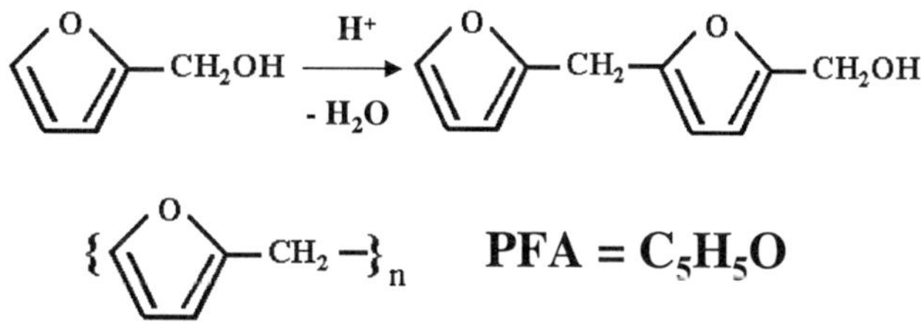

$$\{C_5H_5O_2\}_n \xrightarrow{\blacktriangle}$$

$$1/3\ C_{13}H_9O_3 \xrightarrow{\blacktriangle} 1/3\ C_{12}H_3 + 1/3\ H_2O$$

$$1/3\ C_{12}H_3 \xrightarrow{\blacktriangle} 4\ C + \tfrac{1}{2}\ H_2$$

Figure 1. Furfuryl alcohol, polyfurfuryl alcohol and the likely reactions of pyrolysis.

nearly completely sp^2 and approximately trigonally oriented.[8] Porosity measurements indicate that the skeletal densities of the NPC increased with increasing pyrolysis temperature over a range from ~ 1.5 to just over 2 g/cm^3, both of which are substantially below that of graphite.[1, 2]

3. CHEMICALLY CONSTRAINED NPC STRUCTURAL SIMULATION

Given that there is very little in the way of long range order, it is critical then that each and every measurable piece of chemical and physical property data be utilized in the simulation of the NPC structure. These measurables of the system form a set of constraints which must be imposed upon any of the simulations which we may otherwise consider eligible as structural candidates. To be specific the constraints are:

- All sp^2 or pseudo trigonal carbon
- Density in the range from 0.5 to 0.8 of that of graphite and linearly related to the H/C ratio
- 0.5 nm mean pore size and narrow distribution about the mean
- Lattice energy from 0 to 10 kcal/mol higher than graphite
- Matching XRD, HRTEM and *G[r]*

With the first two of these in mind a simulation methodology which we referred to as "chemically constrained" was developed.[9, 10] A suitable control volume was established and it was populated with multi-six-ring aromatics at random positions and orientations. The ring number and structures were randomized but their individual H/C ratios fit that which was set globally as an initial condition on the simulation. This ratio also fixed the target density for the simulation. Bond making between open sites on the rings were made randomly and in the next step those beyond a realistic bond length were rejected. Bond angles and lengths were minimized and then the density of the new lattice was computed. These same steps were repeated for the remaining lattice sites until the computed density matched the experimental density for the given H/C ratio. With each cycle of bond making the density increased as the reoptimized structure was drawn into itself within a smaller volume. Once complete the file of atom positions could be opened in a computational chemical modeling package such as Cerius.

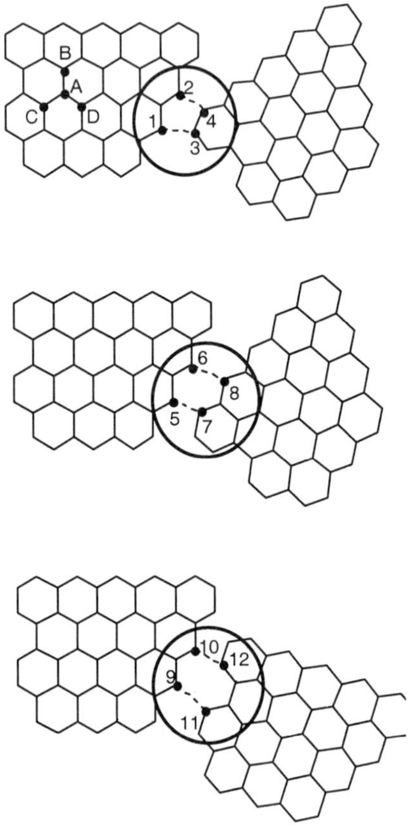

Figure 2. Inclusion of five and seven member rings occurs between all six-ring moieties as a result of random bonding. These defects lead to broken symmetry and non-planarity as the structure continues to develop during minimization (adapted from Ref. 9).

As a result of the random bond-forming processes taking place at the edges of the multi-six-ring structures, included defect structures developed spontaneously with five and seven members (**Figure 2**). Thus with this approach structures which included regions of both positive and negative mean curvature could be generated (**Figure 3**). These aspects of the modeling effort were quite good and it gave us for the first time more valid structural images of the NPC materials compared to those which had been proposed on purely intuitive grounds earlier.

On the other hand the algorithm was a time consuming one. As many as 12 hours of CPU time could be consumed on an SGI workstation dedicated to this simulation when atom numbers close to 1000 were used. Since the atom numbers were necessarily low, the structures were not large enough to enable the computation of a pore size distribution, nor were they suitable for simulations of the HRTEM images associated with NPC. We did compute a PDF for these structures and we found that they were reasonable approximations to those of the experimental materials. On the other hand if *G[r]* in such low order materials is of much less value than it is for materials with more long range order and hence structural uniqueness. Thus a more robust and more computationally efficient algorithm was sought which would allow for the simulation of much larger networks.

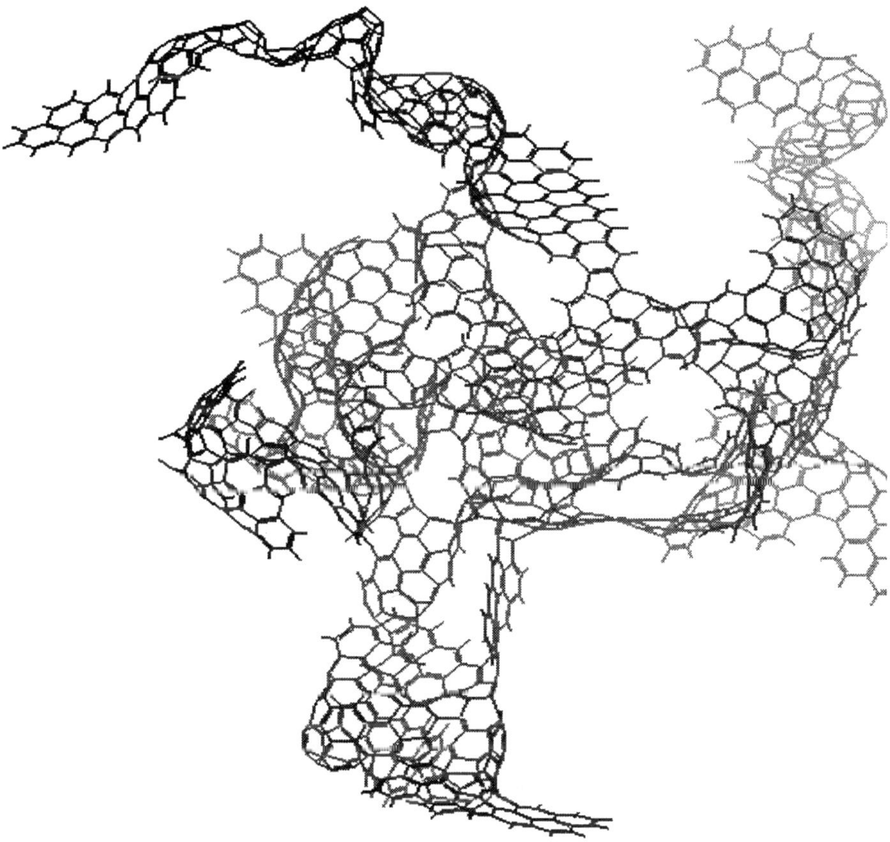

Figure 3. Typical structure from the chemically constrained structural simulation of NPC (adapted from Ref. 9).

4. STOCHASTIC GEOMETRIC MODELING

The essence of the nanoporous carbon is that the local structure around each carbon atom is pseudo-trigonal. In fact all the local and medium range chemistry is locked in by this feature of the covalence and bonding of carbon atoms. It should come then as no real surprise that the distance across C_{60} fullerene, once the atomic radii are included is within 0.1 nm of the average pore size of the NPC **(Figure 4).** So it seemed quite feasible to carry out a reduction in the complexity of the problem by embodying the chemistry and the physics of the carbon in the geometry, albeit stochastic geometry, of the average lattice. We approached this from two perspectives one can be considered to be a defective graphite model and the other one a random ring addition.

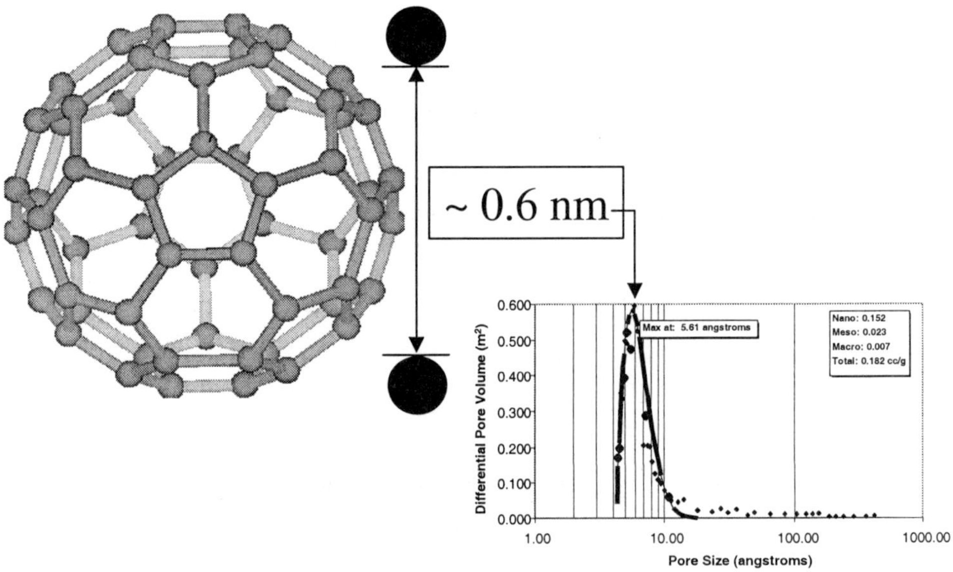

Figure 4. Comparison of the average pore size of NPC and the dimension at the internal surface of C_{60} fullerence.

4.1 Defective Graphite

Deviations away from graphite are taken as the natural result of defects left behind in the NPC structure during pyrolysis which at reaction temperatures do not have time to be annealed out of the solid. These defects lead to the formation of five and seven member rings and hence cause a corruption of the planarity of the graphene sheets and a loss of perfect planar symmetry. By this means much larger simulated NPC structures could be assembled.

One method which we explored along these lines was a model building process which we refer to as proceeding through a defective graphite lattice. In essence the density of the NPC taken in ratio to that of graphite is taken to be a direct measure of the probability of any site being occupied in the graphite lattice. Once this ratio is set it can be used to randomly knock-out atoms from a three dimensional lattice of graphite. Left in this state the system would be unphysical and too high in energy. Therefore a partial minimization can be done to simulate an annealing of the structure. To do this one can form bonds between sites missing a mutual nearest neighbor and then the bond angles and bond lengths can be reoptimized by minimizing the objective function, F, given as follows:

$$F = \sum_{j=1}^{n} (\lambda_g - \lambda_j)^2 + \omega \sum_{j=1}^{n} (\alpha_g - \alpha_j)^2 \tag{1}$$

Where the λs refer to the bond lengths and the αs refer to bond angles. The reference state bond length is taken as 0.14 nm and the bond angle is 120°. This objective function given more explicitly follows:

$$F = (\lambda_g^2 - \xi_1)^2 + (\lambda_g^2 - \xi_2)^2 + (\lambda_g^2 - \xi_3)^2 + \omega\left[\alpha_g - \cos^{-1}(\eta_{12})\right]^2 + \omega\left[\alpha_g - \cos^{-1}(\eta_{13})\right]^2 + \omega\left[\alpha_g - \cos^{-1}(\eta_{23})\right]^2 \tag{2}$$

Where the variables are given in three-space coordinates as:

$$\xi_1 = (x_1 - x)^2 + (y_1 - y)^2 + (z_1 - z)^2 \tag{3}$$

$$\xi_2 = (x_2 - x)^2 + (y_2 - y)^2 + (z_2 - z)^2 \tag{4}$$

$$\xi_3 = (x_3 - x)^2 + (y_3 - y)^2 + (z_3 - z)^2 \tag{5}$$

$$\eta_{12} = \frac{(x_1 - x)(x_2 - x) + (y_1 - y)(y_2 - y) + (z_1 - z)(z_2 - z)}{\sqrt{(x_1 - x)^2 + (y_1 - y)^2 + (z_1 - z)^2}\sqrt{(x_2 - x)^2 + (y_2 - y)^2 + (z_2 - z)^2}} \tag{6}$$

$$\eta_{23} = \frac{(x_2 - x)(x_3 - x) + (y_2 - y)(y_3 - y) + (z_2 - z)(z_3 - z)}{\sqrt{(x_2 - x)^2 + (y_2 - y)^2 + (z_2 - z)^2}\sqrt{(x_3 - x)^2 + (y_3 - y)^2 + (z_3 - z)^2}} \tag{7}$$

$$\eta_{13} = \frac{(x_1 - x)(x_3 - x) + (y_1 - y)(y_3 - y) + (z_1 - z)(z_3 - z)}{\sqrt{(x_1 - x)^2 + (y_1 - y)^2 + (z_1 - z)^2}\sqrt{(x_3 - x)^2 + (y_3 - y)^2 + (z_3 - z)^2}} \tag{8}$$

The objective function is minimized via a multivariate Newton-Raphson method. On a single layer basis the result is one of a symmetry-broken layer (**Figure 5**). Three dimensional results from this type of simulation are quite interesting. We see that the layers retain a graphite-like structure when viewed from the perspective of the edges of the layers (**Figure 6**). The method did not include bond formation between the layers in the form of cross links and as such it is probably a better simulation of non-graphitizing carbons than it is of the highly cross linked NPC materials.

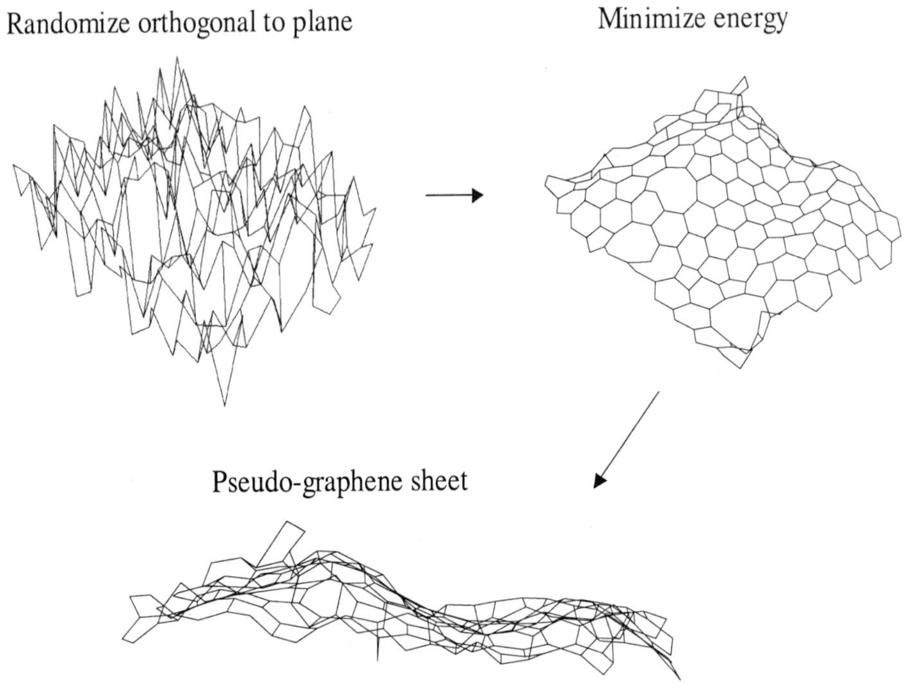

Figure 5. Defective graphite simulation steps: randomize and minimize to produce a symmetry-broken layer.

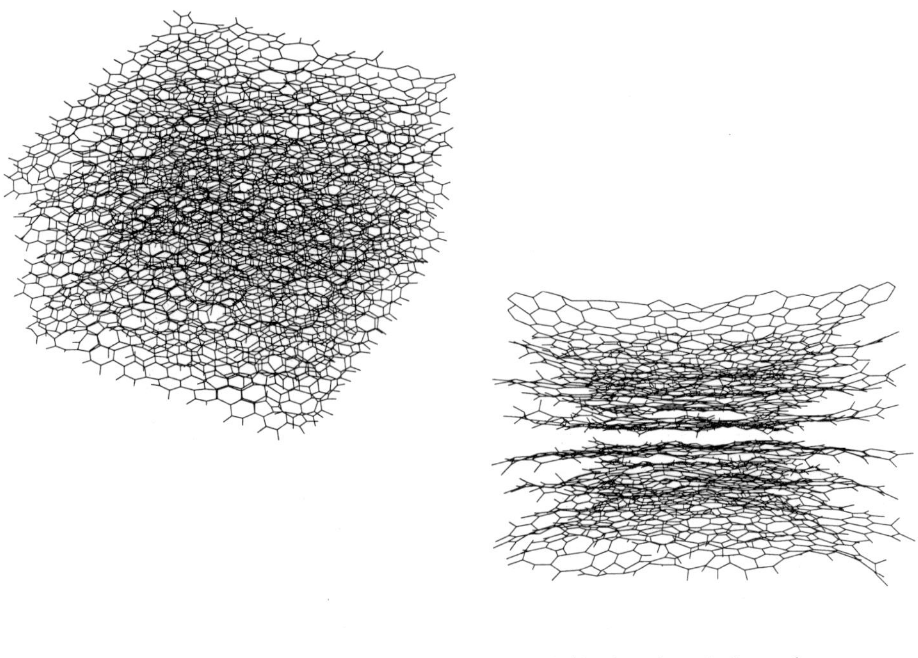

Figure 6. Defective graphite simulations top view and side view along the layer edges.

4.2 Random Ring Addition Simulation

In order to derive more continuous structures with more ribbon-like features an alternative algorithm has been constructed which provided for a *random ring addition simulation* (RRAS) of the growth of the lattice. The objective function of the *defective graphite simulation* was retained. Instead of beginning with graphite and then depopulating it, and reminimizing its structure, a lattice is grown from a seed fragment by this method. In simplest form the steps are as follows:

- Set the simulation box dimension
- Set the number of carbon atoms from by the target density
- Set the probabilities for five, seven and "skipped" (missing) rings
- Create a naphthalenic seed of two fused six rings
- Choose positions and new ring sizes stochastically
- Minimize the new ring and its neighbors on the fly during construction
- Repeat until all atoms are used

Schematics of this process for ring-by-ring build up of the lattice are illustrated in **Figures 7** and **8**. Implemented in Mathematica 4.0, separate modules are called to characterize a locus for new ring addition, to then construct a give ring size and another is used to do a local structural minimization. Snapshots of the growth of such a lattice at various stages of the simulation (**Figure 9**) show that as the pseudo-latticc grows its structural complexity rises. The lattice grows to fill the available space of the simulation box but only to the limit of the present density. In this case the five and seven ring defects were set to 3 and 7%, respectively. We can also see that the structure is essentially continuous and it is highly reticulated.

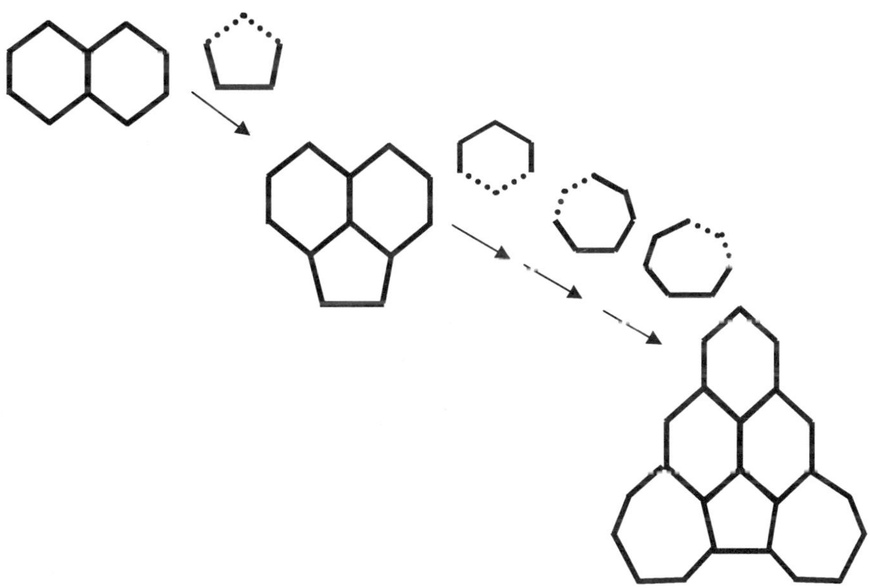

Figure 7. Random ring addition simulation of NPC: seed with fused six-rings, add 5, 6, 7 or a skipped ring stochastically with respect to position and ring number to produce a non-self intersecting continuous lattice.

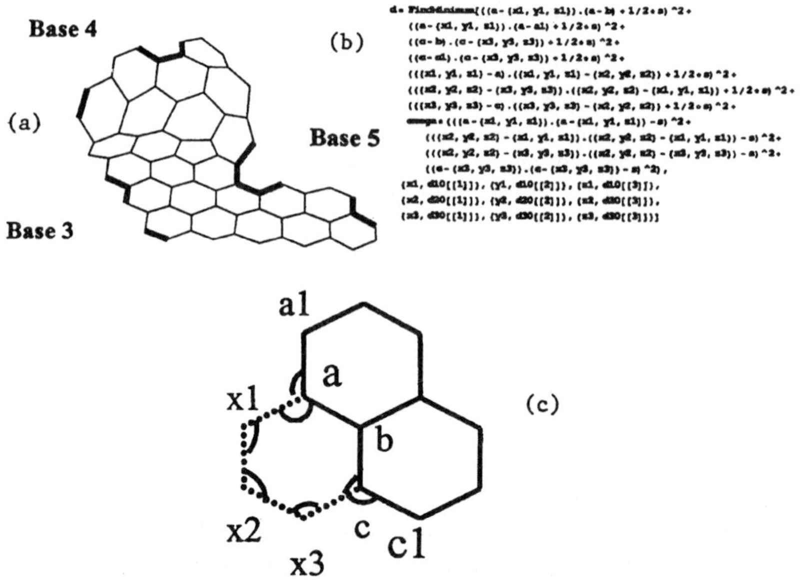

Figure 8. Typical position for ring addition along developing lattice, (b) objective function, (c) three space coordinates used for RRAS.

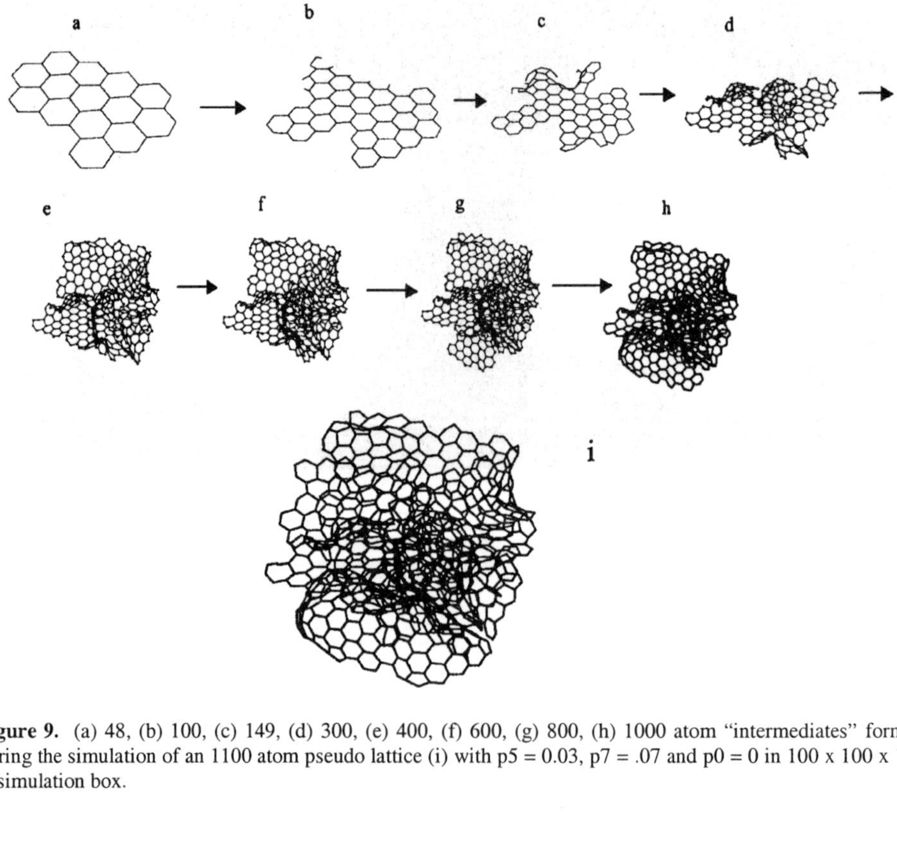

Figure 9. (a) 48, (b) 100, (c) 149, (d) 300, (e) 400, (f) 600, (g) 800, (h) 1000 atom "intermediates" formed during the simulation of an 1100 atom pseudo lattice (i) with p5 = 0.03, p7 = .07 and p0 = 0 in 100 x 100 x 100 Å simulation box.

4.3 Efficiency for Larger Scale Simulations

On of the main goals that we had in going to the stochastic geometric simulation was to be able to construct much larger lattices in considerably shorter times. This new algorithm satisfies that goal. Whereas simulations of 1000 atoms used to take more than 10 hours of CPU time on an SGI workstation using the chemically constrained algorithms, we can now create such lattices in 10 minutes of CPU time on a 1 GHz Pentium III PC. In 5.9 hours of CPU time lattices of 75,000 atoms have been developed (**Figure 10**). This is fast enough for us to make reasonable progress toward the goal of providing lattice files which can be used for molecular dynamics simulations of transport of small molecules in NPC. However, the next challenge is the one of characterizing these lattices *in-silico.*

4.4 *In Silico* Characterization of Super Pseudo Lattices

The super pseudo lattices (SPLs) which we are now able to construct with fairly high computational efficiency represent real progress toward the goal of developing fully chemically and physically constrained structural models of NPC. However, they also represent a very real and new challenge in characterization *in silico*. At this point our ability to create such structures outstrips our ability to characterize them. Once again, as

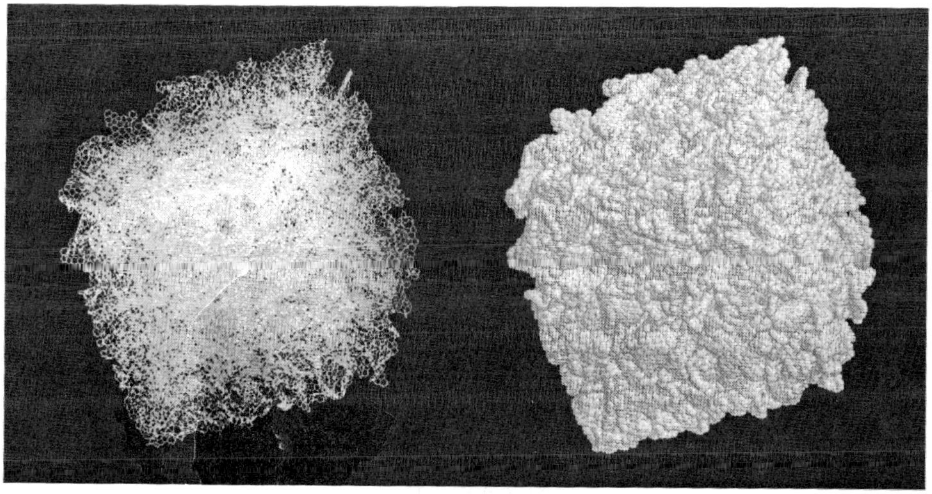

Figure 10. RASMOL images of 75 K-atom super pseudo lattice of NPC constructed via the ring by ring simulation algorithm –(a) wire frame rendering (b) space-filling (p5 = p7 = 0.1, 100 x 100 x 100 Å3 simulation box).

in our earlier work on the CCSM, we have utilized the very powerful algorithms already present in computational chemistry software packages such as Cerius. This package provides powerful sub-routines for the simulation of diffraction patterns, HRTEM images, lattice energies, NMR spectra and similarly critical characterization data.

In this regard at the time of this writing we have made preliminary progress in using Cerius to characterize the super pseudo lattices we have built. To do so we have had to scale down to smaller lattice sizes of up to, but not larger than, 10,000 atoms. One important point, which we had learned earlier, is that doing a global energy minimization as a first step within Cerius, was wholly unrealistic. Beyond about 1000 atoms this energy minimization could not be done in an acceptable length of CPU time (<12 hr). Hence the structures are analyzed as they are after they have emerged from the ring by ring simulation with its local geometric optimization and without any further energy optimization. As a starting point we have begun to simulate HRTEM images of 10,000 atom super pseudo lattices of NPC. Although very preliminary, the results are quite promising in that the typical worm-like structure is at least qualitatively present (**Figure 11**). Deeper and more complete analyses are in the offing.

A major goal for the foreseeable future is to provide reasonable average structures of the NPC so accurate potential functions can be constructed from first principals in order that *a-priori* simulations of small molecule transport dynamics within the nanopores may be accomplished. Ultimately, accurate predictions of diffusive selectivities will be the paramount test of simulated structural similitude with the real systems.

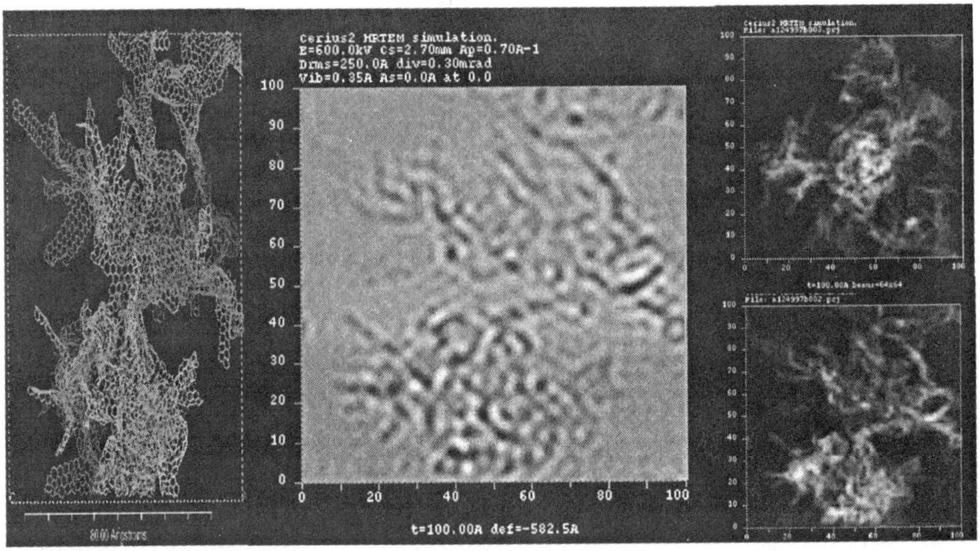

Figure 11. Cerius simulation of an HRTEM image of a 10K atom super pseudo lattice of NPC.

5. CONCLUSIONS

The structural simulation of NPC materials offers a significant challenge in the pursuit of accurate models of average structure. Lacking in long range order but having highly regular and deeply constrained local structures, these materials display a remarkably narrow distribution of pore sizes with a mean value of 0.5 nm. This level of structural complexity requires algorithms which take into account intensive as well as extensive property measures. Explicit chemical constraints offer one approach to this problem, but can be costly in terms of computational efficiency. However, given the rigidity of sp^2 carbon-carbon bonding modes, the chemical constraints can be apprehended in purely geometric rules, which lead to much faster and more efficient construction algorithms. Hence two stochastic geometric simulations have been developed and examined for their efficacy. The reoptimization of a defective layer lattice of graphite leads to structures that are reminiscent of NPC, but which retain a high degree of graphitic character. In contrast stochastic random ring addition from a seed structure leads to continuity and a high degree of internal curvature of the kind deemed necessary for the development of a narrow nanopore size distribution in a globally amorphous solid. Super pseudo lattices of as large as 125,000 atoms have been constructed by this means with high efficiency. Characterization of these lattices for their employ in first principles modeling of small molecule diffusive transport within them is underway.

ACKNOWLEDGMENTS

Work on this project has been funded by grant no. EEC-0085461 ("Molecular Transport in Nanostructured Materials: a Hierarchical Approach to Design Nanostructured Materials") and by the Research Experience for Undergraduates program both provided by the National Science Foundation.

REFERENCES

1. H. C. Foley, ***Micropor. Mats.* 4**, 407–433 (1995).
2. M. S. Kane, J. F. Goellner, H. C. Foley, R. DiFrancesco, S. J. Billinge, and L. F. Allard, Jr., *Chem. of Mats.* **8**, 2159–2171 (1996).
3. M. B. Shiflett, and H. C. Foley, *Science*, **285**, 1902-1905 (1999).
4. M. Acharya, and H. C. Foley, *AIChE Journal*, **46**, 911-922 (2000).
5. H. C. Foley, M. S. Kane and J. F. Goellner, in *Access in Nanoporous Materials*, edited by T. J. Pinnavaia and M. F. Thorpe, (Plenum, New York, 1995) pp.
6. M. S. Kane, H. C. Foley and W. Brendley, *Ind. Eng. Chem. Res.*, **37**, 2416-2425 (1998).
7. R. K. Mariwala, and H. C. Foley, *Ind. Eng. Chem. Res.*, **33**, 2314–2321 (1994).
8. R. K. Mariwala, and H. C. Foley, *Ind. Eng. Chem. Res.*, **33**, 607–615 (1994).
9. V. Petkov, R. G. DiFrancesco, S. J. L. Billinge, M. Acharya, and H. C. Foley, *Philos. Mag., B*, 79, 1519-1530 (1999).
10. M. Acharya, M. S. Strano, J. P. Mathews, S. J. Billinge, V. Petkov, S. Subramoney and H. C. Foley, *Philos. Mag., B*, **79**, 1499-1510 (1999).

THE STRUCTURE OF JAHN-TELLER POLARONS IN THE COLOSSAL MAGNETORESISTIVE MANGANITES

B. J. Campbell, S. J. L. Billinge, J. W. Lynn, R. Osborn and S. K. Sinha*

I. INTRODUCTION

The observation of large negative magnetoresistive effects[1,2,3] has generated tremendous scientific and technical interest in the perovskite manganites and related materials. These phenomena are commonly classified under the heading "colossal magnetoresistance" (CMR) to distinguish them from magnetoresistive effects in metals and metal-multilayers. CMR effects are observed in conjunction with the formation of a ferromagnetic metallic (FM) state at low temperatures and/or high magnetic fields. The close connection between the magnetism and the charge transport in these mixed-valence Mn^{3+}/Mn^{4+} materials is commonly believed to have its origin in the magnetic double-exchange interaction[4,5], in which itinerant e_g electrons mediate the ferromagnetic coupling between neighboring Mn core spins. Recent works have shown that double-exchange alone is insufficient to explain the extraordinary magnetoresistance values observed experimentally, so that electron-phonon interactions must also be invoked.[6,7] Alternatively, a direct Mn-Mn magnetic exchange interaction has been reported to better explain[8] the insensitivity of the Curie temperature (T_C) to the carrier concentration in oxygen-depleted samples.[9] This model also relies on strong electron-lattice coupling, which occurs via the Jahn-Teller (JT) mechanism[10], in which an $Mn^{3+}O_6$ octahedron experiences a spontaneous distortion designed to break the degeneracy of its e_g orbital, and thus lower the energy of its extra electron. If the coupling is strong, the electron may

*Branton J. Campbell, Materials Science Division, Argonne National Laboratory, Argonne IL 60439. S.J.L. Billinge, Department of Physics and Astronomy, Michigan State University, Lansing, MI 48824. Jeffrey W. Lynn, NIST Center for Neutron Research, National Institute of Standards and Technology, Gaithersburg, MD 20899 and University of Maryland, College Park, MD 20742. Ray Osborn, Materials Science Division, Argonne National Laboratory, Argonne IL 60439. Sunil K. Sinha, Physics Department, University of California, San Diego, CA 92093.

From Semiconductors to Protein, edited by S.J.L. Billinge and M.F. Thorpe
Kluwer Academic/Plenum Publishers, 2002

induce a significant local distortion of the lattice, which then follows the electron as it hops from site to site, thus forming a polaron.[8,11,12]

Coherent JT distortions, which indicate a long-range ordering of the occupied e_g orbital configuration, occur in purely Mn^{3+} systems such as $LaMnO_3$, as well as in charge-ordered systems like $La_{0.5}Ca_{0.5}MnO_3$.[13] The incoherent or local distortions associated with hopping JT polarons are also considered to be an important structural feature of the perovskite manganites. This has been amply verified as a wealth of evidence now shows the formation and collapse of the polaronic state to be essential to the large magnetoresistance that accompanies the transition to the FM state, and demonstrates that the spin, charge, orbital, and lattice degrees of freedom are all intimately coupled. JT polarons have been studied in the manganites using a variety of experimental techniques that includes a number of spectroscopies (e.g. UV/IR, Raman, NMR, EPR, μSR, inelastic neutron scattering, XAFS, etc.), thermodynamic and transport measurements (optical conductivity, resistivity, magnetoresistance, Hall effect, thermopower, thermal expansion, etc.), as well as dilatometry, and scanning probe microscopy. Broad summaries of these findings have been given elsewhere.[14]

This article considers the structural characterization of JT polarons, primarily reviewing our use of x-ray and neutron scattering to probe polaronic local structure. Section II of this article considers the coherent JT distortions associated with such long-range charge and/or orbital order. Sections III-V then focus on the incoherent JT distortions associated with the local polaronic structure. Section III discusses deviations from crystallinity and the methods of local structure analysis; section IV reviews the contributions of the PDF and Rietveld methods; section V describes the measurement and importance of polaron-induced lattice strain fields; and finally, section VI reviews short-range polaron correlations.

II. CHARGE AND ORBITAL ORDER IN THE MANGANITES

Early diffraction experiments on the manganite perovskites demonstrated that the MnO_6 octahedra in a number of manganite perovskites experience JT distortions in addition to octahedral rotations. Rotations about pseudo-cubic (100) axes can occur in response to an A-site cation slightly smaller than the corresponding interstitial volume, and tend to yield orthorhombic lattice parameters ($a \sim \sqrt{2}a_{cub}$, $b \sim \sqrt{2}a_{cub}$, $c \sim 2a_{cub}$) such that $c/(\sqrt{2}a) > 1$ (the so-called "O" phase). Cooperative JT distortions appear in some systems, such as $LaMnO_3$, and produce a distinct orthorhombic phase (O′) that usually obeys the condition, $c/(\sqrt{2}a) < 1$, with no additional cell-doubling beyond that dictated by the octahedral tilts. In the O′ phase, each $Mn^{3+}O_6$ octahedron experiences an apical stretch, making one of its three Mn-O bonds longer than the other two by as much as 0.25 Å.[15] The distortions cooperate to minimize the resulting elastic-strain energy so that all of the long bonds lie in the orthorhombic *a-c* plane and form a checkerboard-like pattern of perpendicular in-plane distortion directions as seen in Fig. 1. An early review by Goodenough[16] contains an unusually detailed catalogue of perovskite structures and properties.

A classic series of neutron powder diffraction experiments[17] on $La_{1-x}Ca_xMnO_3$ in the mid-1950's uncovered a wide array of distinct magnetic orderings, that were successfully explained by J. B. Goodenough in terms of ordered sublattices of Mn^{3+} and Mn^{4+} charges together with cooperative JT distortions on the Mn^{3+} sublattice.[13] In addition to corroborating the observed magnetic orderings, these hypothetical charge and orbital

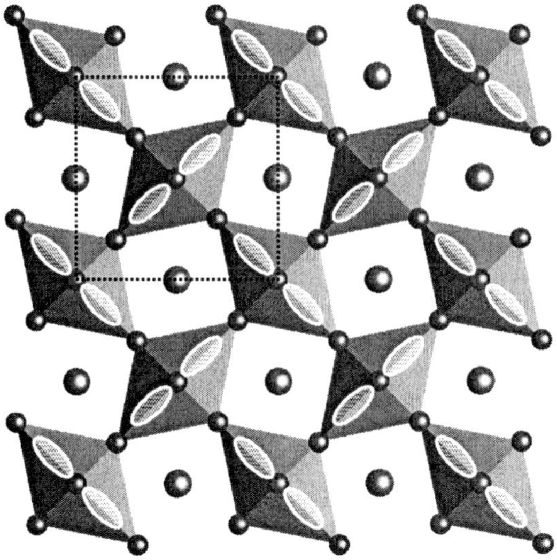

Figure 1. Illustration of the cooperative JT distortions and the inferred orbital ordering in the 100% Mn^{3+} lattice of $LaMnO_3$. These idealized distortions involve a primary apical stretch of one of the Mn-O bond pairs. After figure 4 of Goodenough.[12] Pairs of ovals indicate occupied e_g orbitals.

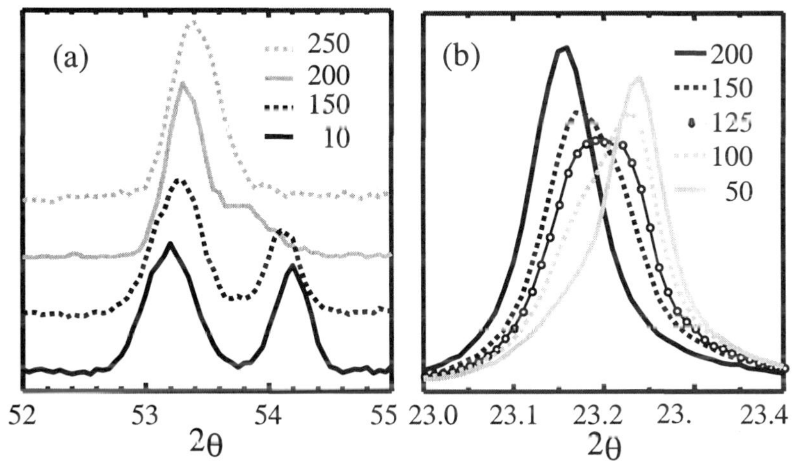

Figure 2. (a) A second-phase (0,3,1) reflection tracks the development of the low-temperature C/O-ordered phase in $La_{0.5}Ca_{0.5}MnO_3$. The peak on the left contains the (0,3,1) of the primary high-temperature phase as well as the unresolved (2,1,1) and (1,1,2) reflections. After Huang *et al.*[20] (b) The evolution of the (0,0,10) reflection through the C/O-AF transition in $LaSr_2Mn_2O_7$. The peak on the left corresponds to the high-temperature phase which develops C/O order at $T_{CO} \sim 210K$. The peak on the right, corresponding to the competing A-type AF phase, appears as a shoulder at $T_N = 170$ K and largely consumes the C/O-ordered phase by 100 K. After Argyriou *et al.*[25]

ordering patterns also provided qualitative predictions for ordering temperatures, lattice parameters, and magnetic sublattice moments that agreed well with observations. However, a charge-ordered sublattice multiplies the unit cell size and also predicts the appearance of non-magnetic superlattice reflections, which were not observed at that time.

The observation of these super-structural signatures has only recently become possible with the development of bright high-resolution x-ray and neutron sources and modern electron diffraction instruments. In $La_{0.5}Ca_{0.5}MnO_3$, for example, which is ferromagnetic above T_C~260K and CE-type antiferromagnetic below T_N~160K, the appearance of weak superstructure reflections has been reported to develop between T_C and T_N.[18,19] A recent neutron powder diffraction (NPD) study shows this superstructure to appear at 230K as a second phase, in competition with the ferromagnetic phase[20], and subsequently hosts CE-type AF order below T_N. A careful Rietveld analysis of the superlattice peak intensities[19], using synchrotron x-ray powder diffraction (PXD) data, confirms Goodenough's earlier hypothesis of long-range charge/orbital (C/O) order, and the resulting zig-zag pattern of JT distortions is now commonly referred to as "CE-type" C/O order by association with CE-type magnetic order. Fig. 2a shows the appearance of a 2nd-phase (0,3,1) reflection to the right of the primary-phase (0,3,1) reflection that tracks the development of the C/O order. This peak arises because the lattice parameter perpendicular to the distortions, *b*, is significantly smaller in the JT-distorted region, whereas the lattice parameters in the *a*-*c* plane of the distortions grow larger. Long-range C/O order has now been observed in many manganite systems,[21,22] including some "charge-stripe" phases.[23,24]

A similar situation occurs in the bilayered $La_{1+2x}Sr_{2-2x}Mn_2O_7$ system. In $LaSr_2Mn_2O_7$ (x = 0.5), CE-type C/O order forms at $T_{CO} \sim 210$K, well above the AF ordering temperature.[25,26,27] Fig. 3 contains an illustration of this zig-zag orbital arrangement and the corresponding transverse displacements within a single perovskite sheet of this material. The oxygen displacements within the JT-distorted octahedra of this system are roughly 0.05 Å, which is about half that observed in $La_{0.5}Ca_{0.5}MnO_3$.[19] While the C/O-ordered phase of $LaSr_2Mn_2O_7$ has been reported to host CE-type AF order within a narrow temperature range[27], the predominant magnetic ordering in this material appears at $T_N \sim 170$K and consists of AF-coupled ferromagnetic sheets (i.e. A-type order). The A-type phase competes with the C/O-ordered phase until the C/O order is overcome below 100K. The (0,0,10) Bragg reflection, shown in Fig. 2b, has a low-angle component associated with the primary high-temperature phase that charge orders at T_{CO}. The right-hand shoulder is a 2nd-phase peak belonging to the A-type phase, which appears at T_N and grows with decreasing temperature until the C/O phase is consumed. The broad phase-coexistence regions seen in $La_{0.5}Ca_{0.5}MnO_3$ and $LaSr_2Mn_2O_7$ are typical of systems that exhibit C/O order. The local-disorder created by the solid-solution of size-mismatched A-site cations may create a microscopically inhomogeneous Jahn-Teller stabilization energy, though such an explanation needs more careful examination.

The $Pr_{0.7}Ca_{0.3}MnO_3$ system, which has a very small A-site cation-size mismatch, is interesting in that it exhibits a relatively sharp transition between a long-range C/O-ordered phase[28,29] and a low-temperature FM phase, accompanied by a very large CMR effect.[30] As the system is pushed from the C/O-ordered state into the FM metallic state by lowering the temperature, applying a magnetic field, or illuminating with an intense x-ray beam, the superstructure peaks diminish, indicating that the e_g electrons delocalize and the C/O order melts.[29,31] Generally, even in the absence of long-range C/O order, the competition between local JT distortions and ferromagnetic correlations persists and proves to be fundamental to CMR, as discussed below.

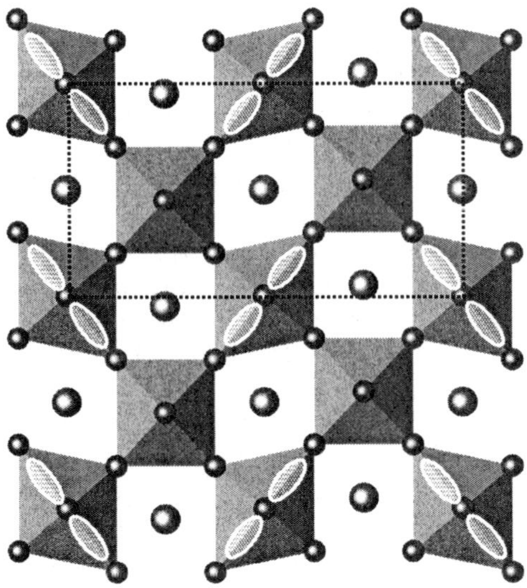

Figure 3. Illustration of charge-ordered sublattices and zig-zag (CE-type) charge and orbital order within a perovskite sheet of $LaSr_2Mn_2O_7$. After figure 4 of Goodenough.[12]

Because Mn^{3+} and Mn^{4+} have identical neutron cross-sections and nearly identical x-ray cross sections, neutron and x-ray diffraction are sensitive to orbital order rather than charge order. The superlattice reflections that we observe in C/O-ordered systems arise from the oxygen displacements induced by cooperative orbital order, while the charge-order is usually inferred. Polarized resonant x-ray diffraction techniques have now been applied to the CMR manganites for the purpose of obtaining independent measures of orbital and charge order.[32,33] Charge-ordered domains were shown to be many times larger than the orbitally ordered domains, both above and below T_{CO}, in the $Pr_{1-x}Ca_xMnO_3$ system (x = 0.4,0.5), demonstrating that the charge and orbital degrees of freedom are closely but not perfectly coupled.

III. LOCAL STRUCTURE FROM SCATTERING DATA

At specific compositions, such as undoped $LaMnO_3$, the JT polarons form a long-range ordered lattice. However, at general values of the composition the structure of JT polarons involves local deviations from perfect crystallinity. In the present context, the atomic displacements induced by the distortions are the deviations of interest. These displacements modify the diffracted intensities and give rise to diffusely-scattered intensity as well. Consider the effect of static local displacements on the elastic scattering distribution. This scattering is proportional to the structure function, which can be expressed as

$$I(\mathbf{Q}) = \left\langle \sum_{m,n} f_m e^{i\mathbf{Q}\cdot(\mathbf{r}_m+\mathbf{u}_m)} f_n^* e^{-i\mathbf{Q}\cdot(\mathbf{r}_n+\mathbf{u}_n)} \right\rangle = I_B(\mathbf{Q}) + I_D(\mathbf{Q}), \quad \text{where} \qquad (1)$$

$$I_B(\mathbf{Q}) = \sum_{m,n} (f_m d_m)(f_n^* d_n^*) e^{i\mathbf{Q}\cdot(\mathbf{r}_m - \mathbf{r}_n)}, \tag{2}$$

$$I_D(\mathbf{Q}) = \sum_{m,n} f_m f_n^* (d_{mn} - d_m d_n^*) e^{i\mathbf{Q}\cdot(\mathbf{r}_m - \mathbf{r}_n)}, \tag{3}$$

$$d_{mn} \equiv \left\langle e^{i\mathbf{Q}\cdot(\mathbf{u}_m - \mathbf{u}_n)} \right\rangle \text{ and } d_n \equiv \left\langle e^{i\mathbf{Q}\cdot\mathbf{u}_n} \right\rangle. \tag{4}$$

Here, the f_n are the x-ray or neutron scattering factors, while the $\mathbf{r}_n$, and $\mathbf{u}_n$ are the respective ideal atomic positions, and displacements from those ideal positions, summed over the whole crystal. The first term in Eq. (1), which we call $I_B(\mathbf{Q})$, is only non-zero at the Bragg points of the reciprocal lattice, and represents the modified Bragg peak intensities. The modifier, $d_n(\mathbf{Q})$, is commonly written as $e^{-w(\mathbf{Q})}$ and referred to as the "displacement" factor or the Debye-Waller (DW) factor. The displacement factor is often used to represent thermal disorder,[34] though it can clearly contain information about other types of defects and disorder. It modifies the diffracted intensities so as to correspond to the new average structure, which has been averaged over any distribution of local displacements. The second term in Eq. (1), which we refer to as $I_D(\mathbf{Q})$, is the diffuse scattering function, and is defined continuously over reciprocal space, rather than discretely. The diffuse scattering term directly probes the atomic displacement-displacement correlation function, which views the local displacements relative to one another rather than individually.[35] Thus, there is information about the nature of the local structure encoded in both the diffuse scattering the Bragg-diffracted intensities.

Static displacements produce only elastic scattering, whereas dynamic displacements either yield scattering features that have a finite quasi-elastic energy width that is inversely proportional to the time scale of the dynamics, or inelastic (phonon) scattering. X-ray scattering measurements typically integrate over a wide range of energies, and therefore produce a "snapshot" of the instantaneous displacements (i.e. broad energy resolution = narrow time range). Neutron scattering measurements, on the other hand, are commonly able to resolve the quasi-elastic width and/or inelastic response, allowing one to study the low-energy dynamics.

Total scattering experiments measure the average and local structure simultaneously, including quasi-elastic structure when it can be resolved. Single-crystal diffuse scattering is a particularly powerful tool because it measures full three-dimensional total scattering data. Some of its more challenging aspects include (1) the tremendous dynamic range required to simultaneously measure Bragg reflections and diffuse features in 3D reciprocal space without saturation, distortion, or loss of weak signals, (2) the collection, reconstruction, and reduction of 3D reciprocal space data from multi-detector hypersurfaces such that the end product is suitable for quantitative analysis, and (3) the computational burden of manipulating complex correlation models and large 3D scattering data sets. While valuable orientational information is lost in 1D scattering data from powder samples, the problems of dynamic range, data reduction, and computer power are much less severe for powder methods. Atomic pair distribution function (PDF) analysis, applied to powder samples, is a total scattering method in which powder diffraction data are Fourier-transformed to real-space coordinates[36] before being analyzed using techniques analogous to Rietveld refinement.[37] Rietveld analysis accesses the average local displacements via the DW factor, whereas the PDF method includes the diffuse-scattering intensity, $I_D(Q)$, thereby providing the potential to study locally correlated displacements.

The analytical tools that have been applied to the study of both average structure and local structure problems[38] tend to be classified according to three common denominators (1) a multi-parameter structural model, (2) an algorithm for navigating the multi-parameter space, and (3) an agreement-factor defined over the parameter space. An effective algorithm is tailored to the application at hand and should be thorough (samples all of the relevant parameter space), efficient (moves quickly towards good solutions), and robust (escapes local minima). A well-designed agreement factor will effectively differentiate a parameter set that approximates the true structure from one that does not over a wide dynamic range of "agreement", and should be sensitive to the model selected. The common use of a least-squares or steepest-descent algorithm, together with an agreement factor that measures only the difference between experimental and calculated data (e.g. diffraction or PDF data), might be termed the "straightforward" approach. However, the choice of model, algorithm, and agreement-factor is bounded only by the imagination. The Monte Carlo algorithm, for example, can be used in conjunction with agreement factors based purely on inter-atomic potentials as well as agreement factors based on a fit to experimental diffuse scattering data.[38] As data collection, reduction, and analysis tools improve, we move steadily towards the ability to apply a suite of fully three-dimensional PDF ("3DPDF") tools to single-crystal total scattering data.

IV. POLARONS FROM POWDERS

Long-range C/O-order is typically observed at special compositions with simple fractional values of x. In $LaSr_2Mn_2O_7$ ($x = 0.5$), for example, the resistivity rises sharply at T_{CO}, but falls off again as the C/O state gives way to a competing magnetic structure consisting of AF-coupled ferromagnetic sheets.[25] Most manganites that exhibit large CMR effects, however, do not exhibit long-range C/O order, despite their strongly analogous physical behavior. The resistivity of $La_{1.2}Sr_{1.8}Mn_2O_7$ ($x = 0.4$) increases rapidly with decreasing temperature despite the absence of a C/O-ordered state and drops precipitously at $T_C \sim 120$ K to make way for an entirely ferromagnetic structure.[39] The state above T_C is clearly polaronic, but very difficult to characterize structurally since the polaronic structure is local rather than crystalline.

Much of our knowledge of polaronic structure has been obtained from scattering experiments on polycrystalline samples. In $La_{1-x}Ca_xMnO_3$, the neutron powder PDF peak near 1.95Å corresponds to the Mn-O bond. At $x = 0$, this peak is split due to the presence of JT distortions[40,41], so that separate peaks appear at 1.92 Å and 2.12 Å. Because the splitting also occurs in the crystallographic-model PDF, it is clear that these are the cooperative or "coherent" JT distortions of the average structure. At $x = 0.25$, however, the PDF peak associated with the JT-elongated bond appears above T_C, even in the absence of any coherent JT distortions in the crystallographic average structure, implying the presence of local "incoherent", and possibly dynamic, distortions.[42] Note the exceptional quality of the fit to the crystallographic model below T_C in Fig. 4, and the significant discrepancies between the crystallographic and data PDFs above T_C. The Mn-O bond distortions disappear below T_C, where the e_g electrons become delocalized, demonstrating that the observed distortions are polaronic in origin.

The PDF peak at 2.75 Å corresponding to O-O bonds is also a sensitive probe of polaronic local structure. While the integrated area of this peak is independent of temperature, it is observed to markedly broaden below T_C at $x = 0.21$ and $x = 0.25$, which causes the peak height to decrease. The evolution of the 2.75 Å peak height vs. temperature appears in Fig. 5. Below T_C, the peak is sharp, indicating a resolution-

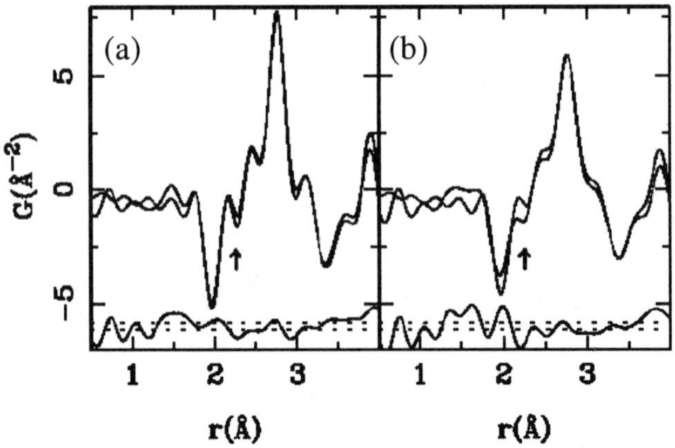

Figure 4. Best fit of the crystallographic model to neutron-powder PDF data for $La_{1-x}Ca_xMnO_3$ at x = 0.25 both below and near T_C. a) 10 K and b) 220K. The darker line is the data PDF, while the lighter line represents the model PDF. The difference PDF appears at the bottom of the figure, where straight dashed lines indicate the 2σ uncertainty level due to counting statistics. After Billinge.[14]

limited O-O bond-length distribution and providing further evidence that the polarons vanish in the FM state. This evidence is corroborated by XAFS results from a series of samples from x = 0 to 1 which show a pronounced anomalous broadening of the Mn-O bond length distribution above T_C for compositions with a FM ground state.[43] The DW factors also contain evidence of local polaronic distortions. In $La_{1-x}Ca_xMnO_3$ (x = 0.21 and 0.33) and $La_{1-x}Pb_xMnO_3$ (x = 0.30), the isotropic oxygen DW parameters obtained from Rietveld refinements are seen to clearly drop below T_C, accompanied by a slight decrease in the lattice parameters (see Fig. 6).[44,45] This again reflects the disappearance of local JT distortions in the FM state and the concomitant decrease in the disorder at the oxygen sites. A similar but weaker effect is also seen in the DW parameters of the A and B-site cations.[45]

The early PDF data from $La_{0.79}Ca_{0.21}MnO_3$ provided a remarkably detailed picture of polaronic distortions in the insulating state above T_C. At these doping levels (x < 0.5) the charge carriers are hole-type Mn^{4+} polarons in an electron-rich Mn^{3+} matrix. Efforts to model the PDF data revealed the presence of small isotropic MnO_6 octahedra (associated with Mn^{4+} sites) coexisting with elongated JT-distorted ones (associated with Mn^{3+} sites), such that the long and short bond lengths differed by about 0.24 Å,[42] which is similar to the difference observed in orbitally-ordered $LaMnO_3$.[15] The octahedra occupied by hole-type polarons were observed to be uniformly contracted (δ ~ 0.12) compared to the average structure. The PDF model, for the sake of convenience only, consisted of a particular C/O-ordering scheme similar to that of Fig. 3, but modified for a doping level of x=0.25, with every second Mn^{4+} site replaced by a Mn^{3+} site with its long JT-distorted Mn-O bond oriented out of the plane.[46] While these details were only explained in the later work[46], the excellent fit to the early PDF data[42] well beyond 10 Å suggested that this C/O ordering scheme had merit. As discussed below, subsequent

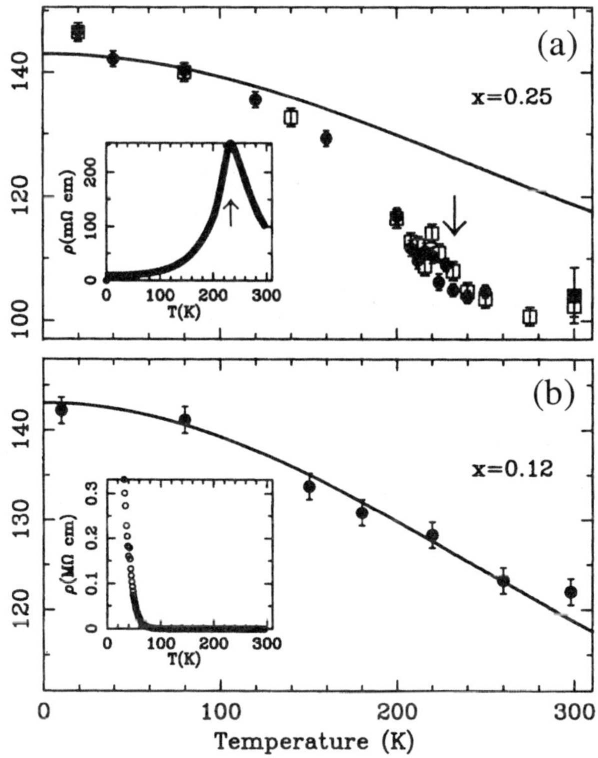

Figure 5. The neutron-powder PDF peak-height (2.75 Å) vs temperature for $La_{1-x}Ca_xMnO_3$ at a) x = 0.25 and b) x = 0.12. At x = 0.25, this peak, which is located at the nearest-neighbor O-O separation distance, anomalously broadens and falls-off above T_C (indicated by arrows). In contrast, the smooth variation of the peak shape at x = 0.12 is consistent with ordinary thermal vibrations. Circles and squares are data collected on cooling and warming, respectively. Corresponding resistivity data appear in the insets. After Billinge *et al.*[42]

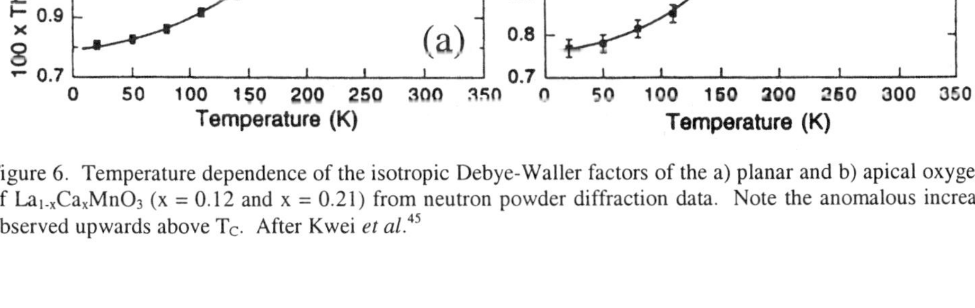

Figure 6. Temperature dependence of the isotropic Debye-Waller factors of the a) planar and b) apical oxygens of $La_{1-x}Ca_xMnO_3$ (x = 0.12 and x = 0.21) from neutron powder diffraction data. Note the anomalous increase observed upwards above T_C. After Kwei *et al.*[45]

single crystal diffuse scattering measurements conclusively demonstrated that short-range CE-type polaron correlations do occur in $La_{1-x}Ca_xMnO_3$ and other related systems.

Powder diffraction studies of $La_{1+2x}Sr_{2-2x}Mn_2O_7$ have shown that as the doping level is varied from x = 0.32 to x = 0.40, there is an unusual structural crossover effect in the coherent octahedral distortion amplitude at the optimum doping level, $x_{opt} = 0.36$.[47,48,49] This is explained in terms of distorted $Mn^{3+}O_6$ octahedra that shrink and $Mn^{4+}O_6$ octahedra that expand when the e_g electrons are delocalized, as in the case of the PDF results[42] from $La_{0.79}Ca_{0.21}MnO_3$. Most importantly, the oxygen DW factors of $La_{1+2x}Sr_{2-2x}Mn_2O_7$ reveal that the amplitude of the incoherent JT distortions is completely disconnected from that of the average coherent JT distortion. Independent of x, these local distortions build up above T_C and vanish abruptly at T_C, demonstrating that the polarons still collapse in the FM state.[48,50]

V. THE RISE AND FALL OF THE POLARON: STRAIN FIELDS AND DIFFUSE SCATTERING

The most detailed information about the three-dimensional structure of JT polarons in the CMR manganites comes from single crystal diffuse scattering (SCDS) experiments. X-ray and neutron SCDS, which has been used to study defects and disorder in a variety of materials[38] ranging from C_{60} to the high-T_C superconducting oxides[51], is also a powerful probe of the correlated displacements at the core of a JT distortion and the cooperative pattern of displacements that comprise any polaron-induced lattice strain fields that develop in response to the core displacements. Due to the nature of the Fourier transform, the sizes of related real and reciprocal space features are inversely related. Diffusely scattered features associated with displacements near the core of a distortion are broadly distributed across reciprocal space, whereas the diffuse scattering due to long-range strain-field displacements far from the core is distributed very close to the Bragg reflections. Obviously, because it is more localized in reciprocal space, the strain-field or "Huang" scattering[52] near a Bragg reflection is easier to measure. However, the core displacements are closely related to the strain-fields, and the physics that couples them is well understood.[35,53] Thus the diffuse strain-field scattering can be also used to obtain both local and extended polaronic structure.

Synchrotron x-ray diffuse scattering data have recently been obtained from single-crystal samples of several CMR perovskites that reveal unmistakable evidence of anisotropic strain-field scattering from polaronic lattice distortions.[54,55,56] Figure 7a contains a two-dimensional ($h0l$) section of reciprocal space that includes the (0,0,8), (0,0,10), and (0,0,12) Bragg reflections of $La_{1.2}Sr_{1.8}Mn_2O_7$.[54] This scattering falls off like $1/q^2$ away from the allowed Bragg reflections, indicative of the typical long-range $1/r^2$ decay of the lattice strain fields away from point defect cores. And in addition to being proportional to the Bragg-peak structure factor, the strain-field scattering tends to be stronger for a Bragg peak far from the origin due to a factor of Q^2. The rod of scattering parallel to (0,0,1) centered at $h = 0$ is due to a low concentration of intergrowths of other members of the related Ruddlesden-Popper series: $(La/Sr)_{n+1}Mn_nO_{3n+1}$, and is unrelated to the polaron scattering. The "butterfly"-shaped anisotropy visible in Fig. 7a contains vital information about the symmetry and orientation of the JT distortions, relative to the perovskite bilayers. Fig. 7b, contains an (h,k,0) section centered at (0,0,10.2), and reveals a three-dimensional scattering pattern that looks more like an hourglass than a butterfly, suggesting a distortion that is symmetric under 90° rotations (e.g. a $d(3x^2\text{-}r^2)$ type distortion). The weak anisotropy apparent at the edges of the figure is actually the result

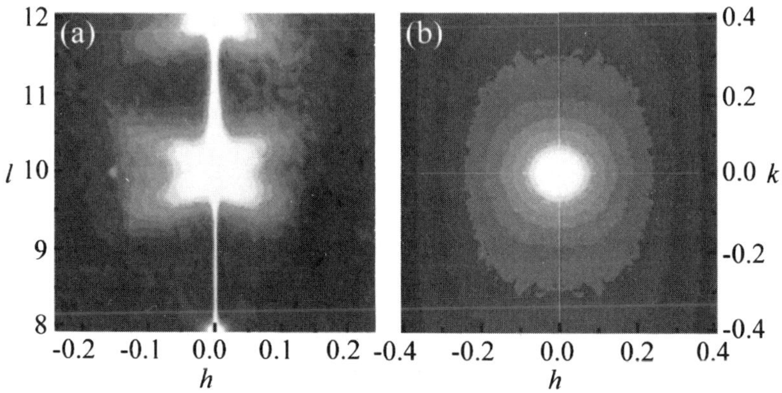

Figure 7. Contour plot showing the measured distribution of x-ray SCDS around the (0,0,10) reflection in $La_{1.2}Sr_{1.8}Mn_2O_7$ at 125 K. Panels (a) and (b) contain the (*h*,0,*l*) and (*h*,*k*,10.2) sections, respectively. After Vasiliu-Doloc *et al.*[54]

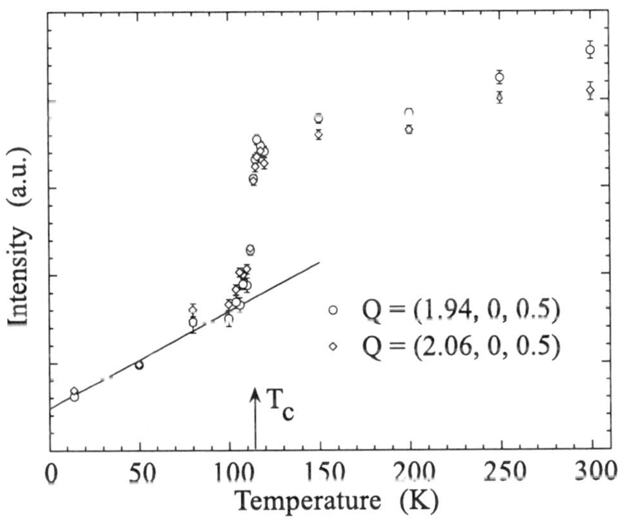

Figure 8. X-ray SCDS data from $La_{1.2}Sr_{1.8}Mn_2O_7$. Temperature dependence of the strain-field scattering measured on the two $l > 0$ lobes at $\mathbf{Q}$ = (2±0.06, 0, 0.5). The straight line indicates the contribution due to thermal diffuse scattering. The remainder of the scattering is associated with polaronic strain fields, which disappear abruptly below T_C, and get slowly damped out at high temperatures (T >> 300 K). After Vasiliu-Doloc *et al.*[54]

of x-ray absorption and the specific geometry of the sample. It should be noted, however, that these data are also qualitatively consistent with an average over $d(3x^2-r^2)$ and $d(3y^2-r^2)$ distortions. Complete three-dimensional SCDS data from a general Bragg position will resolve the true orbital character of these polarons.

In order to interpret the strain-field component of the x-ray diffuse scattering in Fig. 7, one must also take into account the contribution from thermal diffuse scattering (TDS), which arises from the instantaneous atomic displacements associated with thermally populated phonon modes. While the TDS also falls off as $1/q^2$ near allowed Bragg reflections, it can be confidently differentiated from the strain-field scattering via it's characteristic linear temperature dependence.[34] In Fig. 8a, the TDS contribution to the low-temperature butterfly scattering at (2±0.06, 0, 0.5) has been fitted to a straight line. The anomalous loss of intensity at the ferromagnetic transition temperature reveals the remainder of the scattering to be polaronic in nature and marks the sudden collapse of the polaronic state at T_C. The application of a magnetic field above T_C is also seen to destroy the strain-field contribution as the system is pushed into the FM state, providing a strong confirmation of the polaron hypothesis.[54]

In order to calculate the expected strain-field scattering distribution, one must determine the pattern of lattice displacements that arises in response to forces exerted by the defect center upon a finite number of atoms near the core. These displacements (assumed to be small) can then be inserted into Eq (1). Because the displacement field is impossible to calculate in all but the simplest cases, it's Fourier transform is generally calculated instead. This is the well-known method of static-displacement waves.[35,57] Finally, for scattering very near to the Bragg peak, we can employ the elastic approximation, which effectively embeds the details of a realistic JT polaron into an elastic continuum matrix. While there are many ways to group the various terms of the expression, we find it convenient to write the scattered intensity as:

$$I_{strain}(\mathbf{Q}) \propto |F_G|^2 Q_\beta Q_\delta T_{\alpha\gamma} D_{\alpha\beta\gamma\delta}\ , \quad \text{where} \tag{5}$$

$$T_{\alpha\gamma}(\mathbf{q}) = \sum_{\mathbf{R}_1,\mathbf{R}_2} \mathfrak{I}_\alpha(\mathbf{R}_1)\mathfrak{I}_\gamma(\mathbf{R}_2) e^{i\mathbf{q}\cdot(\mathbf{R}_1-\mathbf{R}_2)} \quad \text{and} \tag{6}$$

$$D_{\alpha\beta\gamma\delta}(\mathbf{q}) = \sum_{j_1,j_2} \frac{\hat{e}_\alpha(\mathbf{q},j_1)\hat{e}^*_\beta(\mathbf{q},j_1)\hat{e}^*_\gamma(\mathbf{q},j_2)\hat{e}_\delta(\mathbf{q},j_2)}{\omega(\mathbf{q},j_1)^2\,\omega(\mathbf{q},j_2)^2} \tag{7}$$

Here, **R** is an atomic position vector relative to the defect center, $\mathfrak{I}(\mathbf{R})$ is the force exerted by the defect core on the atom at **R**, $\mathbf{q} = \mathbf{Q}-\mathbf{G}$ is a vector in the Brillouin zone corresponding to reciprocal lattice vector **G**, F_G is the structure factor at **G**, j is a phonon branch index, and $\omega_{\mathbf{q},j}$ and $\hat{e}(\mathbf{q},j)$ are the eigen-frequencies and eigenvectors of the continuum dynamical matrix. The sums over Greek indices are implicit sums over vector and tensor components. To this, we add the TDS contribution, which may be written in the same notation as:

$$I_{TDS}(\mathbf{Q}) \propto |F_G|^2 Q_\alpha Q_\beta \sum_j \frac{\hat{e}_\alpha(\mathbf{q},j)\hat{e}^*_\beta(\mathbf{q},j)}{\omega(\mathbf{q},j)^2} \tag{8}$$

The calculated x-ray SCDS pattern around the (0,0,10) Bragg reflection that appears in Fig. 9 employs reasonable values of the tetragonal elastic constants to obtain the dynamical matrix, and assumes that forces are applied only to the six oxygens of the

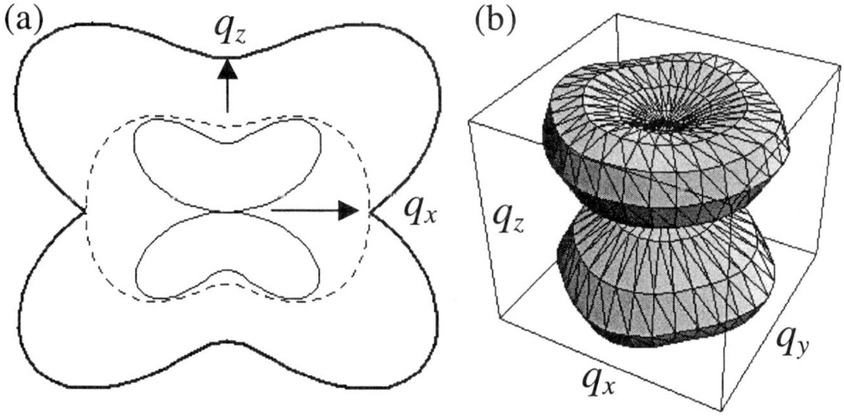

Figure 9. Calculated x-ray SCDS distribution around the (0,0,10) Bragg reflection of $La_{1.2}Sr_{1.8}Mn_2O_7$, based on a local JT stretch || (001) in an elastic medium. (a) Calculated iso-intensity contours for the strain field scattering (thin solid line) and TDS (dashed line) contributions to the total scattering (thick solid line) in the $(h,0,l)$ plane. (b) Calculated iso-intensity surface of the three-dimensional strain-field scattering distribution.

MnO_6 unit that defines the distortion core. Fig. 9b illustrates the relative contributions of the strain-field scattering and TDS to the overall scattering pattern. Note that the strain-field scattering possesses well-defined lobes, whereas the anisotropy of the TDS is more subtle. This fact makes interpreting the x-ray SCDS patterns somewhat easier.

VI. POLARON CORRELATIONS AT THE NANOSCALE

Weak diffuse satellite reflections were recently reported to appear in x-ray SCDS data from $La_{1.2}Sr_{1.8}Mn_2O_7$, indicating the presence of nanoscale spatial polaron correlations above T_C.[34] Fig. 10 contains single-crystal x-ray and neutron diffuse scattering data from this layered system[58] that simultaneously reveals anisotropic strain-field scattering from uncorrelated polarons together with several polaron-correlation peaks. Related features have since been discovered in a number other perovskite manganites[55,56,59,60], suggesting that short-range polaron correlations may be general participants in CMR phenomena.

In $La_{1-x}Ca_xMnO_3$, single-crystal neutron diffuse scattering data reveal broad diffuse peaks at reciprocal-space positions that correspond to propagation vector **q** = (0.25, 0.25, 0) and appear to indicate a short-range incipient form of CE-type C/O order (see Fig. 11a).[59] Like the elastic strain-field scattering associated with uncorrelated polarons, the temperature dependence of the diffuse CE-C/O peaks builds up above T_C and drops sharply to zero at the FM transition. Because this transition is accompanied by a significant CMR effect, one might expect that it can be assisted via the application an external magnetic field. This is indeed the case[51,59], as illustrated in Fig. 11b. The ground state of $La_{0.8}Ca_{0.2}MnO_3$ (x = 20), on the other hand, is not metallic[44], and its polaron correlations persist below T_C.[60] This close relationship between the local structure and the macroscopic transport makes it clear that the charge carriers are polaronic.

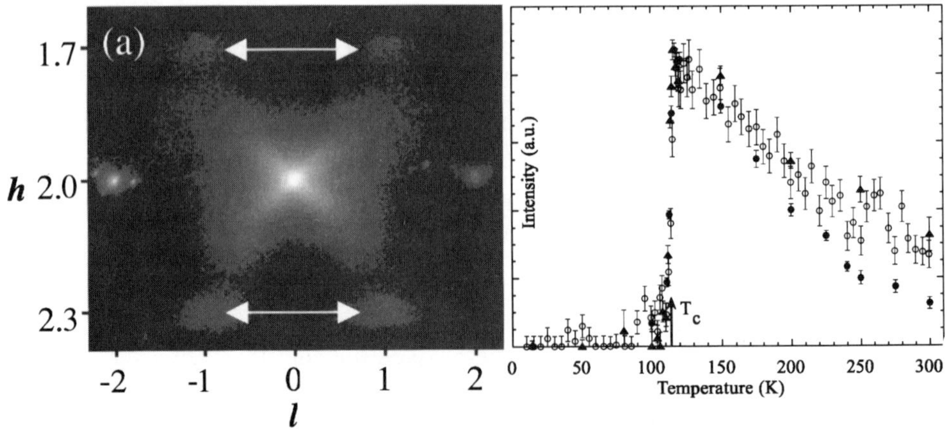

Figure 10. Single-crystal x-ray diffuse scattering data from $La_{1.2}Sr_{1.8}Mn_2O_7$. (a) CCD image of high-energy x-ray SCDS data from the (*h*,0.05,*l*) plane surrounding the (2,0,0) Bragg reflection (logarithmic scale). The diffuse butterfly-shaped scattering pattern is a combination of polaronic strain-field scattering and anisotropic thermal diffuse scattering, while the four diffuse maxima indicated by arrows are evidence of nanoscale polaron correlations. The peaks on the far left and right are tails of the (2,0,±2) Bragg reflections. After Campbell *et al.*[58] (b) Nearly identical trends of the polaron-correlation peak at **Q** = (2.3,0,1) [x-rays, closed circles] and the polaronic strain-field scattering at **Q** = (2±0.06,0,0.5) [x-rays, TDS corrected, closed triangles] and **Q** = (2.05,0,0.25) [quasi-elastic neutron scans, open circles]. After Vasiliu-Doloc *et al.*[54]

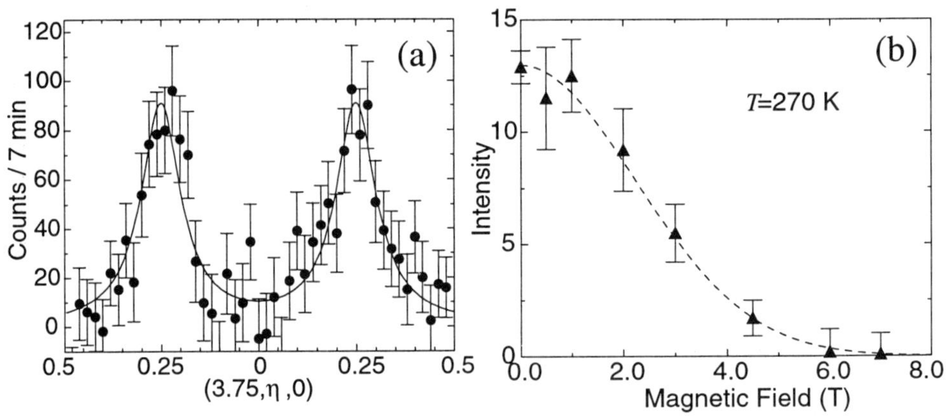

Figure 11. (a) Single-crystal neutron diffuse scattering from $La_{0.7}Ca_{0.3}MnO_3$. This (h,-h,0) scan through a pair of (3.75, 0.25, 0) peaks indicates a short-range analogue of CE-type charge/order. After Adams *et al.*[59] (b) The magnetic field dependence of the (3.75,0.25,0) polaron correlations peak at 270 K. After Lynn *et al.*[59]

At lower doping levels, the propagation vector associated with the short-range polaron correlations **q** = (0.25+ε, 0.25-ε, 0) becomes slightly incommensurate, much like that observed in the case of long-range C/O in $La_{0.5}Ca_{0.5}MnO_3$,[19] with a value of ε = 0.05 being reported[60] for $La_{0.8}Ca_{0.2}MnO_3$. When the doping level is reduced to x = 0.15, however, the diffuse peaks appeared at entirely different positions[60] which appear to be consistent with the more complicated long-range C/O configuration reported for $La_{1-x}Sr_xMnO_4$ (x ~ 0.125).[61] $Pr_{0.7}Ca_{0.3}MnO_3$ provides an interesting combination of incommensurate long and short-range C/O order.[56] Here, critical charge and orbital fluctuations are evidenced by diffuse CE-C/O satellites with an incommensurability that decreases with temperature and finally locks in at $\varepsilon = 0$ when the correlation length diverges at $T_{CO} \sim 230$ K.

Like the C/O fluctuations, the ferromagnetic spin fluctuations in $La_{0.7}Ca_{0.3}MnO_3$ build up as T_C is approached from above.[59] Below T_C, the central quasi-elastic peak of the spin fluctuations then disappears as spin-wave peaks develop in their place.[62,63] The intrinsic q-space widths were simultaneously measured for both the polaron-correlation peaks and the ferromagnetic-fluctuation peaks. The comparable correlation lengths, ~10 Å and 16 Å, respectively, suggest that the quasi-elastic spin fluctuation should be thought of as the spin component of a magnetic polaron. Figure 12 shows the nearly identical trends of the CE-C/O peak at (3.75,0.25,0), the central peak of the magnetic fluctuation spectrum at (1.03,0,0), and the sample resistivity.[59] In light of the competition between C/O order and long-range magnetic order, this surprising connection between the magnetic and polaronic fluctuations above T_C deserves further examination.

While the temperature and magnetic field behaviors of the polaron correlation peaks in $La_{1.2}Sr_{1.8}Mn_2O_7$, relative to the transition to the FM state at T_C, are mirrored by the $La_{0.7}Ca_{0.3}MnO_3$ system, the unusual propagation vector, q = (0.3, 0, ±1), of the polaron correlations in the bilayered $La_{1+2x}Sr_{2-2x}Mn_2O_7$ system immediately set them apart as distinct from the CE-type correlations of the 3D perovskite manganites.[54,58] Though 90° twin domains are represented in equal proportions, these diffuse satellite reflections point to a one-dimensional incommensurate modulation along the (100) axes of the tetragonal I4/*mmm* lattice. A crystallographic analysis of the integrated intensities of over 100 of these diffuse satellites yields the novel longitudinal JT stretch mode[58] shown in Fig. 13, which shall be referred to hereafter as "L"-type.[64] This pattern of distortions points to a JT-coupled charge density wave fluctuation, the structure of which has no known long-range analogue. The real-space correlation lengths are roughly $\xi_{\parallel} \approx 23$Å, $\xi_{\perp} \approx 15$Å, and $\xi_c \approx 10$Å, and triple-axis energy scans demonstrate that the structure is static on a 1 ps time scale.[54] Straightforward strain considerations additionally suggest that further-than-nearest-neighbor interactions must be responsible, while the fact that this arrangement has only been observed in a layered system may be connected to Fermi-surface effects reported to exist in the FM state below T_C.[65]

The predominant feature in Fig. 13 is the JT stretch mode of the Mn-O bonds that lie along the direction of the modulation. The Mn-O bonds that lie along (001) at a given site are compressed when the corresponding bonds along (100) are stretched, and vice-versa. A pattern of octahedral tilts accompanies the alternating expansion and compression of the Mn-O bonds in a physically intuitive way Because the modulation amplitudes of all of the seven unique atoms have the same phase, we are able to represent the displacement directions with the sinusoidal solid and dashed curves shown in the figure. These relative phases of these curves are dictated by the symmetry of the modulation and show how the displacements in one perovskite sheet are related to those in another. Notably, this symmetry stacks the JT-distorted octahedra of a single bilayer together along the (010) direction, giving the underlying charge density wave a striped character as well. It also sets the modulations in adjacent bilayers to be 180° out of phase.[58]

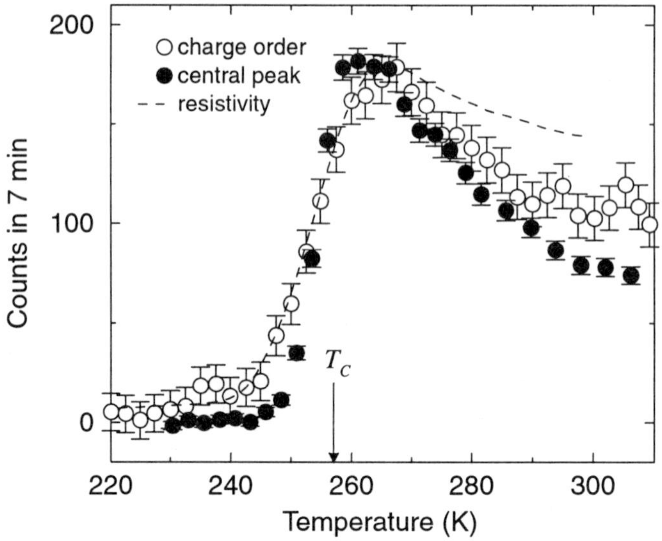

Figure 12. Single-crystal neutron diffuse scattering and transport data from $La_{0.7}Ca_{0.3}MnO_3$ reveals nearly identical temperature trends for the quasi-elastic magnetic fluctuation peak at (1.03,0,0), the polaron correlation peak at (3.75,0.25,0), and the bulk resistivity. After Adams *et al.*[59]

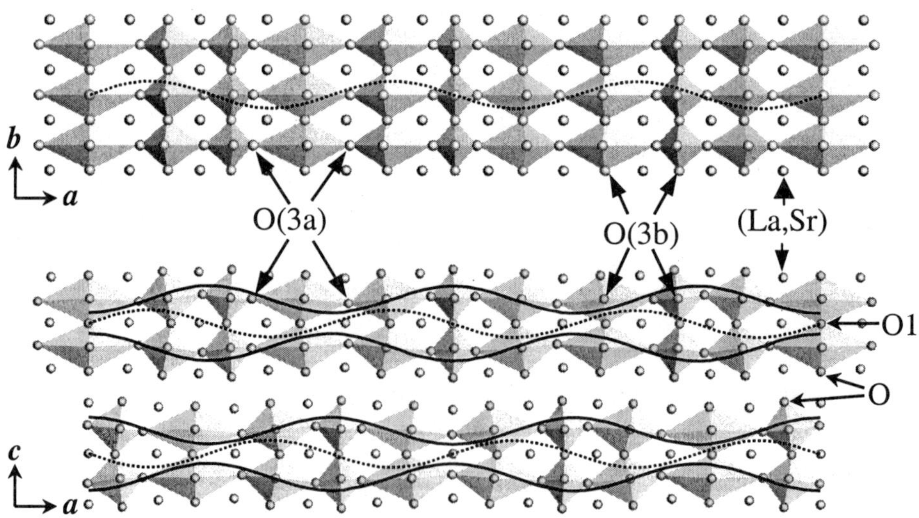

Figure 13. Crystallographic representation of the polaron correlations in $La_{1.2}Sr_{1.8}Mn_2O_7$. The directions of the atomic displacements within each perovskite sheet and bilayer follow the solid (z-component) and dashed (x-component) curves, where peaks indicate $+x$ or $+z$ displacements and troughs indicate $-x$ or $-z$ displacements. The displacements are exaggerated in order to make the more subtle features of the modulation visually apparent, and the modulation period has been approximated by the commensurate value of 10/3. After Campbell *et al.*[58]

VI. CONCLUDING REMARKS

The detailed crystallographic picture of coherent JT distortions, obtained in the limit of long-range C/O order, has added greatly to our understanding of CMR phenomena. The evolution of these coherent distortions and their associated superstructures at magnetic and metal-insulator transitions has firmly established a relationship between spin, charge, orbital, and lattice degrees of freedom at certain special compositions. Now, a growing suite of experimental and analytical scattering tools that are sensitive to local structural features have opened a new eye on the CMR universe. Local JT distortions have recently been observed to correlate strongly with anomalous magnetic and transport properties where no evidence of long-range C/O order could be found, demonstrating that the CMR effect in the perovskite manganites is derived from the existence of a polaronic state and its subsequent collapse under the influence of the double-exchange interaction.

Powder PDF and single-crystal diffuse scattering studies of JT polarons have uncovered a number of distinct structural features, including the local distortions of JT-distorted MnO_6 octahedra and the strain fields that comprise the accompanying lattice response. Furthermore, nanoscale (10 to 25 Å) polaron correlations have been discovered in several manganite systems, and can be viewed as a short-range form of C/O order that can exist over wide regions of the phase diagram that do not permit long-range order. Straightforward powder measurements quickly uncover features and trends of interest, while the more difficult and comprehensive single-crystal measurements are required to build a complete picture. Understanding this complementarity is important to the further development of both powder and single-crystal methods. Relevant issues that continue to intrigue us include the high-temperature fate of local JT-distortions and correlations, the time-scale of polaron dynamics, and the existence of local orientational order. Ongoing efforts to probe the local structure of the CMR manganites should address these and many other questions.

ACKNOWLEDGMENTS

Special thanks to a number of colleagues that have been involved in various aspects of this research, including C.P. Adams, D.N. Argyriou, A.A. Arsenov, A.R. Bishop, C. Booth, J.A. Borchers, E. Bozin, D.C. Dender, R.G. DiFrancesco, T. Egami, R.W. Erwin, A.J. Fedro, K. Ghosh, R.L. Greene, Q. Huang, M.F. Hundley, Z. Islam, S. Krasnicki, G.H. Kwei, Z.Y. Li, C.D. Ling, D. Louca, J.F. Mitchell, Y.M. Mukovskii, J.J. Neumeier, J.L. Peng, V. Petkov, G. Preosti, Th. Proffen, S. Rosenkranz, U. Ruett, A. Santoro, O. Seeck, D.A. Shulyatev, V.N. Smolyaninova, J.D. Thompson, M.F. Thorpe, L. Vasiliu-Doloc, and D. Vermeille. Research at ANL is supported by the U.S. Dept. of Energy, Office of Science under Contract No. W-31-109-ENG-38. Research at UM is supported by the NSF-MRSEC, DMR 00-80008. Research at MSU is supported by NSF grant DMR-0075149. Research at the Manuel Lujan Neutron Scattering Center at LANL was supported by the U.S. Dept. of Energy under contract W-7405-ENG-36.

REFERENCES

[1] G.H. Jonker, J.H. van Santen, Physica **16**, 337 (1950).
[2] R.M. Kusters, J. Singleton, D.A. Keen, R. McGreevy, W. Hayes, Physica B **155**, 362 (1989).
[3] J. Volger, Physica **20**, 49 (1954).
[4] C. Zener, Phys. Rev. **82**, 403 (1951).
[5] P.G. de Gennes, Phys. Rev. **118**, 141 (1960).
[6] A.J. Millis, P.B. Littlewood, B.I. Shraiman, Phys. Rev. Lett. **74**, 5144 (1995).
[7] H. Röder, J. Zang, A.R. Bishop, Phys. Rev. Lett. **76**, 1356 (1996).
[8] D. Emin, Mat. Res. Soc. Symp. Proc. **494**, 163 (1998).
[9] J. Tanaka, M. Umehara, S. Tamura, M. Tsukioka, S. Ehara, J. Phys. Soc. Jpn. **51**, 1236 (1982).
[10] H.A. Jahn, E. Teller, Proc. R. Soc. London A **161**, 220 (1937); J. H. van Vleck, J. Chem. Phys. **7**, 72 (1939).
[11] A. Moreo, S. Yunoki, E. Dagotta, Science **283**, 2034 (1999).
[12] J.B. Goodenough, Annu. Rev. Mater. Sci. **28**, 1 (1998).
[13] J.B. Goodenough, Phys. Rev. **100**, 564 (1955).
[14] See articles by Billinge, Tomioka *et al.*, Noh *et al.*, Zhao *et al.*, and Jaime *et al.*, in *Physics of Manganites*, eds. T.A. Kaplan and S.D. Mahanti, Kluwer Academic/Plenum Publishers, New York (1999).
[15] J.F. Mitchell, D.N. Argyriou, C.D. Potter, D.G. Hinks, J.D. Jorgensen, S.D. Bader, Phys. Rev. B **54**, 6172 (1996).
[16] J.B. Goodenough and J.M. Longo in *Landolt-Börnstein Tabellen*, New Series III/4a, pp. 126-314, ed. K.H. Hellwege, Springer-Verlag (1970).
[17] E.O. Wollan, W. C. Koehler, Phys. Rev. **100**, 545 (1955).
[18] C.H. Chen, S.-W. Cheong, Phys. Rev. Lett. **76**, 4042 (1996).
[19] P.G. Radaelli, D.E. Cox, M. Marezio, S.-W. Cheong, Phys. Rev. B **55**, 3015 (1997).
[20] Q. Huang, J.W. Lynn, R.W. Erwin, A. Santoro, D.C. Dender, V.N.Smolyaninova, K. Ghosh, R.L. Greene, Phys. Rev. B **61**, 8895 (2000).
[21] J.F. Mitchell, D.N. Argyriou, J.D. Jorgensen, in *Colossal Magnetoresistive Oxides*, ed. Y. Tokura, Gordon & Breach (2000).
[22] C.N.R. Rao, A. Arulraj, A.K. Cheetham, B. Raveau, J. Phys. Cond. Mater. **12**, R83 (2000).
[23] S. Mori, C.H. Chen, S.-W. Cheong, Nature **392**, 473 (1998).
[24] P.G. Radaelli, D.E. Cox, L. Capogna, S.-W. Cheong, M. Marezio, Phys. Rev. B **59**, 14440 (1999).
[25] D.N. Argyriou, H.N. Bordallo, B.J. Campbell, A.K. Cheetham, D.E. Cox, J.S. Gardner, K. Hanif, A. dos Santos, G.F. Strouse, Phys. Rev. B **61**, 15269 (2000).
[26] T. Kimura, R. Kimai, Y. Tokura, J.Q. Li, Y. Matsui, Phys. Rev. B **58**, 11081 (1998).
[27] M. Kubota, K. Yoshizawa, Y. Moritomo, H. Fujioka, K. Hirota, Y. Endoh, J. Phys. Soc. Jpn. **68**, 2202 (1999).
[28] Z. Jirak, S. Krupicka, Z. Simsa, M. Dlouha, S. Vratislav, J. Magn. Magn. Mater. **53**, 153 (1985).
[29] H. Yoshizawa, H. Kawano, Y. Tomioka, Y. Tokura, Phys. Rev. B **52**, 13145 (1995).
[30] Y. Tomioka, A. Asamitsu, Y. Moritomo, Y. Tokura, J. Phys. Soc. Jpn. **64**, 3626 (1995).
[31] V. Kiryukin, D. Casa, J.P. Hill, B. Keimer, A. Vigliante, Y. Tomioka, Y. Tokura, Nature **386**, 813 (1997).
[32] M.v. Zimmerman, J.P. Hill, D. Gibbs, M. Blume, D. Casa, B. Keimer, Y. Murakami, Y. Tomioka, Y. Tokura, Phys. Rev. Lett. **83**, 4872 (1998).

[33] Y. Murakami, H. Kawada, M. Tanaka, T. Arima, Y. Moritomo, Y. Tokura, Phys. Rev. Lett. **80**, 1932 (1998).
[34] B.T.M. Willis, A.W. Pryor, *Thermal Vibrations in Crystallography*, Cambridge Univ. Press, Cambridge (1975).
[35] M.A. Krivoglaz, *X-ray and neutron diffraction in non-ideal crystals*, Eng. translation, Springer-Verlag, Berlin/Heidelberg (1996).
[36] See articles by T. Egami and S.J.L. Billinge, in *Local Structure from Diffraction*, eds. S.J.L. Billinge and M.F. Thorpe, Plenum Publishers, New York (1998).
[37] R.A. Young, in *The Rietveld Method*, ed. R.A. Young, IUCr Monograph No. 5, Oxford Univ. Press (1993).
[38] See articles and accompanying references in *Local Structure from Diffraction*, eds. S.J.L. Billinge and M.F. Thorpe, Plenum Publishers, New York (1998).
[39] Y. Moritomo, A. Asamitsu, H. Kuwahara, Y. Tokura, Nature **380**, 141 (1996).
[40] Th. Proffen, R.G. DiFrancesco, S.J.L. Billinge, E.L. Brosha, G.H. Kwei, Phys. Rev. B **60**, 9973 (1999).
[41] D. Louca, G.H. Kwei, J.F. Mitchell, Phys. Rev. Lett. **80**, 3811 (1998).
[42] S.J.L. Billinge, R.G. DiFrancesco, G.H. Kwei, J.J. Neumeier, J.D. Thompson, Phys. Rev. Lett. **77**, 715 (1996).
[43] C.H. Booth, F. Bridges, G.H. Kwei, J.M. Lawrence, A.L. Cornelius, J.J. Neumeier, Phys. Rev. Lett. **80**, 853 (1998).
[44] Q. Huang, A. Santoro, J.W. Lynn, R.W. Erwin, J.A. Borchers, J.L. Peng, K. Ghosh, R.L. Greene, Phys. Rev. B **58**, 2684 (1998).
[45] G.H. Kwei, D.N. Argyriou, S.J.L. Billinge, A.C. Lawson, J.J. Neumeier, A.P. Ramirez, M.A. Subramanian, J.D. Thompson, Mat. Res. Soc. Symp. Proc. **475**, 533 (1997).
[46] S.J.L. Billinge, Th. Proffen, V. Petkov, J.L. Sarrao, S. Kycia, Phys. Rev. **62**, 1203 (2000).
[47] J.F. Mitchell, D.N. Argyriou, J.D. Jorgensen, D.G. Hinks, C.D. Potter, S.D. Bader, Phys. Rev. B **55**, 63 (1997).
[48] M. Medarde, J.F. Mitchell, J.E. Milburn, S. Short, J.D. Jorgensen, Phys. Rev. Lett **83**, 1223 (1999).
[49] D.N. Argyriou, H.N. Bordallo, J.F. Mitchell, J.D. Jorgensen, G.F. Strouse, Phys. Rev. B **60**, 6200 (1999).
[50] The term "collapse" here refers to the disappearance of the JT polaron, whereas in Ref. [8], it is used to describe small polaron formation.
[51] S.T. Johnson, P.D. Hatton, A.J.S. Chowdhury, B.M. Wanklyn, Y.F. Yan, Z.X. Zhao, A. Marshall, Physica C **219**, 61 (1994).
[52] K. Huang, Proc. Royal Soc. A **190**, 102 (1947).
[53] P.H. Dedhrichs, J. Phys. F Metals Phys. **3**, 471 (1973).
[54] L. Vasiliu-Doloc, S. Rosenkranz, R. Osborn, S.K. Sinha, J.W. Lynn, J. Mesot, O.H. Seeck, G. Preosti, A.J. Fedro, J.F. Mitchell, Phys. Rev. Lett. **83**, 4393 (1999).
[55] S. Shimomura, N. Wakabayashi, H. Kuwahara, Y. Tokura, Phys. Rev. Lett. **83**, 4389 (1999).
[56] S. Shimomura, T. Tonegawa, K. Tajima, N. Wakabayashi, N. Ikeda, T. Shobu, Y. Noda, Y. Tomioka, Y. Tokura, Phys. Rev. B **62**, 3875 (2000).
[57] For a detailed presentation of the phonon formulation, see A.A. Maradudin, E.W. Montroll, G.H. Weiss, I.P. Ipatova, Theory of Lattice Dynamics in the Harmonic Approximation, 2nd ed., Academic Press, New York (1971).

[58] B.J. Campbell, R. Osborn, D.N. Argyriou, L. Vasiliu-Doloc, J.F. Mitchell, S.K. Sinha, U. Ruett, C.D. Ling, Z. Islam, J.W. Lynn, Phys. Rev. B, *in press*, cond-mat/0106577 (2001).
[59] C.P. Adams, J.W. Lynn, Y.M. Mukovskii, A.A. Arsenov, D.A. Shulyatev, Phys. Rev. Lett. **85**, 3954 (2000); J.W. Lynn, C.P. Adams, Y.M. Mukovskii, A.A. Arsenov, D.A. Shulyatev, J. Appl. Phys. **89**, 6846 (2001).
[60] P.-C. Dai, J.A. Fernandez-Baca, N. Wakabayashi, E.W. Plummer, Y. Tokura, Phys. Rev. Lett. **85**, 2553 (2000).
[61] Y. Yamada, O. Hino, S. Nohdo, R. Kanao, T. Inami, S. Katano, Phys. Rev. Lett. **77**, 904 (1996).
[62] J. W. Lynn, R. W. Erwin, J. A. Borchers, Q. Huang, A. Santoro, J. L. Peng, and Z. Y. Li, Phys. Rev. Lett. **76**, 4046 (1996); J. W. Lynn, R. W. Erwin, J. A. Borchers, A. Santoro, Q. Huang, J. L. Peng and R. L. Greene, J. Appl. Phys. **81**, 5488 (1997).
[63] For a recent review of the spin dynamics see J. W. Lynn, J. Superconductivity and Novel Magnetism **13**, 263 (2000).
[64] This label was suggested by D.N. Argyriou.
[65] Y.-D. Chuang, A.D. Gromko, D.S. Dessau, T. Kimura, Y. Tokura, Science **292**, 1509 (2001).

ENERGY DISPERSIVE X-RAY ABSORPTION SPECTROSCOPY; A TOOL FOR TIME RESOLVED STRUCTURAL STUDIES OF CHEMICAL REACTIONS IN SOLUTION

Sofía Díaz Moreno*

1. INTRODUCTION

Most of the chemical processes that occur in nature take place in liquid media. For this reason, chemical reactions in solution have been widely studied for many years. By and large, these sorts of studies have been focused on the mechanism of the reactions, i.e. the paths that the starting reactants follow to form the chemical products, whilst the structures of the species involved in the reaction have received considerably less attention. The main reason for this state of affairs has arisen from the difficulty to experimentally determine, in detail, the structures of chemical species in liquid media. A key question in this field of research has been the detection of possible intermediate chemical species. To date, this important question has largely been addressed by the most standard techniques used for the study of chemical reactions in solution, such as uv-vis and infrared (IR) spectroscopies. Unfortunately these techniques are seldom structurally specific so the determination of the detailed chemical form of any such metastable species has awaited the arrival of a fast and more structurally focused technique.

In this chapter the relatively new technique of Energy Dispersive X-ray Absorption Spectroscopy, EDXAS, will be introduced as a novel means by which it is increasingly becoming possible to study chemical reaction kinetics from a more structurally insightful perspective. Here the term *chemical reaction kinetics* will be taken to refer to the study of the mechanism by which a reaction proceeds from a macroscopic point of view. This is in contrast to the distinctly different field of *chemical reaction dynamics*, which focuses on the study of the individual collisions of atoms and molecules that form chemical products i.e. a microscopic perspective.

---

* European Synchrotron Radiation Facility, E.S.R.F., 6, rue Jules Horowitz, B.P.220, 38043 Grenoble Cedex 09-France.

From Semiconductors to Protein, edited by S.J.L. Billinge and M.F. Thorpe
Kluwer Academic/Plenum Publishers, 2002

The key strength of energy dispersive X-ray absorption spectroscopy as it has evolved at the new third generation synchrotron sources such as the European Synchrotron Radiation Facility and that make the technique a potentially powerful tool for the study of in-solution chemically active species, is its ability to collect spectra on timescales as fast as a millisecond. This limiting timescale coming not from the experimental probe but rather from the time for the mixing process of the reactants to take place. This ability to collect data at such a high rate is combined with all the characteristic information content that can be obtained by conventional x-ray absorption spectroscopy i.e. chemically specific local structure information centered on a selected atomic species such as inter-atomic distances, number and type of neighbouring atoms and degree of thermal and structural disorder.

2. HISTORICAL PERSPECTIVE

The initiation of chemical reactions to obtain a new product from a combination of starting components has been utilized for many years[1]. For example in ancient times the pseudo-science of alchemy evolved through the mixing of reactants to produce products and was driven by the quest to discover a means to transmute base metals into gold. To this day, chemical reactions underpin our knowledge of all chemical and biological processes and this knowledge can very easily be argued to be the *gold* of the new millenium. Throughout history the commonly acknowledged indicators for the occurrence of reactions were changes of color of solutions, production of bubbles suggesting the formation of gas etc. and the earliest known attempt at a quantitative study of a chemical reaction was made in 1850 by Wilhelmy[2]. He studied the reaction of the inversion of sucrose and for the first time used the concentration of remaining reactant and concentration of product at any given moment to infer a speed of reaction. From this date, many different reactions began to be studied, for example, the famous experiment performed by Berthelot and St. Gilles in 1862[3]. They studied the equilibrium between ethanol and acetic acid, and concluded that the rate of the reaction was proportional to the products of the two concentrations. In all of these early studies the method of starting the reaction was what has become known as the bath method. This method consists of mixing the reactants at time "zero", followed by introducing the mixture into a thermostatic bath to keep the temperature constant, and collecting aliquots for analysis at different times. The concentrations of the reactants and/or products were determined by chemical and physical methods applied to these aliquots, methods such as titration, dilatometry, conductivity, etc. To freeze the reaction while determining the concentration, different strategies were used, such as lowering the temperature or adding a quenching agent such as another chemical species to react with one of the initial reactants and to eliminate it from the reaction mixture.

Although useful to study chemical reaction kinetics for those processes that occur on timescales ranging from minutes and hours, the bath method is unsuitable for reactions that go to completion on timescales less than a few minutes due to the dead time necessary for the manual mixing of the reactants. This restriction on the available temporal window of investigation provided the incentive for the development of the flow methods for starting and following the majority of reactions. In 1923, Hartridge and Roughton used the continuous flow method for the first time to follow the reaction of haemoglobin with oxygen and carbon monoxide[4]. When using this method, the reactants are made to flow through a mixing chamber, and the concentrations of products and/or reactants are monitored at different distances from the mixing chamber. In consequence the distance from the mixing point to the observation point controls the temporal resolution and timescales better than milliseconds can be achieved. Unfortunately the continuous flow method has the considerable disadvantage that the amount of solution that is needed to follow a reaction is quite high. As an example, in the first reaction followed by Hartridge and Roughton, seven liters of solution were necessary. Although today this amount has been reduced dramatically due to the new design of mixing chambers, this method is still not particularly appropriate if one wishes to follow reactions in which the reactants are very expensive or difficult to synthesize. In order to solve this difficulty, in 1940 Chance developed a new method for starting a reaction that is called the stopped flow method[5]. In this case the reactants are mixed in a mixing chamber, and then the flow is stopped. The concentration of the species that has been chosen for study is determined at a fixed distance from the mixing chamber with the temporal resolution governed by the rate at which this mixture can be passed to an observation cuvette. The amount of solution that is needed in this case is much less as after mixing the flow of reactants to the mixing chamber is stopped. The limiting dead time of the system that can vary from seconds to milliseconds depends on many factors, principally the distance from the mixing chamber to the observation cuvette, the flow rate of the reactants, and the volume that the observation cuvette is able to contain.

The flow methods were able to evolve thanks to the parallel development of new methods to determine the concentrations of the species in solution, such as the spectroscopic methods: uv-vis and IR spectroscopies, NMR, EPR, etc. These methods of monitoring the reaction allowed the experimental scientist to determine the concentration without stopping the reaction and thus opened the door to the study of reactions *in situ*.

Within this framework, Bazulin was the first to perform, in 1936, an experiment in which a reaction was started by what is called the relaxation method[6]. Although originally proposed by Nernst in 1910, the experiment failed because sufficiently fast and sensitive equipment did not exist at that time[7]. This method is based on the equilibrium state of the reactions. When a reaction is in equilibrium, any small perturbation of this situation results in a response that will bring the reaction back to the equilibrium state i.e. the famous *Le Chatelier's Principle*. By studying the way that a reaction returns to this state, it mathematically becomes possible to extract the kinetics of the original reaction. The perturbations that can be applied can be of various different sorts, but are always very small, so that one does not complicate the primary goal, which is the extraction of information about the kinetics of the main reaction. Too large a perturbation complicates the necessary mathematical approximations utilized in the formal analysis of the results. If the perturbation originates from a sudden change in the temperature, the method is called temperature jump whilst if pressure is the variable that it is suddenly changed, we talk about a pressure jump. More complex perturbations such as magnetic or electric fields or ultrasound etc. can be applied. Generally, relaxation methods cover reactions that occur on timescale less than a millisecond. Unfortunately due to the fact that the principal requirement is that the reaction under study should be reversible so as to give rise to an equilibrium state, the application of these powerful methods is not always suitable.

Several years after the relaxation methods were applied, in 1949 Norrish and Porter received the Nobel Prize for the development of a new and very fast method to start a reaction - Flash Photolysis[8]. By this method, a fast and powerful light pulse, coming from a lamp or a laser is employed to start an appropriate reaction. This method opened a new window of opportunity for gaining insight in the study of very fast chemical reactions, covering timescales ranging from the picosecond up to the second. In some cases it can be seen as a substitute method for the flow methods though this is not completely accurate due to the fact that the two methods study different sorts of reactions. While the flow and the relaxation methods study reactions where the reactants are in the ground state, in Flash Photolysis the reaction is initialized with a pulse of light that places one of the reactants in an excited state. So if for example, the process under study is a substitution reaction in a metallic complex, in which a ligand of the complex is substituted by another molecule of reactant, the Flash Photolysis method is not particularly suitable due to the fact that once the complex is in the excited state, it could react with the molecules of the solvent or any other molecule that is present in the solution.

As in the previous case of the flow methods, the monitoring methods most commonly used to determine the concentration of the reactive species in the case of the relaxation and the Flash Photolysis methods are mainly spectroscopic, since they are the fastest. However it was not until 1990 when the X-ray Absorption Spectroscopy method, in the energy dispersive variant, was used to monitor a reaction[9]. In this case, the source used was a rotating anode, obtaining a time resolution of five milliseconds per spectrum.

This method of monitoring the reaction, in conjunction with stopped-flow techniques, is the subject of the present chapter. We are going to present here how the high photon flux of the third generation synchrotron sources provides a new tool for the study of fast reactions from a structural viewpoint. This aspect is the main point that differentiates the EDXAS technique from other more widely used and easily accessible spectroscopies. Uv-vis is perhaps the most widely used due to its availability in the average chemistry laboratory, but the structural information that can be extracted by the method is minimal. The case is similar for IR spectroscopy, although in some specific instances this method is able to provide information about the structure of different functional groups such as CO or CN. Unfortunately it does not provide general structural information and in particular it has a most considerable shortcoming in that it is poorly applicable to the case of aqueous solutions - water is the most common and universal solvent. The case of NMR spectroscopy has some parallels with EDXAS in the sense that it is able to provide structural information about the species involved in the reaction, but for this spectroscopy to be applied the concentrations of the reactants needs to be quite high. The technique of electronic paramagnetic resonance, EPR, is another that has been applied to the study of chemical reactions. It is especially suitable to detect short life intermediates, but unfortunately can only be used in the case of paramagnetic species.

The above historical perspective is summarized in Figure 1.

3. INSTRUMENTAL ASPECTS

3.1 The X-ray spectrometer

The main characteristic of Energy Dispersive X-ray Absorption Spectroscopy is its ability for collecting complete spectra in as little time as a microsecond. This is considerably faster than the traditional way of performing an X-ray Absorption spectrum. In the latter technique, each spectrum is collected in minutes as traditionally in this mode, the intensity of the X-ray radiation absorbed by the sample is monitored in function of the incident energy and the scan proceeds in a stepwise manner i.e. change the energy by ΔE then measure the intensity and repeat[10].

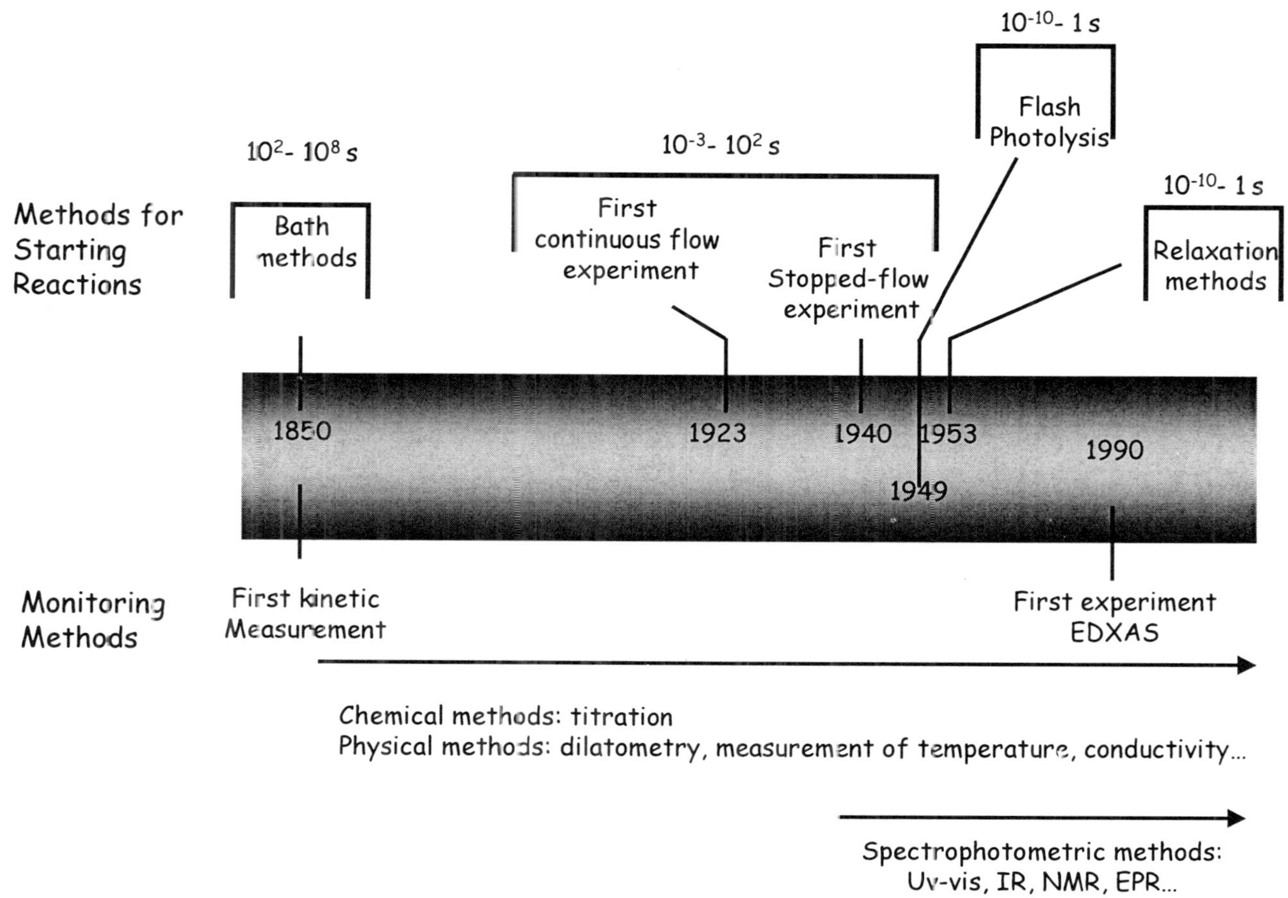

Figure 1. Diagram giving a historical overview of the evolution of the study of chemical reaction kinetics in solution.

This means that at any given time the sample is illuminated with a monochromatic beam, with two detectors, one before and one after the sample. This experimental geometry allows us to measure and normalize to the incident beam the intensity of transmitted and hence absorbed light. In the dispersive variant of the technique the sample is illuminated with a polychromatic beam of X-ray photons. This is called a pink beam. By using a bent crystal polychromator instead of a traditional scanning monochromator, the different energies in the incident beam are separated spatially, following the Bragg law. The whole fan of energies is then focused on the sample, and a position sensitive detector is then used to measure the spectrum in a single shot by being placed in the divergent beam after the sample. Thus, this mode of data collection has the advantage that there is no need to mechanically scan any component on the X-ray beamline. This is a very important advantage as this intrinsic instrument stability makes the energy dispersive setup especially suitable for studying very small changes in the spectrum. The caveat to this advantage is that because all the energies are taken at the same time, the intensity of the X-ray beam without the sample has to be measured by moving the sample out of the beam. This means that the measurements of the intensity of the beam with and without sample are consecutive, and not simultaneous. This in general is the most significant challenge in performing the technique as good data normalization and low signal to noise ratios are the desired goal of any spectroscopic technique. Though difficult, this challenge is not insurmountable, provided that both the X-ray source and beamline are working optimally. A schematic diagram on an energy dispersive X-ray absorption beamline is shown in Figure 2.

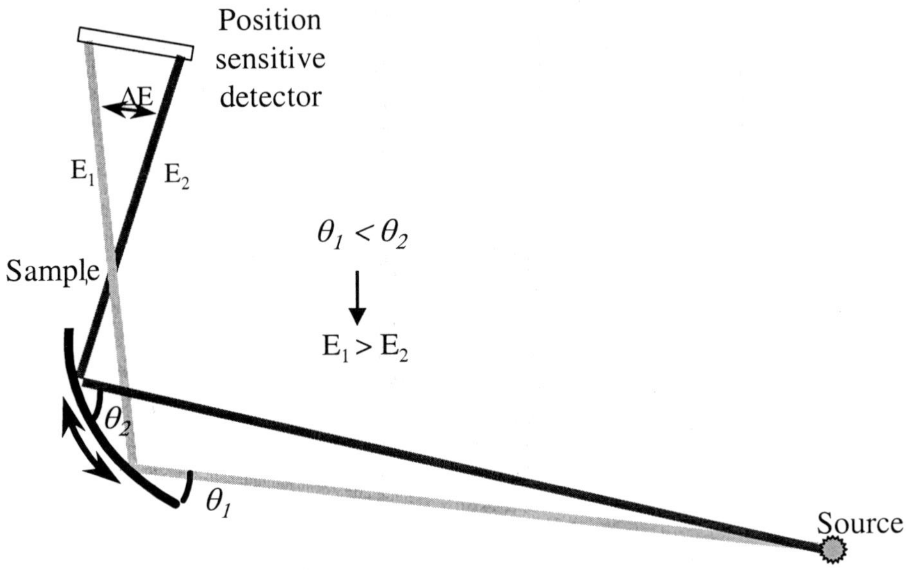

Figure 2. Schematic diagram of an energy dispersive X-ray absorption spectroscopy beamline such as ID24.

The work that will be presented here has all been performed at the European Synchrotron Radiation Facility at beamline ID24[11]. This instrument utilizes a pair of matched undulators as its X-ray source. The chosen energy for the experiment is tuned by changing the gaps of the undulators. The small X-ray beam produced by these undulators is optimized for energy dispersive spectroscopy using two coated mirrors placed in the

Kirkpatrick-Bayes geometry, i.e. vertically focused and horizontally diverged. This beam is then incident upon a bent crystal polychromator that can operate either under Bragg or Laue conditions, corresponding to energies from 6 keV up to 14 keV for the Bragg geometry, and from this value up to 26keV for Laue. The polychromator focuses the beam at the sample position, with a spot size of typically 50-100 μm horizontally by 300 μm vertically. This value delimits the smallest sample size that can be used.

The key element of the beamline for time resolved experiments is the position sensitive detector, the CCD camera[12]. This camera consists of a fluorescent screen on to which the X-ray beam is incident and this in turn is monitored by a 1242 pixels (horizontal) by 1152 pixels (vertical) CCD array. The CCD chip is not completely exposed but rather only 64 lines of the 1152 are exposed, and these are treated as a single spectrum. The rest of the chip is masked and acts as an analog buffer, which can store up to 17 rapidly acquired spectra, before the read out cycle of the camera. Each spectrum of the series of 18 can be taken in 100 microseconds. The shifting time of the charges from different fast spectra is of the order of 10 microseconds and the readout time of the camera is 250milliseconds. These very different times give rise to two different modes of collecting spectra that are called the *fast acquisition mode* and the *slow acquisition mode*.

In the fast mode, each group of 64 lines is treated as a single and individual spectrum, and the time resolution is given by the time to take each spectrum, with no time in between two consecutive spectra. Since the integration time for any spectra is very short, the process has to be repeated several times and averages made to get a sufficient signal to noise ratio for subsequent analysis of the spectra. For this reason only processes that can be repeated in a highly reproducible manner can be followed by this method. The main disadvantage is that only 18 spectra can be taken in a row, so the reaction under study has to be quite fast, of the order of seconds or less. A schematic of this acquisition mode is shown in Figure 3.

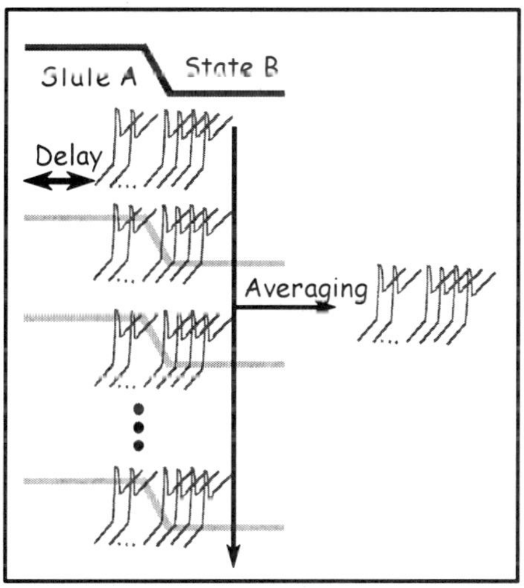

Figure 3. Scheme of the fast acquisition mode, illustrating the parallel averaging.

The slow mode is appropriate for the study of slower reactions than in the previous case. Each spectrum is an average of the 18 spectra taken by the camera before the readout. Then, the camera takes another spectrum that is again the result of an average of the 18, and so on. So each spectrum is taken in 18 times the selected integration time for an individual 64 line spectrum, and each averaged set of spectra are separated from the previous set by 250 milliseconds, the camera readout time. In this operation mode the number of averaged spectra that can be taken is unlimited, but the time resolution is worse than in the fast mode due to the readout time of the camera. However, this limitation is not a great handicap since through careful use of both the two acquisition modes, it is possible to "zoom in" on any region of interest identified in the slow acquisition by offsetting the starting time of the fast acquisition mode relative to the starting time of the reaction. A schematic diagram of the slow acquisition mode is shown in Figure 4.

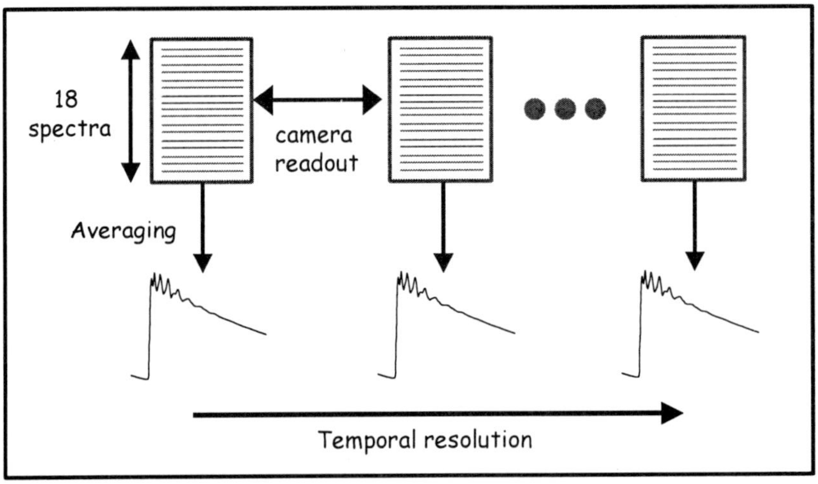

Figure 4. Scheme of the slow acquisition mode. The time resolution of the mode is limited by the averaging process.

3.2 The sample environment

Although several different methods can be used for starting the process under study, as described in section 2, the most widely utilized method with the Dispersive variant of X-ray Absorption spectroscopy is the stopped-flow method. There are several reasons for this. The first is that the amount of solution that is necessary for the technique is smaller than for the other flow methods. This is a great advantage if the reactants are expensive or difficult to synthesize. The second reason is the easy availability of relatively cheap commercial instruments that can be readily adapted for use on the X-ray spectrometer.

The stopped-flow device that has been used for this experiment has three vertical syringes to avoid any bubbles that can be tricky to avoid if the syringes are horizontal. Each syringe is moved by microstep motors allowing fine control of the quantity of reactants delivered to the reaction cell, typically each reactant can be delivered in quantities as small as 20μl. The three syringes are filled with different reactants required for the process to be studied, if only two reactants are required one can either not use one of the three syringes or simply fill two syringes with the same solution. In the first step of the mixing process, the reactants from the first two syringes are mixed in a premixing device. Subsequently this mixture is then driven to a second mixer where the third reactant is added and from here the now reacting mixture is driven to the observation cuvette. This cuvette is equipped with X-ray transparent windows. Most commonly in the case that only two reactants are necessary for the reaction, only the second mixing device tends to be used. The first syringe would then be filled with the solvent used in the reaction and this can then be used to wash the stopped-flow device once the reaction is finished. A scheme of this device is shown in Figure 5.

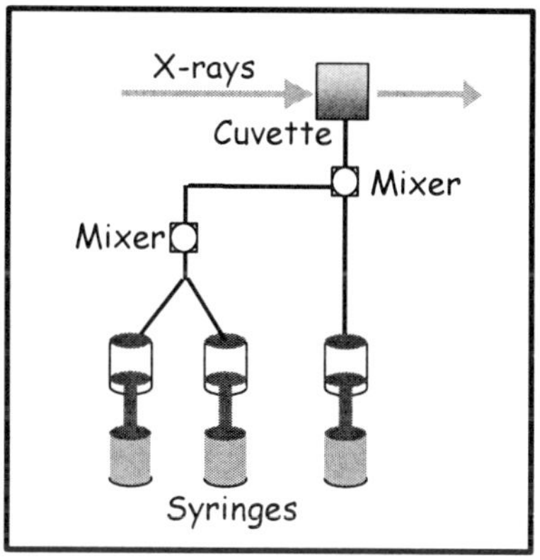

Figure 5. Schematic view of the stopped-flow apparatus as used at ID24.

The distance between the second mixer and the cuvette is the parameter that defines the dead time for the apparatus. This distance has to be as small as possible so as to make this time short, but still has to be big enough to allow the reacting mixture to homogenize. The dead time also considerably depends on the flux of the mixture, i.e. the faster the mixture travels, the shorter the time is, as well as on the volume of solution in the cuvette. Before starting the reaction for detailed study, these parameters must first be optimized to take account of the viscosity and miscibility of the various components. In the commercial apparatus used in this experiment, the maximum flow rate for each syringe motor is 6ml/ms that gives typical dead time values of the order of 1ms.

The observation cuvette is the key element of the whole apparatus. The windows have to be transparent to the X-ray beam, even at low energies, but yet must be sufficiently rigid to avoid the changes in the X-ray path length seen by the beam due to the pressure produced by the flow of reactants into the cell. For the example that we will present in this chapter, a cuvette made of PTFE has been used, with windows made from pyrocarbon of 200μm thickness. The carbon is transparent to the X-rays at the iron K edge (7.111 keV), and inert to the majority of the reactants that can be used. A picture of the cuvette is shown in Figure 6.

3.3 Matching the spectrometer and the sample environment

The key to time resolved studies is the synchronization of the data collection via the spectrometer with the sample environment that controls the timing of the initiation of the chemical reaction under study. This is done with a triggering system that assures that the reaction mixture reaches the observation cuvette at the time the detector starts collecting data.

To avoid the saturation of the detector during the readout time, in both the fast and slow acquisition modes, a fast mechanical shutter has to be used to stop the X-ray beam during this period of time. This shutter is also synchronized with the other elements, detector and sample environment, to assure that it is open at the correct time. This fast shutter is placed in front of the sample because it also serves a second purpose. As the volume of solution in the cuvette is very small so that the dead time is short and the amount of solution used is small, there is the risk that the absorption of the intense X-ray beam by the sample induces density changes in the solution by beam heating. Placing the shutter before the sample prevents this from occurring as the sample is only subject to the incident X-rays when the camera is collecting data.

Figure 6. Photograph of the cuvette used for the Iron K-edge. The carbon windows can be observed in the image.

4. ELECTRON TRANSFER REACTION OF HYDROQUINONE

The redox reactions of hydroquinone (H_2Q) with various oxidizing agents to form quinone (Q) are highly complex. This is because they involve both electrons and protons coming from the hydroquinone itself. In concentrated solutions of H_2Q with Fe^{III} as an oxidizing reagent (0.1-0.6M), the formation of a blue intermediate species has been observed by uv-vis spectroscopy and is easily corroborated by direct observation[13]. The concentrated solution is seen to transform from an initial state with a pale purple colour to a final state that is brown, via an intermediate state that is seen to be blue. By spectroscopy it has been shown that the formation of this intermediate is very fast (<5 ms) though its decomposition and formation of the final products is slower, (between 1s to 2s). The scheme of the reaction is shown in Scheme 1[14,15].

This reaction has been followed by time resolved X-ray absorption spectroscopy in order to detect the structural changes in the coordination sphere of the iron cation arising from the formation of the intermediate complex, followed by any changes involved in its subsequent decomposition to the products. The tracking of the formation and decomposition process will further allow us to extract information about the kinetic parameters that control the reaction, and of particular interest are the parameters characterising the first step of the process.

The reaction was followed by the changes that occur in the XAFS spectra collected at the Fe K-edge. The energy of the Fe K-edge is near the lower limit of ID24, and the energy range that it is possible to collect is thus quite small, the spectra being limited to the XANES region. To follow this reaction, the chosen oxidizing agent was iron perchlorate and additional perchloric acid was added to avoid hydrolysis of the iron complex. The concentration of both reactants in the observation cuvette is 0.1M, while the concentration of protons (the solution pH) has been made variable in order to study the influence of this species on the reaction.

$$Fe^{3+}(aq) + H_2Q \xrightarrow{\text{fast}} [(Fe^{III}\text{-}H_2Q)^{n+}{}_{aq}] \xrightarrow{\text{slow}} Fe^{2+}(aq) + Q + 2H^+$$

where

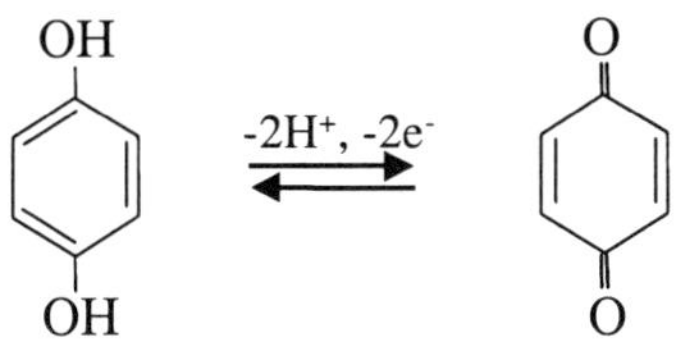

H_2Q Q

Scheme 1. Oxidation reaction of hydroquinone, showing the two different steps proposed.

A first and a very important baseline study that needs to be performed before undertaking a time resolved energy dispersive X-ray absorption study is the measurement of the X-ray absorption spectra of the initial and the final states of the reactant and products that comprise the reaction under study. To obtain this information for the reaction presented here, two static spectra were collected:

(1) the initial state - a 0.1M iron perchlorate solution measured in a static solution 2mm path length cell and

(2) the final state - a mixture of a 0.2M hydroquinone solution in a 50/50 water/ethanol mixture with a 0.2M solution of iron perchlorate (in 1 to 1 proportion by volume).

Both spectra are energy calibrated to the known spectra of an iron foil and are shown in Figure 7.

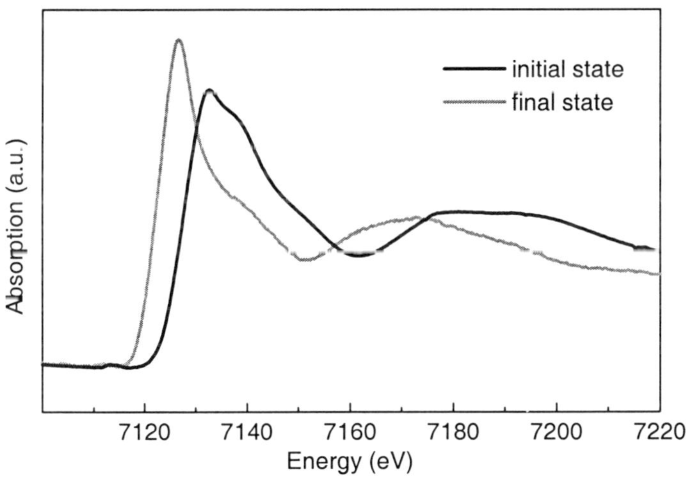

Figure 7. Initial and final state corresponding to the reaction scheme shown in Scheme 1.

As can be seen in the figure, the main difference in the spectra of both species is the position of the white line, separated by 6.1eV. The position of the edge is related with the oxidation state of the absorbing element and this difference matches with the fact that the iron has two different formal oxidation states, 3+ and 2+. There are further differences seen in the XANES features and also in the limited EXAFS range that is visible. These differences largely come from the different distances and potential bonding geometries for water molecules that are located around the central cation. Though not performed here, this structural information can potentially be accessed in a full detailed XANES analysis and for elements that allow data collection over a larger energy range, through a conventional EXAFS analysis.

Following the characterisation of the initial and final states the reaction was then initiated in the stopped flow apparatus and followed at different values of the pH. This allowed us to study the influence of this parameter on the rate of the reaction. Four different pH values were used, and refer to the mixture in the observation cuvette. They were pH = 1.0, 1.3, 1.6, and 1.9. For each pH value, 15 spectra were collected as the reaction proceeded and each spectrum was collected in 50ms. In Figure 8 we show the four series of spectra.

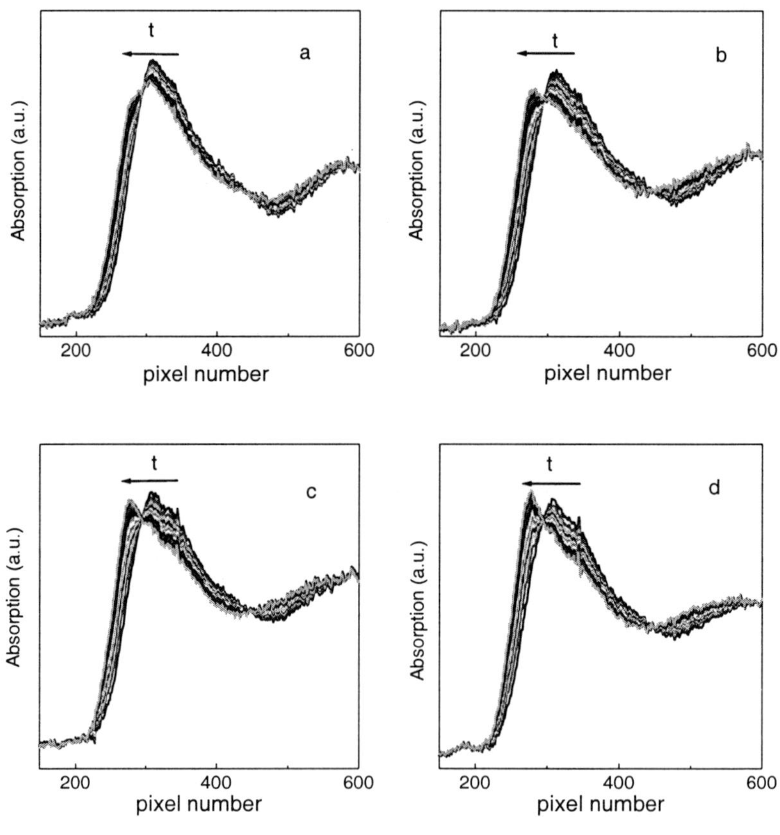

Figure 8. Series of XANES spectra taken in 50ms each at different values of pH. From a to d corresponding to pH values of 1, 1.3, 1.6, and 1.9. The energy scale corresponds to the distribution of the fan of X-ray energies across the CCD camera used to collect the data.

These series all show a smooth transformation as the reactants are converted into products. We note that the first spectrum in any of the series is not the same as that measured for the initial state. This indicates that the formation of the products starts sometime in the 50ms that are necessary to take the first spectrum, and very soon after the mixing of the reactants. Also, the last spectrum does not correspond to that of the final state and this tells us that the reaction is not yet finished 0.75s after mixing.

However, by this first comparison of the four series of spectra we can conclude that the reaction is faster when the concentration of protons is lower. This is in good agreement with the mechanism proposed above, and with previous observations for the same reaction.

The series of spectra included in figure 8 show the formation of the final product. However, with the time resolution shown in the figure, the first step of the reaction, i.e. the formation of the intermediate, cannot be seen. In order to study the formation of the proposed intermediate we must increase the time resolution and hence a new experiment was performed. The concentration of the reactants was chosen to be the same as in the previous cases with the pH value fixed at 1.6. A series of 15 spectra were collected, in which each spectrum was taken in 10ms. This first and the last spectrum of the series is shown in Figure 9. The spectrum of the initial state has been included for comparison.

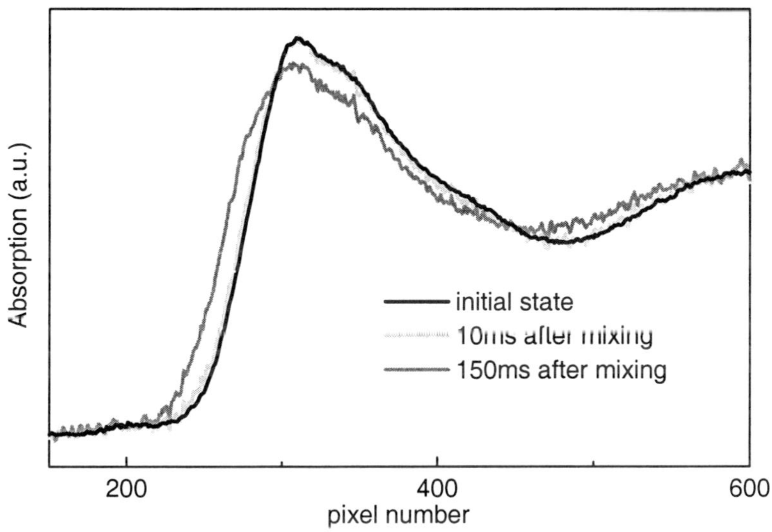

Figure 9. XANES region of the spectra indicated in the figure.

In this case the spectrum corresponding to the initial state is identical to the first one measured in the series of the 10ms spectra. This seems to indicate that the reaction did not start in the first 10ms after the mixing of the reactants. However, by uv-vis techniques, two steps of the reaction have been identified, and the formation of a blue intermediate is detected by eye. Measurements with 2 and 1 ms per spectrum have been performed, with no changes at all.

To explain these facts we have to have a deeper insight to the proposed structure of the blue intermediate. In the proposed reaction scheme the Fe^{3+} ion should substitute one of the water molecules that are forming its hydration shell by one of the oxygen atoms coming from the hydroquinone molecule. However, the length of this bond should be quite similar to the one formed with the solvent water molecules. From the EXAFS point of view, the new ligand is coordinated through the same sort of atom (oxygen) so the phase shift and the backscattering amplitude of the X-ray absorption signal are very likely to be the same. In addition, there is not expected to be any chemically induced change in the oxidation state so only with very high quality spectra and a detailed analysis of the EXAFS region of the spectra would we expect it to be possible to find any differences. Unfortunately due to the limiting signal to noise ratio of the current data it is not possible to find any evidence in this case.

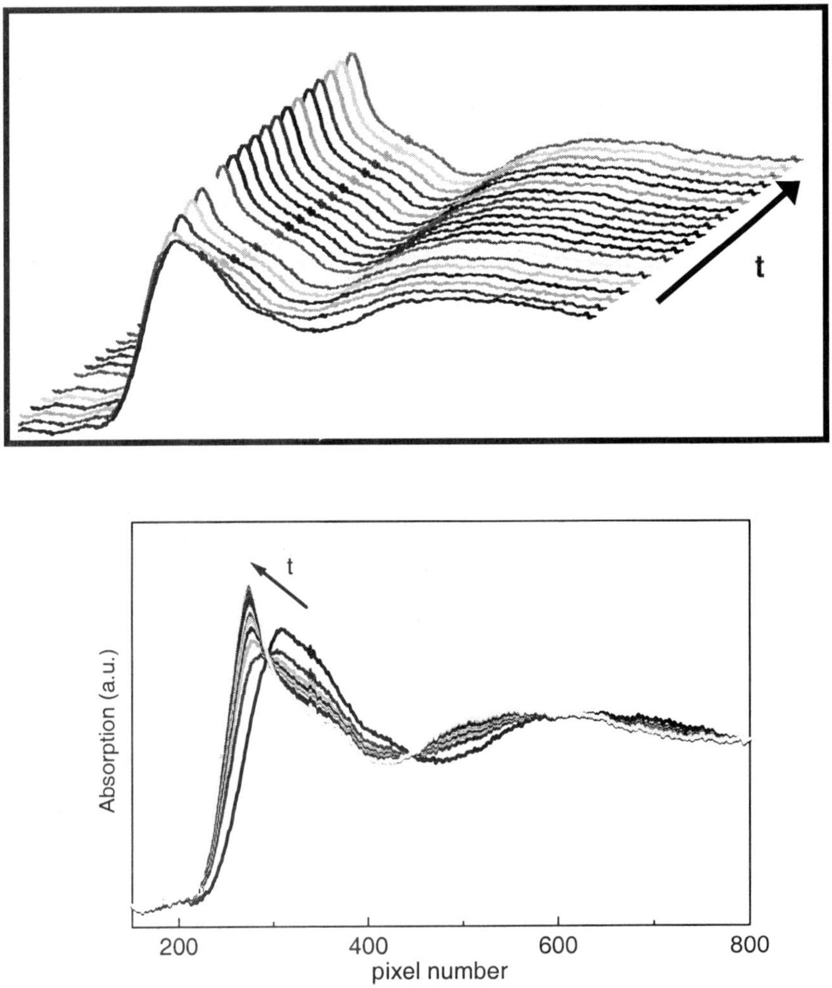

Figure 10. XANES region of the spectra collected by using the slow acquisition mode, in 3D view and 2D. The arrow indicates the direction with advancing time.

In a final step to study the reaction until completion, the slow acquisition method has been used. The series of spectra for the reaction performed at a pH of 1.0 is represented in Figure 10. Each spectrum in the series was collected in 270ms and the time between two consecutive spectra is 500ms. By a careful analysis of this series we can conclude that the reaction is finished approximately 13s after the mixing of the reactants.

4. CONCLUSIONS

In this chapter we have introduced the concepts and basic methodology that is required for studying in solution chemical reactions with the relatively new technique of Energy Dispersive X-ray Absorption Spectroscopy. Within the current limitations imposed by third generation synchrotron radiation sources, it is possible with this technique to study reactions on time scales from millisecond to minutes, provided that the reactants contain a suitable atomic site for the X-ray absorption probe. At the current time, with the facilities available, the concentration of the X-ray active element in the reaction mixture should be of the order of 20-50mM, depending on the photo-absorbing element and the balancing elements that compose the mixture. The outlined method has been illustrated to show that it is sensitive to both the local electronic and structural environments of the photo-excited atomic sites and thus provides us with a unique and highly complementary probe to the more traditional techniques used to study chemical reaction kinetics.

An example of a redox reaction has been presented. The process has been studied by two different data collection modes. These modes allowed us to follow the very fast (1ms per spectrum) as well as the slower processes. The influence on the rate of the reaction of different concentrations of protons has been shown and the necessity for there to be a clear change in the structure of the complexes under study during the reaction has also been pointed out in the chosen example. Without such changes, the study of such processes by X-ray absorption spectroscopy techniques is very challenging and not ideal. Though we were unable to observe the formation of the first intermediate by the EDXAS technique in this particular case, the lesson we have been taught is a useful one. It highlights clear lines for future developments that are still needed if this technique is to become a widespread and useful chemical tool, and points us to improvements that can readily be made. Such improvements reflect the fact that often a series of complementary techniques will be necessary to better understand a reacting chemical system and examples that are currently under consideration and development are combinations such as simultaneous *in-situ* EDXAS and optical spectroscopies.

5. ACKNOWLEDGMENTS

I would like to thank all the staff of ID24 who have contributed to the realization of the time resolved capabilities of the beamline. Without all their technical and scientific support this project would have been impossible. I would like thank Dr Thomas Neisius for his key contributions to the development of the time-resolved study capabilities of beamline ID24. Many thanks also to Prof. John Evans for his inestimable help during the collection of the data presented in this chapter, and to Dr. D.T.Bowron for his help during the preparation of the manuscript.

REFERENCES

1. For a complete reference of chemical reactions see K. J. Laidler, *Chemical Kinetics* (McGraw-Hill, London, 1965) and H. Strehlow, *Rapid Reactions in Solution* (VCH Publishers, New York, 1992).
2. L. Wilhelmy, *Pogg. Ann.*, **81**, 413 (1850).
3. M. Berthelot and L. P. St. Gilles, *Ann. chim. et phys.*, **63**, 385 (1862).
4. H. Hartridge and F. J. W. Roughton, *Proc. Roy. Soc. (London)*, **A104**, 376 (1923).
5. B. Chance, *J. Franklin Inst.*, **229**, 455, 613, 737 (1940).

6. P. Bazulin, *C. R. Acad. Sci. USSR*, **3**, 285 (1939); *J. Exp. Theor. Phys.*, **9**, 1147 (1938).
7. W. Nernst; see F. Keutel in: *Dissertation* (Berlin, 1910).
8. R. G. W. Norrish and G. Porter, *Nature*, **164**, 658 (1949).
9. N. Yoshida, T. Matsushita, S. Saigo, H. Oyanagi, H. Hashimoto and M. Fujimoto, *J. Chem. Soc., Chem. Commun.*, 354 (1990).
10. D.E.Sayers, E.A.Stern and F.W.Lytle, *Phys. Rev. Lett*,. **27**, 1204 (1971).
11. M. Hagelstein, A. Fontaine and J. Goulon, *Jpn. J. Appl. Phys.*, **32**, 240 (1993) ; M. Hagelstein, A. San Miguel, T. Ressler, A. Fontaine and J. Goulon, *J. de* Physique *IV*, **1**, C2-303 (1997).
12. A. Koch, M. Hagelstein, A. San Miguel, A. Fontaine and T. Ressler, *IS&T / SPIE Proc. Of Symposium on Electronic Imaging: Science and Technology*, **85**, 2416 (1996).
13. J. H. Baxendale, H. R. Hardy and L. H. Sutcliffe, *Trans. Faraday Soc.*, **47**, 963 (1951).
14. B. S. Brunschwig, C. Creutz, D. H. Macartney, T-K. Sham and N. Sutin., *Faraday Discuss. Chem. Soc.*, **74**, 113 (1982).
15. A. Haim, *Acc. Chem. Res.*, **8**, 264 (1975).

EXPERIMENTAL DETERMINATION OF THE STRUCTURES OF COMPLEX LIQUIDS: BEYOND THE PDF

J. L. Finney and D. T. Bowron*

1. INTRODUCTION

Interesting things happen in solution. Proteins fold into their native, active conformations. Enzyme substrates bind to the proteins that operate on them, and drug molecules bind to their targets. Much of the chemical industry works with solutions in both manufacture and separation. Understanding – and hence tailoring – these processes depends on a detailed understanding of the structures in solution, and how they change as external conditions – temperature, pressure, concentration, cosolvents – are changed. Yet our understanding of structures in solution remains poor.

The reason for this poor understanding is not difficult to see. Since von Laue observed X-ray diffraction spots from crystals in 1916, we have exploited the use of crystals to arrange molecules in the regular arrays that maintain phase relationships between the X-ray, neutron, or electron waves scattered by atoms in the system, and that enables us to solve the structures of extremely complex molecules. Much of the present drive for structural genomics is based on the exploitation of X-ray scattering from protein crystals to determine their structures, and hence, hopefully, improve our understanding of the biological processes they catalyse. In a liquid, however, we have no lattice. We lose the ability to maintain phase relationships between different scattered beams. The diffraction experiment thus produces a distribution of distances between pairs of molecules – the pair distribution function, or PDF. Whereas looking at crystalline systems using the PDF has a number of advantages over straight Bragg scattering used in crystallography in that it gives essentially *local* information – as is described elsewhere in this volume – the liquid scatterer is handicapped in that this local information is the *only* structural information he has to work on. Worse still, in the simplest case, we don't even know the kinds of atoms at the ends of each pair distance that the diffraction experiment reports to us – we have to average over atom types, resulting in information that, except in the simplest case of monatomic liquids, is not very helpful in understanding structural

* John L. Finney, Department of Physics and Astronomy, University College London, Gower Street, London WC1E 6BT, UK
Daniel T. Bowron, ISIS Facility, CLRC Rutherford Appleton Laboratory, Chilton, Didcot, Oxon OX11 0QX, UK, and Department of Physics and Astronomy, University College London, Gower Street, London WC1E 6BT, UK

details. If we are to progress to understanding structures in more interesting, more complex liquids, we need to find ways of moving beyond the simple PDF.

This is possible. Developments over the last thirty years have allowed liquid scatterers to indeed move beyond the simple PDF to the stage that we can now obtain structural information at a quite detailed level. In solutions, we can see how solute molecules interact with each other, and how these interactions change with changed external conditions. We can begin to see how adding ions to an aqueous solution of an amphiphile modulates the interaction between the solute molecules, bringing us towards an understanding of the salting out process central in much of chemistry and biology. And, as we shall see in some detail in this chapter, we can begin to see how the hydrophobic interaction – thought to be a major driving force in the folding and interactions of proteins – may operate.

This chapter proceeds by setting out the basic descriptions used for describing liquid structures, and shows how, by the use of neutron scattering with isotope substitution, we can indeed move beyond the PDF – beyond the 'average structure' of the liquid state scientist. Using as an example recent work on aqueous solutions of an alcohol with a large non-polar headgroup – tertiary butanol – we look at the kind of information that we can now obtain experimentally at several levels below the average structure described by the PDF, and how that information can give us insight into solvent-modulated processes in solution.

2. STRUCTURE FACTORS AND RADIAL DISTRIBUTION FUNCTIONS

2.1 A Single Component Liquid

The useful quantity used to describe the structure of a single component liquid is the ***radial distribution function*** $g(r)$ (the rdf) of the system. This quantifies the ratio between the local density of atoms at a distance r from an atom at the origin to the average number density of atoms in the system. The ***pair distribution function*** (PDF) used in many of the chapters of this book is simply defined as $4\pi\rho r g(r)$, where ρ is the average number density of the system.

A diffraction experiment uses the known wavelength of the (usually X-ray or neutron) scattering probe to obtain the information on pair distance distributions contained in the radial distribution function. The scattering experiment measures the intensity scattered as a function of the scattering vector **Q**, defined simply as the difference between the scattered and incident wave vectors, with a magnitude $|\mathbf{Q}| = 4\pi\sin\theta/\lambda$, where 2θ is the scattering angle and λ the wavelength of the radiation used. From the scattered intensity, the structurally-significant ***structure factor*** $S(Q)$ can be extracted which can then be related to the radial distribution function through a Fourier transform relationship:

$$[S(Q)-1] = 4\pi\rho\int_0^\infty r^2 dr\,(g(r)-1)\frac{\sin Qr}{Qr} \tag{1}$$

The above account includes only the structurally-significant parts of the functions commonly used. Fuller information on both the basic descriptions and the processing of scattering data can be found elsewhere[1-3].

2.2 A Multicomponent Liquid

For any solution, there must be more than one different atomic component, so the above functions need to be generalised. We consider a two component system, with components α and β, with N_α, N_β (where $N_\alpha + N_\beta = N$) atoms of each, with the atomic fractions of each component defined as $c_\alpha = N_\alpha / N$, $c_\beta = N_\beta / N$. The structure factor can then be split into three terms – ***partial structure factors*** $S_{\alpha\beta}$ – each relating to different pairs of interacting atoms αα, ββ, and αβ. Thus, ***partial radial distribution functions*** $g_{\alpha\beta}(r)$ are defined similarly to the radial distribution function, but with the added identification of the type of atom both at the local origin and at the distance *r*. In essence, the partial radial distribution function $g_{\alpha\beta}(r)$ tells us the relative probability of finding an atom of type β at a distance r from an atom of type α and, of course, *vice-versa*. We can then rewrite equation (1) as

$$[S_{\alpha\beta}(Q)-1] = 4\pi\rho\int_0^\infty r^2 dr(g_{\alpha\beta}(r)-1)\frac{\sin Qr}{Qr} \tag{2}$$

Thus, for a two component system, we require 3 partial radial distribution functions to describe the structure of the system. For a more complex *n* component system, we will require in general $n(n+1)/2$ partials. Each of these partials will be related to a partial structure factor that is embedded in the measured diffraction data, or the total structure factor

$$F(Q) = \sum_{\alpha,\beta}(2-\delta_{\alpha\beta})c_\alpha c_\beta b_\alpha b_\beta[S_{\alpha\beta}(Q)-1] \tag{3}$$

where b_α, b_β are the scattering lengths of the components α and β.

2.3 Getting Beyond the Average Structure: the Principle

A single diffraction experiment will yield only the total radial distribution function, which is the (concentration and scattering length) weighted sum of the partial radial distribution functions. As we stated in the introduction, this information tends to be not very helpful, especially in the relatively complex solutions in which we are particularly interested. An example is given in figure 1, where the total radial distribution function of a 0.06 mole fraction solution of tertiary butanol in water is given for two temperatures. The patterns for the two temperatures are very similar indeed, and the short distance behaviour is characteristic of that of water – which is of course the major constituent. To try to draw conclusions from these functions about the detailed structure of the solution, or how it changes with temperature, would clearly be rather foolish. To understand structures at a useful level, we need to go beyond this, and look deeper, and try to move towards extracting information at the partial structure factor level.

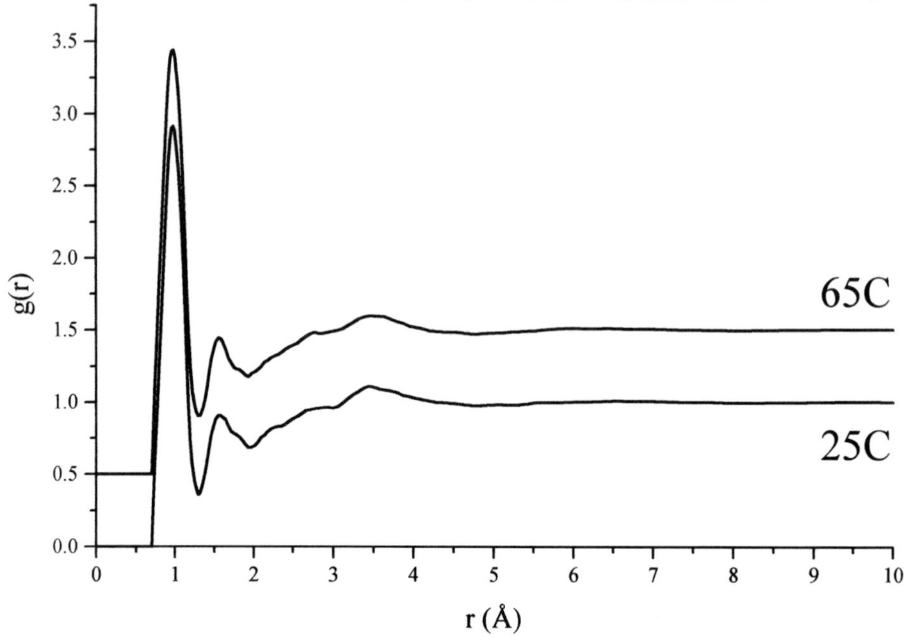

Figure 1. The total rdfs of 0.06 mole fraction solutions of t-butanol at 25C and 65C.

The key to achieving this is given in equation (3). A single diffraction experiment can yield no more than the total structure factor, a weighted sum of the partial structure factors that we would really like to get at. The fact that it is a weighted sum gives us a way forward. One of the weighting factors is the scattering length of each component. If we could perform an experiment on the same system, but somehow change the scattering length, we could perform more than one experiment on chemically similar systems, and begin to extract the kind of more detailed information we really want.

That we can do this is made possible by the fact that the neutron is scattered by the nucleus, and that many elements have different isotopic forms. If these isotopic forms had different neutron scattering lengths, then we could perform an additional experiment on a system in which we change only the isotope of one of the components. Referring to equation (3), we could then obtain from a second scattering experiment a different structure factor $F'(Q)$ in which, say b_{α}, is replaced by $b_{\alpha'}$. Now performing a third experiment replacing the isotope of the second component, b_{β} is replaced by $b_{\beta'}$. Alternatively, isotope α could be replaced by either a third isotope α'' (or what is in effect the same, a mixture of isotopes α and α'). We now would have three equations with the three partial structure factors as the three unknowns. In principle these could be solved for, and the corresponding partial radial distribution functions obtained.

For relatively simple two component liquids and glasses, such 'full' partial structure factors – and hence 'full' partial radial distribution functions – have been extracted[4, 5]. For the relatively complex systems in which we are here particularly interested, performing the much larger number of isotopically distinct scattering experiments that would be necessary to extract all the partial rdfs looks a tall order. Even a simple case such as methanol in water, where there are six chemically distinct atoms in the system (methyl carbon, methyl hydrogen, alcohol oxygen, alcohol hydroxyl hydrogen, water oxygen, water hydrogen), would require $6 x 7 / 2 = 21$ isotopically distinct experiments to yield the 21 partial radial distribution functions (or 'partials' for short). Setting aside whether an appropriate set of isotopically distinct solutions could be prepared (not all elements have isotopes with significantly different scattering lengths), the requirements for neutron beam time – which is not cheap – would be extremely high. As we shall see below, there are other ways round the problem that enable us to obtain not only all the partials for a system such as this, but also to go beyond even the partials to obtain even more detailed structural information on relatively complex solutions.

3. AN ILLUSTRATIVE SYSTEM: AQUEOUS SOLUTIONS OF TERTIARY BUTANOL

In exploring how we can now go beyond the average structure in complex aqueous solutions, we concentrate in the rest of this chapter on the structures of aqueous solutions of tertiary butanol[6, 7]. This molecule, $(CH_3)_3COH$, was chosen for detailed study for a number of reasons. First, it is an amphiphile, having a polar 'tail' – an alcoholic hydroxyl group – that can hydrogen bond to other polar groups, and a non-polar 'head'. Secondly, this non-polar head is of particular interest in that it gives the molecules thermodynamic behaviour that is typically 'hydrophobic'[8] – i.e. its interactions in aqueous solution are driven by the hydrophobic effect that is thought to be a major – some would say dominant – driving force in important biological processes such as protein folding, protein-protein and protein-substrate interactions. It is also a central factor in a range of self-assembly processes, for example the formation of lipid-based superstructures that are both biologically and chemically important. Examining the structures of this system thus gives us the opportunity to study the structural basis of the hydrophobic interaction, the origin of which is still unclear (see below). Thirdly, despite its large non-polar headgroup, the molecule is miscible in water in all proportions, so there are no complications of phase separation – a larger headgroup would result in restricted miscibility. Finally, it is a reasonably symmetrical molecule, a factor that is important in interpreting the results – i.e. it simplifies matters, always a good thing when working with difficult and relatively new procedures.

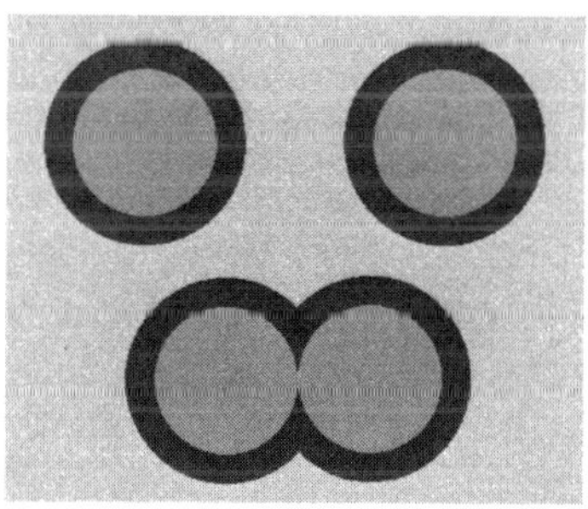

Figure 2. Schematic representation of the association of two non-polar molecules in aqueous solution. The solvent in the shaded hydration region is thought to be in some way more ordered than in the bulk. On association, some of this water is displaced to the bulk with a consequent gain in entropy to the system.

Before moving to consider the structural data that has been obtained on this system, it is worthwhile to set down what is considered to be the conventional explanation of the hydrophobic interaction. In this process, two non-polar molecules come together in aqueous solution (figure 2). The thermodynamics of this interaction clearly shows that there is a gain in entropy in the process – it is entropically driven. The conventional view of the origin of this entropic driving force was set down over 50 years ago by Frank and Evans[9] who argued that the water in the hydration shell of a non-polar group was perturbed to a structure of 'greater order' than is found in bulk water. On bringing together two such molecules (see figure 2), the two hydration shells overlap, releasing previously hydrating water molecules to the bulk. It is the entropic gain of these released water molecules that is argued to be the source of the entropic driving force of this solute association interaction.

The nature of this solvent 'ordering' remains controversial. Frank and Evans themselves suggested that the water structure might be perturbed by the presence of the non-polar surface in the direction of a structure corresponding to that of water at a lower temperature, or perhaps the hydrogen bonds between the water molecules in the

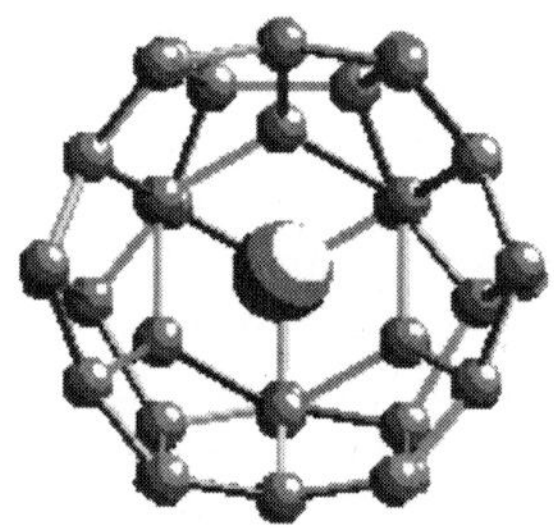

Figure 3. A clathrate cage containing a guest molecule. The one shown here is the large cage of the type II clathrate.

hydration shell were stronger than in the bulk, an increase in strength that might be seen through a reduced hydrogen bonding distance. Furthermore, from their seminal paper comes a picture of the hydration shell of a non-polar group that has dominated thinking about the structural nature of hydrophobic hydration. In their paper, they say[9]: "When a ... non-polar molecule dissolves in water at room temperature, it modifies the water structure in the direction of greater 'crystallinity' – the water, so to speak, builds a microscopic iceberg around it.... This 'freezing' of the water ... causes heat and entropy to be lost, beyond what would otherwise have been expected." In using the word 'iceberg' – which has been much used and misused since – they were quite careful to explain it as "... a microscopic region in which water molecules are tied together in some sort of quasi-solid structure. .... not implied that the structure is exactly ice-like...". Bearing in mind that this was written before we had a realistic picture of structure in liquids, it seems that they are trying to describe some kind of structural ordering that is not to be considered to look structurally like ice – a crystal – itself, but some other semi-structured arrangement. Perhaps partly because of this need to find some kind of locally ordered arrangement that is different from the local order in crystalline ice, the 'canonical' view of the 'ordered' hydration shell has become that of the clathrate ('cage') shells found surrounding molecules such as inert gases and methane in the so-called clathrate hydrates. In these, (see figure 3 for a typical clathrate cage), the water molecules all are capable of forming their full complement of four hydrogen bonds, and do not have a directed attractive interaction with the central non-polar molecule. Looked at in terms of the water molecule's dipole moment, these molecules in a clathrate shell tend to be aligned with their dipole moments tangential to a sphere centred on the central molecule. Similarly, most of the water molecule OH bonds tend also to be tangential to this shell, and certainly do not point towards the central molecule. We will make use of these orientational properties later when the structure of non-polar hydration shells obtained experimentally is compared to this canonical model.

4. BEYOND THE PDF: EXPERIMENT DESIGN.

4.1. Solute and Solvent Terms in Solution.

In trying to understand interactions – and hence structure – in solutions, we can usefully consider three aspects.

First, we would like to know how the solute molecules are solvated. In the case of our t-butanol, questions that are relevant include how the water is organised around the non-polar headgroup, and how this organisation relates to the conventional view of some sort of clathrate cage structure. This means we need information on the *solute-solvent* structure.

Secondly, what is the source of the solvent ordering that is thought to drive the hydrophobic interaction when this hydration shell is partially destroyed by the coming together of two non-polar surfaces? To answer this, we would like to know what the structure is of the solvent in the near-neighbourhood of the non-polar group. How does it differ from bulk water at the same temperature? Are the hydrogen bonds stronger in the hydration shell? Does the water structure look like that of water at a reduced temperature? To try to answer these questions, we need information on the *solvent-solvent* pair distributions.

Finally, we are really interested in how the solvent affects the way in which the solute molecules interact in the solution. How are the solute molecules organised at a given temperature and concentration? What are the dominant interactions between the molecules? Are the non-polar groups – as some theoretical work has implied[10] – separated by a layer of solvent or are the molecules in direct contact? Do the alcohol molecules hydrogen bond to each other as the standard chemical view to date of alcohol-water solutions would have us believe? How do the solute-solute interactions change with temperature – classical hydrophobic interaction ideas suggest there should be a stronger attraction between non-polar groups in water as the temperature is raised to around 60-70C[11]? To try to answer these questions, we need information on the *solute-solute* pair distributions.

4.2. The 'Standard' Set of Experiments

For our example of t-butanol in water, we have the option of changing the hydrogen isotopes on both the water and the methyl groups of the alcohol headgroup. For each substitution, we work with (a) fully deuterated; (b) fully hydrogenated; (c) a 50:50 hydrogen/deuterium mixture. This gives three values of the neutron scattering length of the substituted sites, and hence, for each triplet of substitutions, enables us to obtain three sets of (partial) radial distribution functions. We will call these $g_{HH}(r)$, $g_{HX}(r)$, and $g_{XX}(r)$. Here the subscript H refers to a site whose hydrogens are substituted, while the X subscript refers to all the non-substituted sites. Thus, for a set of substitutions on the methyl hydrogen sites of t-butanol, H refers to all the substituted methyl hydrogen sites, while X refers to all other sites.

Our standard set of experiments can now be specified as follows.

1. Solvent-solvent distribution functions are probed through substitution on the water hydrogen sites to yield the (partial) radial distribution functions $g_{HH}(r)$, $g_{HX}(r)$, and $g_{XX}(r)$ for the solvent-solvent correlations. The three experiments are made with (a) $(CD_3)_3COD$ in D_2O, (b) $(CD_3)_3COH$ in H_2O, and (c) 50:50 $(CD_3)_3COH$ /$(CD_3)_3COD$ in 50:50 H_2O/D_2O.
2. Solute-solute distribution functions are probed through substitution on the alcohol methyl hydrogen sites to yield the (partial) radial distribution functions $g_{HH}(r)$, $g_{HX}(r)$, and $g_{XX}(r)$ for the solute-solute correlations. The three experiments are made with (a) $(CD_3)_3COD$ in D_2O, (b) $(CH_3)_3COD$ in D_2O, and (c) 50:50 $(CD_3)_3COD/(CH_3)_3COD$ in D_2O.
3. Solute-solvent distribution functions are probed through substitution on the alcohol methyl hydrogen sites and the water hydrogen sites. In combination with the solute-solute and solvent-solvent distribution functions, this yields the (partial) radial distribution functions $g_{HH}(r)$, $g_{HX}(r)$, and $g_{XX}(r)$ for the solute-solvent correlations. The three experiments are made with (a) $(CH_3)_3COH$ in H_2O, (b) $(CD_3)_3COD$ in D_2O, and (c) 50:50 $(CD_3)_3COD/(CH_3)_3COH$ in 50:50 H_2O/D_2O.

Thus we appear to have specified nine isotopically-distinct experiments. However, we note that samples 1(a), 2(a), and 3(b) are identical. We therefore perform only 7 experiments to yield the 9 (partial) radial distribution functions specified above, namely $g_{HH}(r)$, $g_{HX}(r)$, and $g_{XX}(r)$ for each of the solvent-solvent, solute-solute, and solute-solvent cases considered.

5. BEYOND THE PDF: SOME RESULTS

5.1 Level 1: Composite Partial Radial Distribution Functions.

Figure 4 shows the nine functions specified in the above section for a 0.06 mole fraction solution of t-butanol in water at 25C and 65C, obtained from a seven-sample experiment[7]. Although we see much more structure than in the total radial distribution functions of this system (figure 1), interpreting these results is not straightforward. Focussing for example on the solute-solute functions, the top left-hand panel shows $g_{HH}(r)$, i.e. the distribution of distances between the methyl hydrogens on two neighbouring t-butanol molecules. $g_{HX}(r)$ shown in the second panel from the left tells us the distribution of distances between methyl hydrogens on one t-butanol and all the other sites on the neighbouring molecule. Thus, although these functions do contain information on solute-solute structure, it is not – in these ***composite partial radial distribution functions*** – easy to interpret these data in terms of specific site-site interactions. Although we have labelled sites on each molecule, the distance distributions we get are still averaged over a number of different sites.

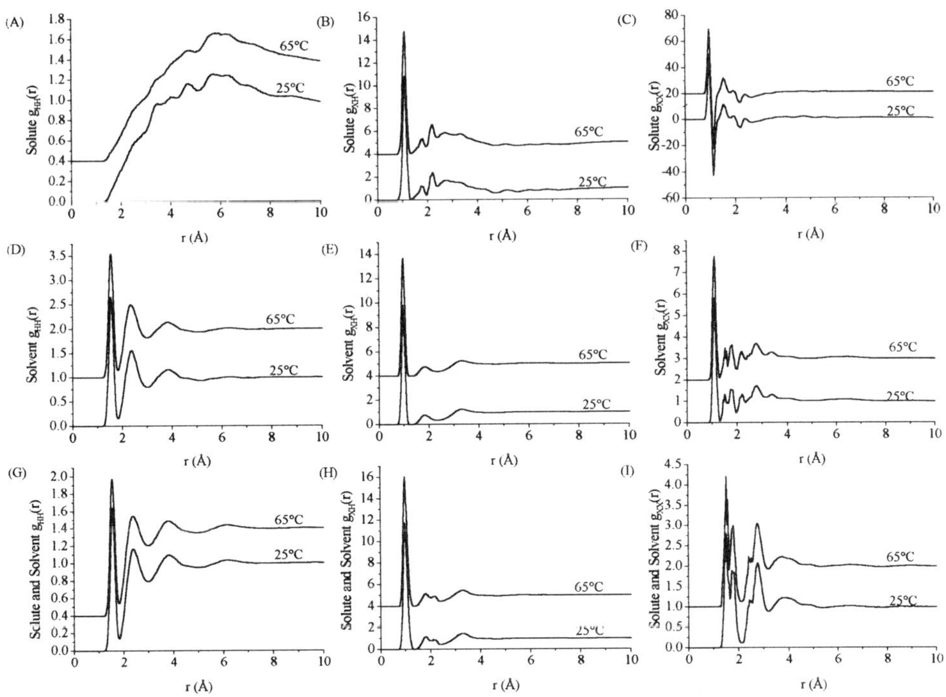

Figure 4. The composite partial rdfs for a 0.06 mole fraction aqueous solution of t-butanol at 25C and 65C. A, B, and C correspond to solute-solute functions, D, E, and F to solvent-solvent, and G, H and I to solute-solvent.

In favourable cases – for example for single site solutes such as ions, or solute molecules of high symmetry – it is possible to obtain useful data from these composite partial distribution functions. For example, much of the early isotope substitution work on ionic solutions obtained results that modified in a major way our ideas on electrolyte solutions[12]. Moreover, earlier work on solutions of both polar molecules such as urea and simple amides[13, 14] worked successfully within this framework, and the first work on alcohol solutions likewise, after some further manipulation of the data to in essence centre the labelled solute site onto the central carbon[15]. Furthermore, we should note that in the case of aqueous solutions, the solvent-solvent composite functions are in fact true partial radial distribution functions, as the two hydrogen sites on the water molecule are equivalent and the water X site is the only other site on the molecule, namely the oxygen. In general, however, to proceed further to understand the detailed structure of this system, we need a greater level of detail than is given in these nine composite partial radial distribution functions.

5.2 Level 2: Partial Radial Distribution Functions.

The t-butanol/water system has 7 different atomic sites methyl hydrogen, methyl carbon, central carbon, oxygen and hydrogen on the solute molecule, hydrogen and oxygen on the solvent. The structure of this system can therefore be specified by $8 \times 7/2 = 28$ partial radial distribution functions. In principle this need 28 isotopically distinct experiments. We have the results of only seven.

At this point, it is worth comparing our situation with that of the protein crystallographer. In solving the structure of a protein containing say 1000 non-hydrogen atoms (using X-rays, the hydrogens are only weakly observed and will be ignored in the initial structure determination), the structure is determined by 3000 (x,y,z) co-ordinates. Except in the very highest resolution experiments, the number of sets of data obtained will be less than this. But the structure is still solved. In this 'underdetermined' situation, use is made of other chemical information to constrain both the initial structure determination and the final structure refinement. The justification of this is that the final structural model agrees with both the experimental data and known chemical constraints, and so this constraining is regarded as a perfectly acceptable process. The liquid scatterer can therefore similarly look for ways of making use of known chemical constraints to compensate for the underdetermined nature of the problem, and obtain by some kind of refinement process structural models that are consistent with the experimental data. Like the protein crystallographer, the liquid state scientist can then interrogate the structural models – which are consistent with the experimentally determined scattering data – to begin to understand the processes that occur in solution.

The procedure increasingly used to extract 'full' sets of partial radial distribution functions was developed by Soper[16] and described also in a previous volume in this series[17]. This Empirical Potential Structure Refinement (EPSR) procedure produces model ensembles that are consistent with the experimental scattering. These can then be interrogated to see the detailed geometries of the various intermolecular interactions in a given solution.

In summary, the procedure starts by setting up a Monte Carlo simulation of the system using a set of standard potential functions $U^0_{\alpha\beta}(r)$. The Monte Carlo simulation is then run to equilibrium, from which the various partial radial distribution functions $g_{\alpha\beta}(r)$ are estimated. This in general (in all cases we are aware of so far) fails to give adequate agreement with experiment, indicating in passing that the starting potentials are inadequate to reproduce the experimental data. So the initial potential energy function is then modified by adding sets of potentials of mean force between the various different sites. These are derived from comparing the experimental $g^D_{\alpha\beta}(r)$ and the simulated $g_{\alpha\beta}(r)$ radial distribution functions. Thus we obtain a modified potential set:

$$U^N_{\alpha\beta}(r) = U^0_{\alpha\beta}(r) + kT \ln\left[g_{\alpha\beta}(r) / g^D_{\alpha\beta}(r)\right].$$

This new potential is fed into the system and a further Monte Carlo simulation is performed. The above procedure is then repeated until $U^0_{\alpha\beta}(r) \approx U^N_{\alpha\beta}(r)$ and $g_{\alpha\beta}(r) \approx g^D_{\alpha\beta}(r)$ for all r and for all pairs of atoms α, β. The resulting experimentally consistent ensembles can then be interrogated to obtain site-site partial radial distributions to examine the detailed structure of the system. In passing, we can note that the way in which the potential function is modified to bring the simulation into agreement with the experiment may be able to guide us in improving the potential functions originally used in the system – or at least give us an idea of what aspects of the starting potentials need modification.

What is described above is an earlier implementation of EPSR which operated in real space – i.e. the comparisons were made between partial radial distribution functions. The normal implementation is now in reciprocal space – i.e. using the partial structure factors

$S_{\alpha\beta}(Q)$ to modify the potential. Typical fits of the measured partial structure factors are shown in figure 5. The discrepancies between the model and the experimental data at the lower Q end can be shown by Fourier transform to relate to inadequacies in the intramolecular model. In terms of the intermolecular structural information that we are aiming for, the model ensembles give excellent agreement with experiment.

Now that we have these ensembles of molecules, we can extract whatever structural information wc want. Thc 28 partial radial distribution functions for t-butanol water at three concentrations (0.06, 0.11, 0.16 mole fractions)[6] are shown in figures 6 and 7. The various site labels are: CC: central carbon; C: methyl carbon; M: methyl hydrogen; O: alcohol oxygen; H: alcohol hydrogen; O_w: water oxygen; H_w: water hydrogen. There is clearly a tremendous amount of information in these functions – a comparison with the almost total lack of useful information in the total distribution functions of figure 1 is instructive. We now proceed to look at some of this information.

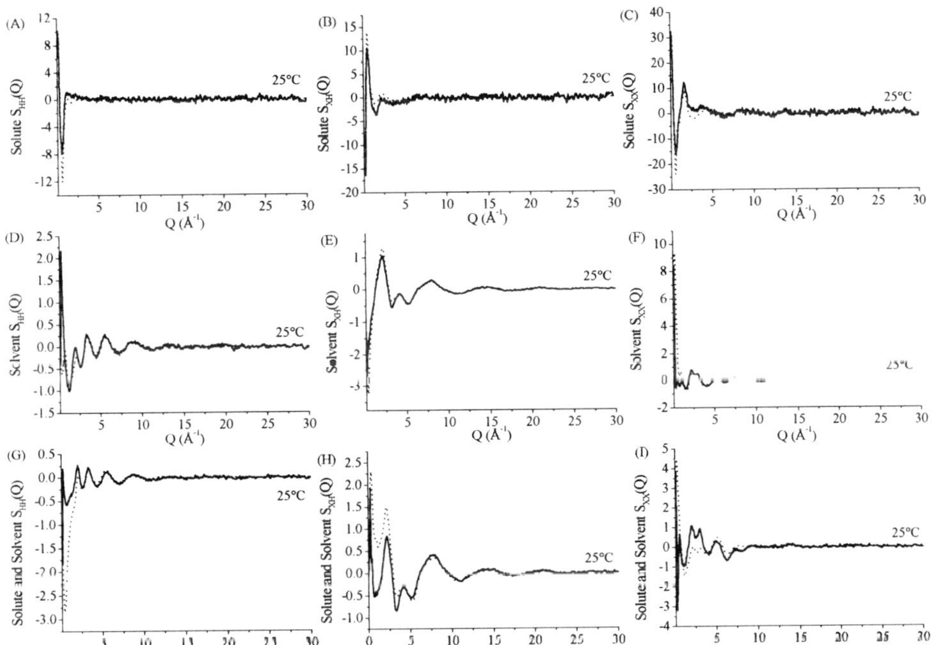

Figure 5. Experimental data (solid line) and EPSR fits (broken lines) for a 0.06 mole fraction aqueous solution of t-butanol at 25C. Labels A-I as in figure 4.

6A

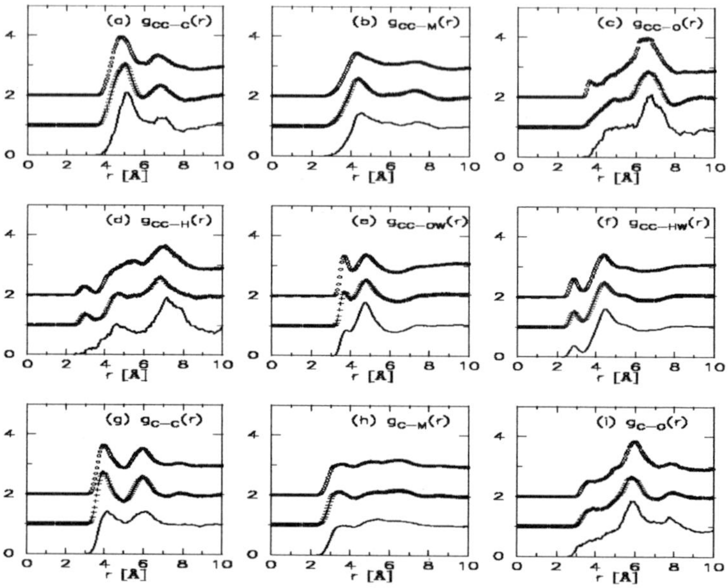
(a) gCC—C(r)
(b) gCC—M(r)
(c) gCC—O(r)
(d) gCC—H(r)
(e) gCC—OW(r)
(f) gCC—HW(r)
(g) gC—C(r)
(h) gC—M(r)
(i) gC—O(r)
r [Å]

6B

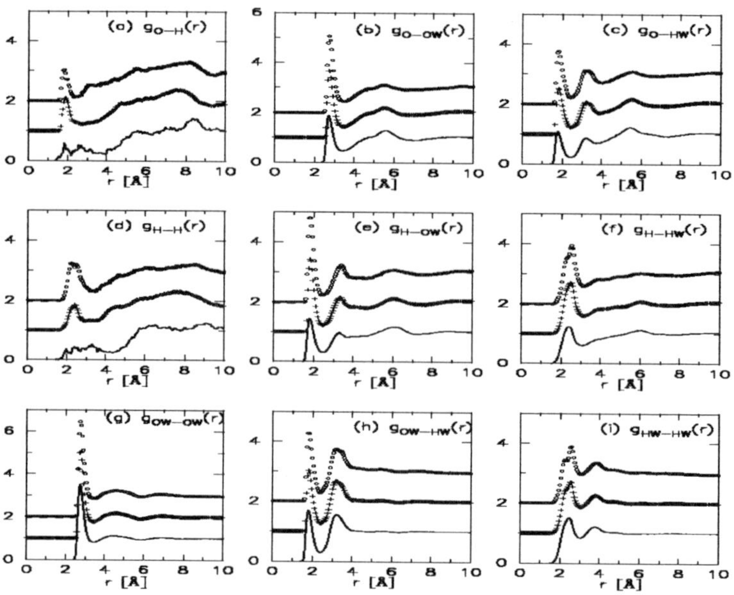
(a) gO—H(r)
(b) gO—OW(r)
(c) gO—HW(r)
(d) gH—H(r)
(e) gH—OW(r)
(f) gH—HW(r)
(g) gOW—OW(r)
(h) gOW—HW(r)
(i) gHW—HW(r)
r [Å]

6C

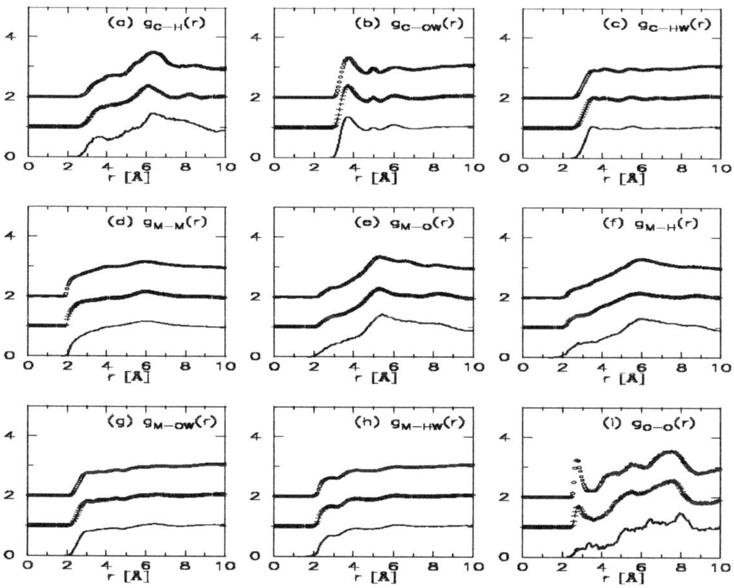

Figure 6A, B, C. Intermolecular partial rdfs for 0.16 (O top), 0.11 (+ middle), and 0.06 (— bottom) mole fraction t-butanol in water.

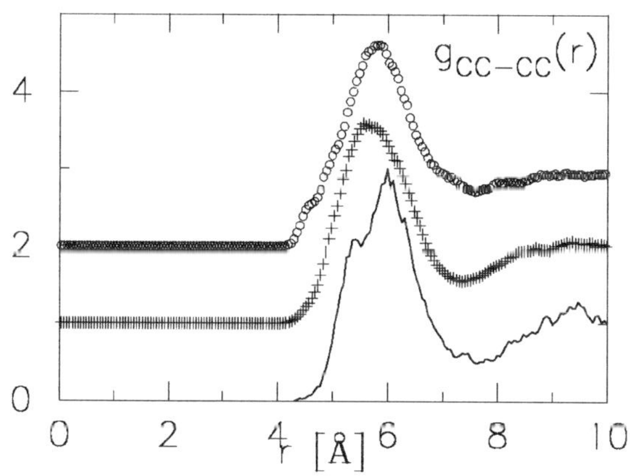

Figure 7. The molecular centres function for 0.16 (O), 0.11 (+), and 0.06 (-) mole fraction t-butanol in water.

First, looking at the central carbon-central carbon radial distribution function (figure 7), we see for all three concentrations a broad peak centred between 5.5 and 6.0Å. The peak position tells us immediately that even at the lowest concentration, there is direct contact between neighbouring t-butanol molecules: the molecules are not – as suggested in earlier models of the hydrophobic interaction[10] – separated by an intervening solvent layer. The area under this peak tells us how many molecules on average surround a central solute molecule. This and other co-ordination numbers are given in table 1: it rises from 2.8 ± 0.6 at 0.06 mole fraction, through 4.4 ± 0.6 at the intermediate concentration, to 5.8 ± 0.6 at 0.16 mole fraction. The fact that significant solute association is found at the lowest concentration was perhaps unexpected, suggesting some tendency to microscopic phase separation. There is an interesting shoulder on the low *r* side of the peak at 0.06 mole fraction, a feature which is explored further below (section 5.3).

Table 1. Selected co-ordination numbers for aqueous t-butanol solutions

Correlation	R_{min}/Å	R_{max}/Å	0.06 mole fraction	0.11 mole fraction	0.16 mole fraction
CC-CC	4.0	7.0	2.8 ± 0.6	4.4 ± 0.6	5.8 ± 0.6
CC-O_w	3.0	4.0	2.1 ± 0.4	2.2 ± 0.3	2.0 ± 0.2
	4.0	6.2	21.0 ± 1.2	16.4 ± 0.9	12.8 ± 0.6
O-O_w	2.3	3.3	2.2 ± 0.4	2.3 ± 0.3	2.1 ± 0.3
O-H_w	1.4	2.5	1.2 ± 0.3	1.3 ± 0.2	1.2 ± 0.2
H-O_w	1.4	2.5	0.8 ± 0.2	0.9 ± 0.2	0.9 ± 0.2
O-O	2.4	3.25		0.1 ± 0.1	0.2 ± 0.1
O-H	2.5	3.5		0.1 ± 0.1	0.1 ± 0.1
H-H	1.6	3.0		0.2 ± 0.1	0.3 ± 0.1

The CC-O_w partial in figure 6A gives us information on how the water molecules are arranged around the solute molecule from the point of view of the t-butanol's central carbon. We note two peaks at around 3.7Å and 4.8Å respectively. The shorter distance is consistent with a water oxygen hydrogen bonding to the hydroxyl group of the alcohol, while the broader peak is at a distance consistent with waters surrounding the non-polar head group – i.e. those in the hydration shell of the non-polar head. Integration under this latter peak tells us (see table 1) that the number of water molecules in this hydration shell falls from about 21 at the lowest concentration to about 13 at the highest – water molecules are, as expected, displaced from the hydration shell as the solute-solute co-ordination increases with concentration in the way described above.

Integration under the first – hydrogen bonding – peak gives an essentially concentration invariant figure of 2 (table 1 line 2). This tells us that each alcohol molecule hydrogen bonds to two water molecules through the polar tail. We can compare this with the O-O_w, O-H_w and H-O_w co-ordination numbers in table 1: not only are the values fully consistent with the alcohol hydroxyl group making two hydrogen bonds to water, but they also demonstrate that one of these waters donates a hydrogen to the alcohol oxygen, while the second water accepts a hydrogen from the alcohol.

The bottom three lines of table 1 give information on the solute-solute hydrogen bonding interactions through the polar tails. For the lowest concentration, as is shown in the appropriate O-O, O-H, and H-H partials in figure 6, there is not even a peak visible at the appropriate hydrogen bonding distance, showing effectively zero intermolecular hydrogen bonding. Even at the two higher concentrations, the peak areas are very small, suggesting that solute-solute hydrogen bonding is extremely limited. These results seem to be telling us that the t-butanol alcohol group preferentially bonds to water molecules rather than to other t-butanols. Thus, we can conclude quite clearly that the experimental data is not consistent with a significant amount of alcohol-alcohol hydrogen bonding at these concentrations. As the initial Monte Carlo simulations performed as part of the EPSR refinement process suggested more intermolecular hydrogen bonding, there are indications here that the model potentials in use for this system overemphasise the strength of the intermolecular hydrogen bonding potential at the expense of the water-alcohol interaction[18]. As the stability of a protein is a delicate balance between protein-protein and protein-water hydrogen bonds, this result indicates that work needs to be done on these hydrogen bonding potentials if they are to be seriously used in modelling biomolecular interactions.

5.3 Level 3: Spatial Density Functions

Although all the information we should need to understand structure might be argued to be included in the full set of partial radial distribution functions, these functions are still averages over orientations of both the central molecule at the origin of the function, and the neighbouring molecule at a distance *r*. As we have complete data on the co-ordinates of all the sites on all the molecules in the model ensembles we are interrogating, we can construct more useful functions which enable us to look at the relative orientations of molecules.

First, we imagine ourselves sitting on a site at the origin and looking at the various neighbours at different distances. We begin to note that there is a preference for a certain atom not only at a given radial distance corresponding to a peak in the partial radial distribution function, but also that that peak tends to be in a particular direction. We thus can construct a further pair distribution function that tells us the orientational preference of a particular site as a function of distance from the central site. This function, the ***spatial density function***, gives us orientational information on the nature of the interactions between the various molecules in our solution. For example, referring again to figure 7, we have a peak which tells us that there are neighbouring t-butanols (specified by the positions of their central carbon atoms) at a distance centred between 5.5 and 6Å. We don't know from this orientational average the directions relative to the geometry of the central t-butanol in which these molecules are to be found. Do they

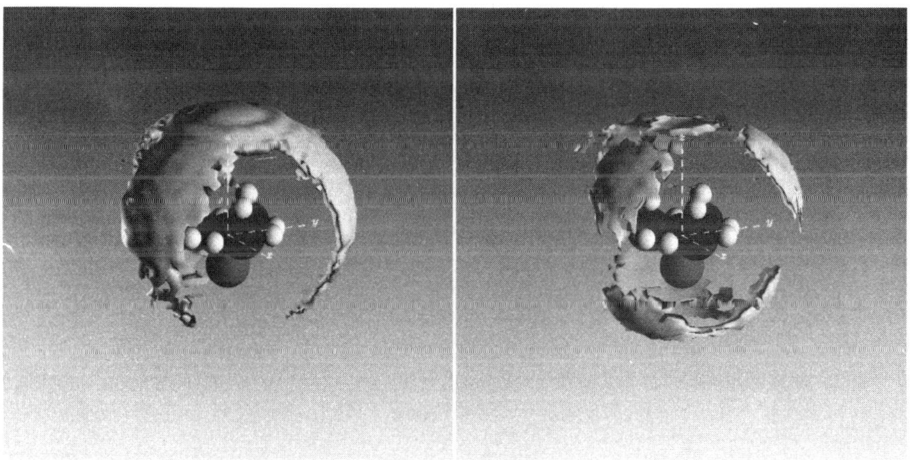

Figure 8. The spatial density function of 0.06 mole fraction aqueous t-butanol at 25C (left) and 65C (right).

cluster around the headgroups, or are there some contacts through the tails? What we really need is to be able to assign a location on a surrounding sphere – in essence a latitude and a longitude – for each of the neighbouring molecules included under the peak shown in figure 7. This is what the spatial density function $g_{\alpha\beta}(r, \Omega)$ does[19]. It represents a three dimensional map of the density of β sites as a function of radial distance, *r*, and orientation Ω about a central site α.

The spatial density function of t-butanol molecules around a central t-butanol molecule, defined with reference to the central carbon sites of each, is shown in the left hand panel of figure 8 for the 0.06 mole fraction concentration at 25C. The continuous density above the central t-butanol molecule and the three sections curving down show the angular distributions around the central molecule of the molecules under the broad peak in figure 7. We thus have more information than in that orientationally-averaged partial radial distribution function: in addition to the preference for the 'capping' position, there is greater density of neighbouring molecules in three orientations separated by about 120°. Interestingly, these directions correspond to the positions between the three methyl groups on the headgroup of the central molecule, which enables the neighbouring molecule to approach slightly closer than if it were approaching along a direction in which a methyl group was located. In fact, if we consider the concentration dependence of the peak position in figure 7, we see a slight outward move as the concentration is reduced. This outward move – which separates out into a low-*r* shoulder at the lowest concentration – appears to be related to an apparent compressive driving force at the higher concentrations which 'pushes in' the neighbouring t-butanol molecule. To anticipate the following section, the methyl groups on neighbouring molecules seem to be engaging with each other like two cogwheels.

5.4 Level 4: Orientational Distribution Functions

In the discussion of the previous paragraph, we are able to comment only on the positional distributions of the β sites around the central α site. We do not know how the neighbouring molecule is oriented with respect to the central one. This information would obviously be useful in understanding the chemical nature of the interactions between the molecules in the various orientations around the central molecule: for example, are the non-polar groups of the neighbouring t-butanol in direct contact with the non-polar groups of the central t-butanol, or are other orientations involved? Moreover, if we increase the temperature to 65C, how would an enhancement of the hydrophobic interactions driving the t-butanol association show up? Does the nature of the solute-solute interactions change? The spatial density function for 65C in the right hand panel of figure 8 already shows that there is a shift in solute-solute interaction away from the headgroup to the polar tail: what is the chemistry of these interactions in this case?

Just as when moving from the partial radial distribution function to the spatial density function we added orientational information to the molecules under the one-dimensional peak of the partial, so here we now need to add orientational information to the molecules in the lobes of the spatial density functions such as those in figure 8. Again, as we have the detailed co-ordinate data in the ensembles generated by the EPSR refinement, this information can be accessed. The orientational distribution function $g(\mathbf{r}, \omega_1, \omega_2)$ relates the relative position vector $\mathbf{r}$ of two molecules 1 and 2 with their orientations ω_1, ω_2 in the laboratory reference frame. As this is strictly a function of nine variables (three positional, six angular), there are major computer memory limitations to storing this function even when the number of variables is reduced by symmetry[20]. An

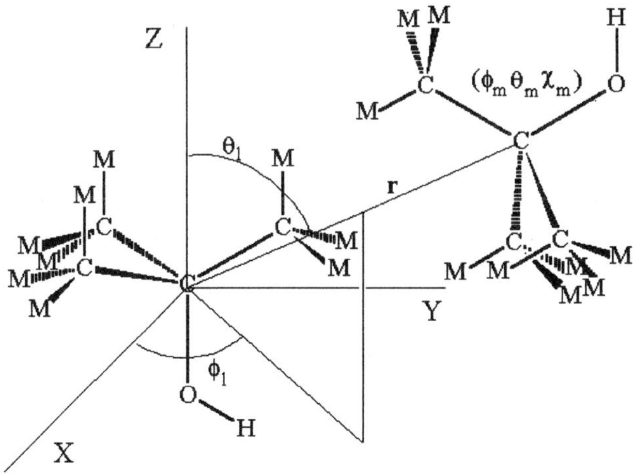

Figure 9. Schematic diagram of the co-ordinate system used to define the orientational correlations between two t-butanol molecules.

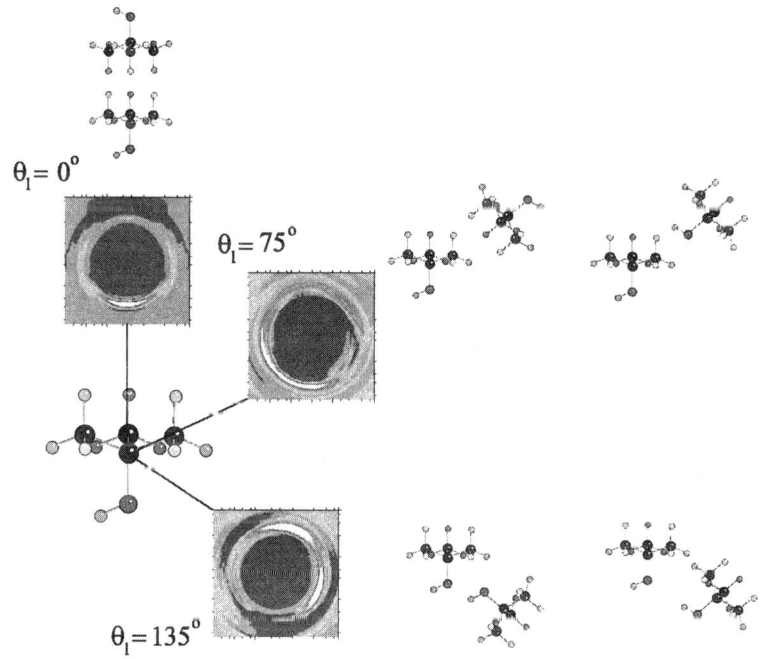

Figure 10. Sections through the orientational correlation function (correlation maps) of two t-butanol molecules in 0.06 mole fraction aqueous solution at room temperature. The molecular arrangements shown are the dominant ones corresponding to the highest contour levels in the plots. For each plot shown, the angular variable plotted is θ_m.

efficient way of dealing with this problem is to expand the orientational correlation function as a spherical harmonic expansion. The correlation function can then be stored as a series of spherical harmonic coefficients[21] $h(l_1l_2l;n_1n_2;r)$, where l_1, l_2, n_1, n_2 relate to the orientational co-ordinates ω_1 and ω_2, while l relates to the spatial distribution of molecule 2 about molecule 1.

Figure 9 shows the definition of the co-ordinate system used for both the orientational position of the neighbour with respect to the geometry of the central molecule, and for the orientation of the neighbouring molecule with respect to the first. The two angles θ_l and φ_l describe the direction of the neighbour (the central molecule being rotationally averaged about the *Z*-axis shown), while the three Euler angles φ_m, θ_m, and χ_m define the orientation of the neighbouring molecule. As a first example of its use, we show in figure 10 correlation maps of θ_m for the 0.06 mole fraction concentration at 25C for three values of θ_l. The $\theta_l = 0°$ plot shows high density for θ_m around 180° - i.e. the dominant interaction in this direction is with the two t-butanol molecules making contact through their non-polar head groups. This is classically what we would expect for a system subject to a hydrophobic driving force. Moving to the $\theta_l = 75°$ plot, we see the highest intensity between the 6 o'clock and 9 o'clock positions, again consistent with a dominant non-polar to non-polar contact as shown in the figure. Here however there is some – though small – intensity around the 1 to 2 o'clock position, indicating some contribution from the polar to non-polar contact again indicated by the molecules in the figure. Finally, moving to the $\theta_l = 135°$ plot, we note bright lobes in the two molecular orientations shown – clearly these correspond to intermolecular hydrogen bonding. We have already noted however that these contacts occur only rarely in this solution. What the orientational correlation functions tell us is that, even if the intermolecular contacts in this region of the central molecule are rare, when they do occur, they are of a typically hydrogen bonding nature.

A second example of the information in the orientational correlation function refers back to the spatial density functions shown for the 0.06 mole fraction concentration at the two temperatures of 25C and 65C (figure 8). What can we learn about the change in the nature of these intermolecular interactions as the temperature is raised in such a way as to (we expect) enhance the hydrophobic driving force? We have already seen that at 25C the lobes in the spatial density function relate to non-polar to non-polar contacts through the headgroups of neighbouring molecules; at the higher temperature (figure 8), these lobes shrink, and there is significant growth in contacts at the polar end of the molecule. What is the chemical nature of these new contacts?

We can answer all these questions by referring to the plots in figure 11. The left-hand panels relate to the 25C solution, the right hand column to 65C. The top two panels are plots through the two spatial density functions of figure 8, and show that the dominant head-to-head contact indicated by the bright lobe around 12 o'clock for 25C is reduced in favour of increasing intensity around 5 and 7 o'clock in the 65C solution. The second panel shows orientational correlation function plots of θ_m. The dominant intensity around 180° again indicates a dominance of non-polar contacts through the headgroups of contacting molecules. Moving right to the 65C panel, we find two interesting points. First, the 6 o'clock feature is a little 'tighter' indicating reduced disorder in the head-group to head-group interaction, an interesting consequence of the increased temperature. Secondly, there is a significant feature around 12 o'clock which is indicative of a polar to non-polar interaction – i.e. the headgroup of one molecule is interacting with the polar tail of a neighbour. A further point to note is that this lobe is shifted out to a distance of

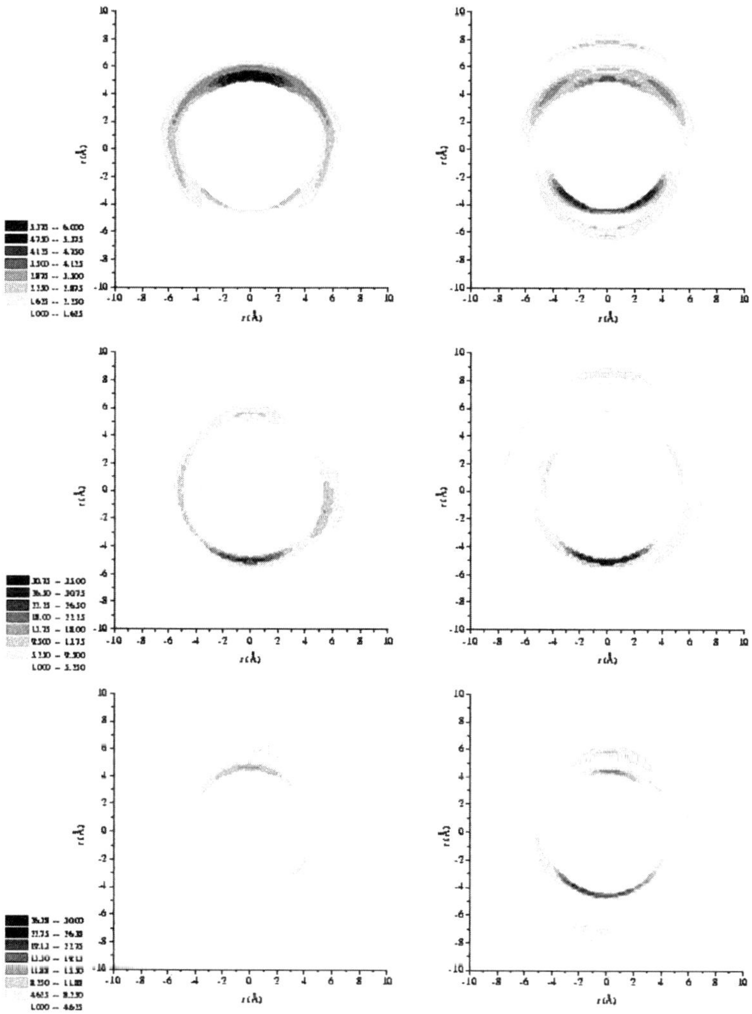

Figure 11. The intermolecular orientation correlation maps corresponding to the solute-solute interactions as a function of molecular centres separation in 0.06 mole fraction t-butanol-water solution. The left-hand column represents the correlations prevalent at 25C, the right hand column those at 65C. The top panels represent the probability of finding a neighbouring alcohol molecule about an arbitrarily chosen central alcohol molecule as a function of the angle θ_l, i.e. the result is averaged over all orientations of the neighbouring alcohol molecule. The middle panels illustrate the dominant orientation as a function of θ_m of a neighbouring alcohol molecule in the direction above the methyl groups of the central molecule, i.e. where $\theta_l=\phi_l=\phi_m=\chi_m=0°$. The lower panels show the relative orientations of alcohol molecules θ_m, in the direction of the hydroxyl group interactions where $\theta_l=135°$, $\varphi_l=60°$and $\varphi_m=\chi_m=0°$.

about 8Å, so this appears not to be a direct contact between the hydroxyl group and the methyl head groups. Rather there is space for an intervening water molecule. We can therefore perhaps consider this as a neighbouring t-butanol hydrogen bonding to one of the water molecules in the non-polar hydration shell of the central molecule.

Finally, the bottom panel shows the orientational plot in the direction of the hydroxyl group of the central t-butanol molecule. The most obvious change to note is the growth of intensity around the 6 o'clock position at 65C. This lobe corresponds to t-butanol to t-butanol contacts through direct hydrogen bonding, an occurrence that, we have already noted, is much rarer at 25C. A second point to note is a splitting of the lobe around 12 o'clock that denotes polar to non-polar contacts; at 65C, the inner lobe comes in closer, the rise in temperature appearing to make easier a close, apparently chemically unfavourable, contact.

Raising the temperature in this system thus seems to do several interesting things. First, the straight non-polar contacts seem to become more ordered. Secondly, there is an increase in the degree of direct hydrogen bonding between t-butanol molecules. And thirdly, the temperature increase seems to make classically unfavourable interactions between polar and non-polar groups easier. The 'strengthening' of the non-polar contacts might be expected from classical ideas of the temperature dependence of the hydrophobic effect. The other two effects seem less easy to relate directly to consequences of a hydrophobic driving force.

6. WELL BEYOND THE PDF: THE KEY TO THE HYDROPHOBIC INTERACTION?

We conclude by discussing further results on this system that impinge directly on the nature of the solvent ordering that is thought to be the driving force for the hydrophobic interaction.

We recall first the discussion at the end of section 3 concerning the classical view of the nature of the hydration shell of a non-polar group. This holds that the water in the hydration shell is somehow 'more ordered' than in the bulk, a popular picture of this ordering being a structure something like the clathrate cage shown in figure 3. Can we use the liquid structure determination techniques discussed in this chapter to justify – or otherwise – this picture?

6.1 The Non-Polar Hydration Shell.

The left hand panel of figure 12 shows the solute-solvent spatial density function for 0.06 mole fraction t-butanol, the solute molecular centre being the central carbon and the water positions being reported through their oxygen positions. Two regimes are apparent: a broad diffuse region around the headgroup, stretching down to and below the molecular equator, and a region around the lower part of the molecule that is closer in than the shell surrounding the headgroup. The lower region relates to waters hydrogen bonded to the hydroxyl tail. The upper region is the non-polar hydration shell of interest. We can now look at the orientation of water molecules in this upper lobe by investigating the appropriate orientational correlation function, and compare what we find with the orientations expected in the classical clathrate cage. Referring again to figure 3, we see that the water molecules in this cage make good hydrogen bonds to other

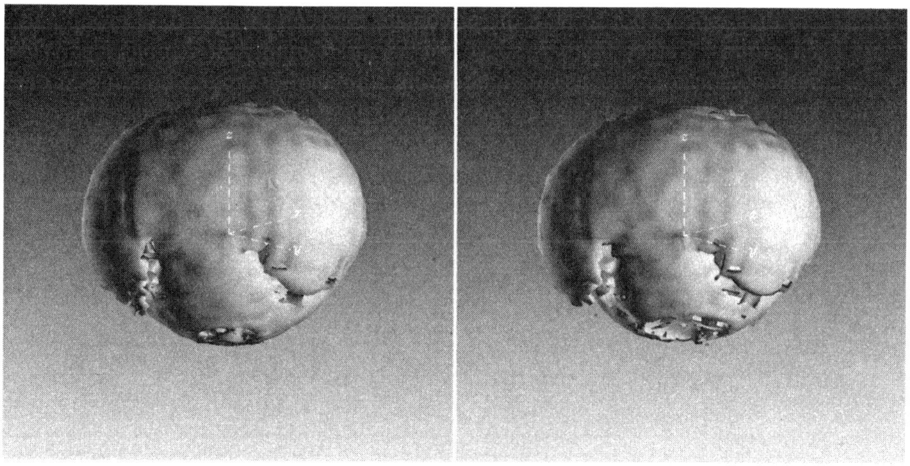

Figure 12. The t-butanol – water spatial density function at (left) 25C and (right) 65C.

molecules in the shell: this suggests that a dominant water molecule orientation is with its dipole moment tangential to the shell. Those waters that hydrogen bond with the surrounding solvent will have their dipole pointing in directions away from the central molecule; there should be none that point towards the cage centre.

Figure 13 gives the required orientational information on the hydrating waters. Concentrating first on the 25C data in the left hand panels, the top panel is essentially the averaged spatial density function of the left hand panel of figure 12, and shows clear lobes around 5 and 7 o'clock that relate to the waters hydrogen bonded to the alcohol hydroxyl group. The waters surrounding the non-polar head group form a continuous diffuse band between about 9 o'clock and 3 o'clock. The bottom left hand panel shows the dominant orientations of the water molecules in this shell: the two lobes around 6 o'clock and 11 o'clock are indeed consistent with a dominant orientation of the water dipole tangential to the shell – as expected in the classical clathrate model. We should note however that there is significant disorder in the water shell which indicates that it would be an oversimplification to think of this shell as an ordered cage structure: a more realistic picture would show the clathrate cage as an idealised version of the actual structure observed. We are after all considering the average structure in a liquid, where significant spatial variability should be expected.

Before leaving this discussion, we examine the changes in the non-polar hydration shell as we increase the temperature to 65C – as mentioned above, we expect some enhancement of the hydrophobic driving force as we raise the temperature. What we see initially from comparing the two panels in figure 12 is that there is little difference in the spatial density functions as a result of this rise in temperature. The plots of fig 13 do however show some interesting changes. First, comparing the top two panels suggests a fractional structural *enhancement* of the upper non-polar hydration shell *and* the lower hydrogen bonded region. Secondly, the non-polar shell water orientations at 65C shown in the lower right panel are more diffuse than at 25C. The temperature rise appears to be spreading intensity in the direction of aligning the water molecule hydrogens in a methyl group straddling position. Further details on these interesting changes can be obtained from inspecting other orientational correlation function sections[7], underlining further the fine structural detail that these methods allow us to obtain.

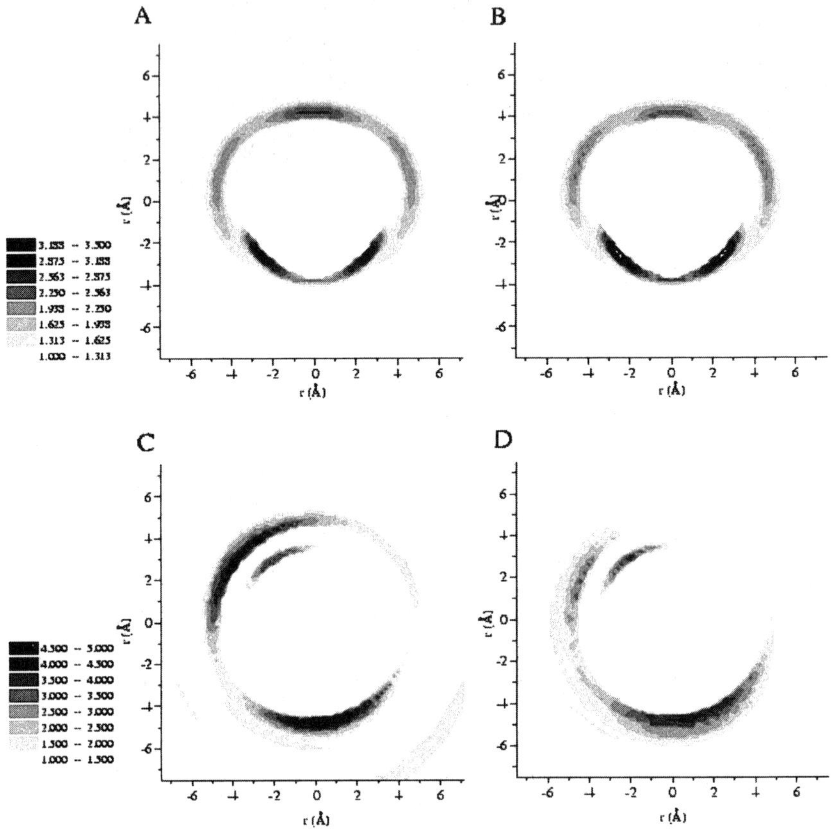

Figure 13. The intermolecular orientation maps corresponding to the first hydration shell, i.e. the solute-solvent interactions as a function of molecular centres separation in 0.06 mole fraction t-butanol-water solution. The left-hand column represents the correlations prevalent at 25C, the right hand column those at 65C. The top panels show the probability of finding an arbitrarily oriented water molecule about any tertiary butanol molecule in the system as a function of the angle θ_l. The lower panels show the dominant correlations of a water molecule as a function of θ_m in the direction defined by the centre of the alcohol molecule and the non-polar methyl groups, i.e. where θ_l=60°, φ_l=0°, φ_m=0° and χ_m=0°.

6.2 The Water's Viewpoint

Is the water in the non-polar hydration shell somehow 'more ordered' than water in the bulk liquid, as the standard model of the hydrophobic interaction would argue? We can answer this by looking at the appropriate water-water partial radial distribution functions, and comparing these with those for the bulk liquid.

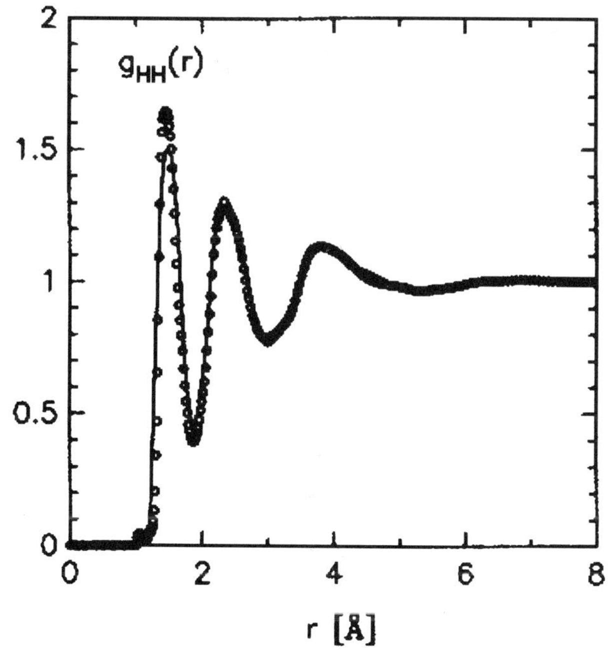

Figure 14. The hydrogen-hydrogen partial rdf for a 1:9 mole fraction solution of methanol in water (line) compared to the same function for pure water (circles).

Returning to the discussion of composite partial radial distribution functions ('level 1', section 5.1), we noted there that even at that modest level of structure determination, for aqueous solutions the solvent-solvent $g_{HH}(r)$ term was a true partial radial distribution function of the solvent. This function for aqueous methanol – the first system for which the structure of the hydration shell of a non-polar group was experimentally tackled[15] – is compared with the same function for bulk water at the same temperature in figure 14. The conclusion is both obvious and unexpected: there is no detectable difference between the two functions. As far as the water is concerned, when its structure is measured through the hydrogen-hydrogen radial distribution function, a sensitive structural measure that includes all-important orientational information, there is no observable perturbation of the structure in the direction of increased – or decreased – order. Similar conclusions have been drawn since in studies of a number of different systems containing non-polar head groups[6, 7, 22]. Thus, there appears to be no structural evidence to support the traditional view that the water in the first hydration shell of a non-polar group is more ordered than in the bulk.

The clue to the actual source of the entropic driving force may be in figure 15, which shows the spatial density functions of the first (top panels) and second (bottom panels) neighbour shells *of water* in (from left to right) bulk water at 25C, 0.06 mole fraction t-

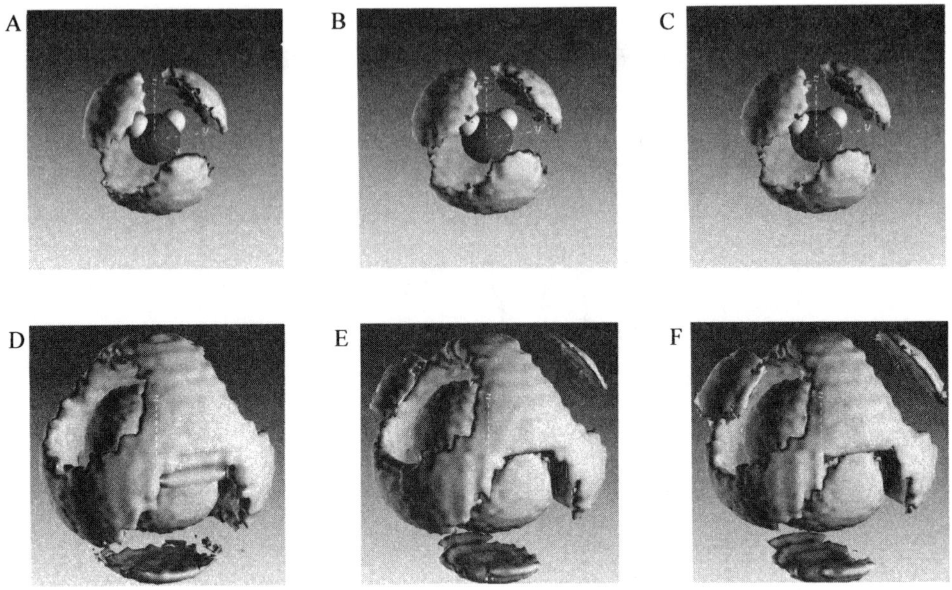

Figure 15. The spatial density functions of water molecules around a central water molecule for (top row) first neighbour shell water molecules and (bottom row) first and second shell waters. In turn the panels show the distributions for (left) pure water at 25C, (centre) 0.06 mole fraction t-butanol – water solution at 25C and (right) 0.06 mole fraction t-butanol – water solution at 65C. The first shell of neighbours (top row) illustrates the invariance between the three systems of the short-range correlations that are of a predominantly tetrahedral local water co-ordination. The appearance of the third neighbour shell in the direction of the central molecule O-H bonds for the alcohol-water mixtures (bottom centre and bottom right) visibly illustrates the solute-induced compression of the solvent molecular density which appears to be enhanced by increasing temperature.

butanol at 25C, and 0.06 mole fraction t-butanol at 65C. We emphasise that we are looking here at the neighbourhood *of the water*, *not* of the non-polar group itself.

Focussing first on the first shells in the top panels, we see the standard spatial density function for the first neighbour shell in water. The two lobes above the central water relate to neighbouring waters that accept hydrogens from the central water, while the broad lobe beneath indicates waters donating hydrogens to the lone pair region of the central water. This first neighbour shell appears unchanged in both t-butanol solutions.

When we look at the bottom panels, where the contour levels have been lowered to show up the second neighbour shell, we see some clear differences. Although the basic geometry of the second shell is similar in all four cases, there are very significant differences: the second shell is progressively 'pulled in' toward the central molecule as we move to the 25C solution, and then to the 65C case. Consequently, lobes of spatial density corresponding to the third shell begin to appear in the two solution plots.

This drawing in of the second shell is accompanied by a significant ordering of the second shell. This is indeed visible in the progressive narrowing of the band in the centre of the bottom panels of figure 15 as we move from left (bulk water) to right (65C 0.06 t-butanol). It is more obvious however in the two-dimensional spatial density maps of figure 16, which are obtained by rotationally averaging $g_{OO}(r,\Omega)$ about the z-axis (shown in figure 15). The second shell structural tightening induced by the presence of the alcohol molecule can be seen in the increasing breaks of continuity in the 10 o'clock and

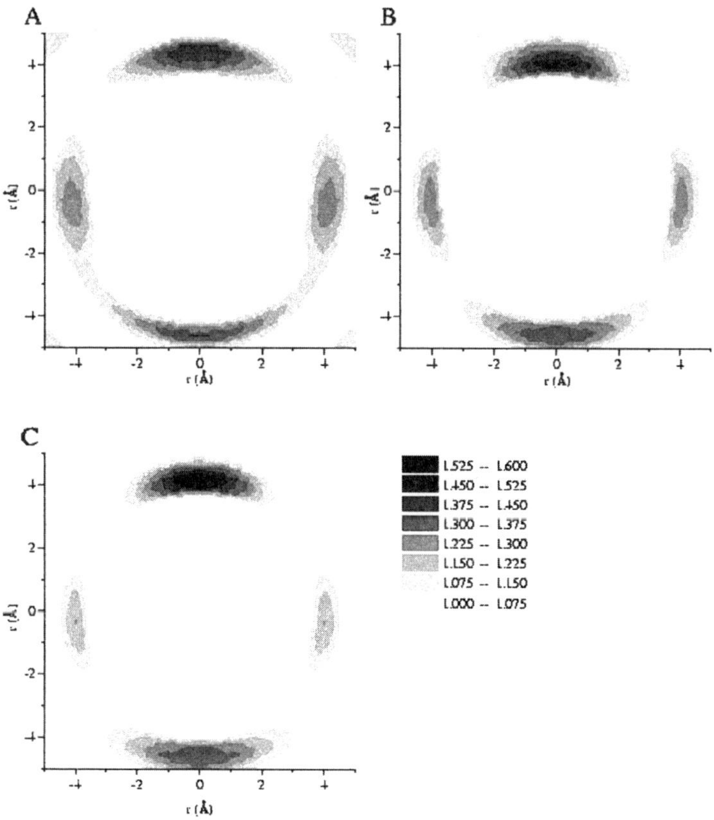

Figure 16. The upper panels are two-dimensional maps of the second water shell spatial density functions of figure 15. These are obtained by rotationally averaging $g_{OO}(r,\Omega)$ about the z-axis, i.e. about the ϕ co-ordinate. The map in (A) corresponds to pure water at 25C, in (B) to 0.06 mole fraction t-butanol – water solution at 25C and in (C) to the alcohol solution at 65C. The disruption of the second shell water in the presence of the alcohol molecule can be seen in the two o'clock direction: this disruption is also marked by the inward migration of the shell, in particular along the twelve o'clock direction.

2 o'clock directions as we move from bulk water to the two solutions. Also visible is the small but significant inward shift of the lobes, which is particularly evident in the 12 o'clock direction. Taking sections through the appropriate lobes allow us to quantify both these inward shifts and the narrowing of the shell widths[7]. In terms of the water molecule's second neighbour shell, we have clear evidence that the presence of t-butanol increases structural order in the system, an ordering which is enhanced with the rise in temperature to 65C. Perhaps it is here – in the *water's second* shell rather than the *alcohol's first* hydration shell – that the source of the entropic driving force for the hydrophobic interaction is to be found.

7. CONCLUSIONS

In 1993, the physical chemist David Feakins observed (personal communication) that "(our understanding of solutions is) in the state that our knowledge of molecules would be if we only had, say, the thermodynamics and kinetics of reaction to guide us". Similar comments could have been made in the 1920s about our knowledge of the structures of molecules in crystals, prior to the development of X-ray crystallography. We all know the developments of the latter technique which now allows us to solve the atomic-level structures of, not only proteins and viruses, but also of macromolecular assemblies such as the ribosome, achievements which even 40 years ago we would have thought fantastic. We have perhaps begun to make similar movements forward in understanding the structures of solutions of chemical and biological importance. Without the assistance that the presence of a regular lattice gives to crystallography, we can, by exploiting both the experimental and computational techniques now available, begin to understand the detailed three-dimensional structures of quite complex liquids. This 'liquid state crystallography' is now very much a practical proposition. It is already giving us new and important information about the way things happen in solution.

REFERENCES

1. P. A. Egelstaff, *An Introduction to the Liquid State*, (Clarendon Press, Oxford, 1994).
2. J. P. Hansen and I. R. McDonald, *Theory of Simple Liquids* (Academic Press, London, 1986).
3. A. K. Soper, W. S. Howells and A. C. Hannon, *ATLAS – Analysis of Time-of-Flight Diffraction Data from Liquid and Amorphous Samples*, Rutherford Appleton Laboratory Report RAL-89-046 (1989).
4. F. G. Edwards, J. E. Enderby, R. A. Howe and D. I. Page, *J. Phys.C* **8**, 3483-3490 (1975).
5. J.E.Enderby, D.M.North, P.A.Egelstaff, *Phil Mag.* **14**, 961-970 (1966).
6. D. T. Bowron, J. L. Finney and A. K. Soper, *J. Phys. Chem.* B **102**, 3551-3563 (1998).
7. D. T. Bowron, A. K. Soper and J. L. Finney, *J. Chem. Phys.* **114**, 6203-6219 (2001).
8. F. Franks and J. E. Desnoyers, *Water Sci. Rev.* **1**, 171-232 (1985).
9. H. S. Frank and M. W. Evans, *J. Chem. Phys.* **13**, 507-532 (1945).
10. L. Pratt and D. Chandler, *J. Chem. Phys.* **67**, 3683-3704 (1977).
11. N. T. Skipper. C. H. Bridgeman, A. D. Buckingham and R. L. Mancera, *Faraday Discuss.* **103**, 141-150 (1996).
12. J. E. Enderby and G. W. Neilson, *Water A Comprehensive Treatise*, edited by F. Franks (Plenum Press, New York), **6**, 1-46 (1979).
13. J. Turner, J. L. Finney, J. P. Bouquiere, G. W. Neilson, S. Cummings and J Bouillot, in *Water and Aqueous Solutions*, edited by G. W. Neilson and J. E. Enderby (Adam Hilger, Bristol, 1986), pp. 277-286.
14. J. Turner, J. L. Finney, J. P. Bouquiere, G. W. Neilson, S. Cummings and J Bouillot, *Physica* **136B**, 260-264 (1986).
15. A. K. Soper and J. L. Finney, *Phys. Rev. Lett.* **71**, 4346-4349 (1993).
16. A. K. Soper, Ch*em. Phys.* **202**, 295-306 (1996).
17. A. K. Soper, in: *Local Structure from Diffraction*, edited by S. J. I. Billinge and M. F. Thorpe (Plenum Press, New York, 1998), pp. 59-83.
18. D.T.Bowron, J.L.Finney and A.K.Soper, *Mol. Phys.* **93**, 531-543 (1998); *Mol. Phys.* **94**, 249-251 (1998)
19. I. M. Svishchev and P. G. Kusalik, *J. Chem. Phys.* **99**, 3049-3058 (1993).
20. A. K. Soper, *J. Chem. Phys.* **101**, 6888-6901 (1994).
21. C. G. Gray and K. E. Gubbins, *Theory of Molecular Liquids Vol. I: Fundamentals* (Oxford University Press, New York, 1984).
22. J. Turner, A. K. Soper and J. L. Finney, *Molec. Phys.* **70**, 679-700 (1990).

X-RAY ABSORPTION SPECTROSCOPY STUDIES OF NON-POLAR SOLUTE HYDRATION STRUCTURE

D.T.Bowron and J.L.Finney*

1. INTRODUCTION

A question of considerable topical interest in the area of solution physics, chemistry and biology relates to the structural nature of non-polar or largely non-polar, atomic or molecular solute hydration. An oft-quoted, yet poorly understood phenomenon that is conventionally thought to drive many important in-solution chemical and biological processes is the hydrophobic effect[1]. This term refers to the tendency of non-polar or largely non-polar solute species to associate when they are placed into an aqueous environment. Thermodynamic observations of model mixture systems where this "force" of association is manifest, show that a fine balance between entropic and enthalpic components exists, with the intriguing twist that the entropic contribution to mixing is negative[2, 3]. In consequence, the conventional wisdom has developed around the postulate of an ordering of the structural configurations of non polar solute hydration shell water, in the direction of greater "crystallinity" [4]. This is the classic "iceberg" model[4], that has dominated a great deal of the literature on hydrophobic phenomena during the latter half of the twentieth century, and in particular shaped many explanations of the mechanisms that drive assembly and molecular motion in biological systems[5].

Until relatively recently, experimental methods that could explicitly probe the structural correlations in the hydration shell of solutes with considerable hydrophobic character remained to be developed. Now, with the arrival of advanced neutron sources and synchrotron radiation facilities, coupled with new computational techniques for data analysis, it has become possible to transform this situation. These new experimental methods allow us to study the fine details of this complex phenomenon and to test the

* Daniel T. Bowron, ISIS Facility, CLRC Rutherford Appleton Laboratory, Chilton, Didcot, OXON, OX11 0QX, UK. and Department of Physics and Astronomy, University College London, Gower Street, London, WC1E 6BT, UK
John L. Finney, Department of Physics and Astronomy, University College London, Gower Street, London, WC1E 6BT, UK

From Semiconductors to Protein, edited by S.J.L. Billinge and M.F. Thorpe
Kluwer Academic/Plenum Publishers, 2002

models and predictions of more structurally indirect experimental and computer simulation methods.

In this paper, we will discuss the development of X-ray absorption spectroscopy methods for liquid systems, and their application to the study of noble gas hydration structure and its perturbations. The noble gases are considered the archetype of a hydrophobic solute, and as such the results obtained relate closely to the fundamental chemical physics that underlies this entire subject area. A simple illustration of the dislike of these solutes for the aqueous environment can been seen in their solubility which, under moderate conditions (room temperature and pressures less than 100bar), is found to be of the order of 10^{-3} mole fraction[6]. This low solubility has been the most significant challenge for local structure characterisation of extremely hydrophobic moieties and hence it is little surprise that advances in their structural characterisation have developed hand in hand with methods to apply compensating pressure to the experimental samples. The application of pressure produces a sufficient increase in solubility to enable more direct structural probes to be utilised. Technical developments in this area for X-ray absorption spectroscopy will also be discussed.

2. X-RAY ABSORPTION SPECTROSCOPY AT 3rd GENERATION SYNCHROTRON RADIATION FACILITIES: STUDIES OF NOBLE GAS HYDRATION

X-ray absorption spectroscopy experimentally came of age with the arrival of synchrotron radiation sources in the mid-1970s. The technique requires an energy tuneable source of X-rays to make the spectroscopic measurement. Though measurements are feasible on laboratory sources they tend to be severely constrained by the limited X-ray flux delivered to the sample. Synchrotron sources increased the intensity of X-rays delivered to the sample by several orders of magnitude, and consequently the technique and its range of applications developed apace with their increasing availability.

Many of the experimental challenges that had to be overcome in the course of this particular project only became manageable with the arrival of 3rd generation synchrotron radiation sources in the mid-1990s. The first of these was the European Synchrotron Radiation Facility (ESRF) in Grenoble, France. These advanced X-ray sources are the culmination of a world wide incremental development programme in electron storage ring technology. Such facilities now routinely provide very stable high intensity X-ray radiation across a wide range of energies, ranging from a few to many thousands of electron volts. These X-ray energies are ideally matched to the study of both atomic and molecular structure as well as the intrinsic electronic interactions that are present in these systems. In contrast to many of the earlier synchrotron sources, these new facilities typically deliver radiation continuously, with a source reliability in excess of 95%. As such they have transformed both the type and degree of difficulty of the experiments that can now be attempted. Experiments requiring considerable applied pressures have been one of the greatest beneficiaries of these developments.

All the results presented here were obtained on the BM29 experimental station[7] at the ESRF. This spectrometer is based on an ESRF bending magnet which provides a versatile energy tuneable radiation source over an energy range of 5-75keV and with a typical spectral energy resolution of the order of $\Delta E/E \approx 5\times10^{-5}$. The typical X-ray flux that reaches the sample ranges between 10^{11} to 10^{9} photons per second as the incident energy is increased across the spectrometer's range. This is sufficient to enable spectra to

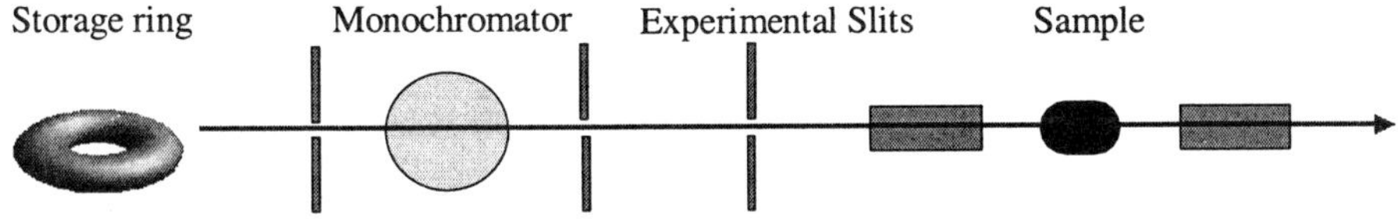

Figure 1: Schematic diagram of a simple transmission geometry X-ray absorption spectrometer. The X-ray absorption signal is measured as $\mu x = \ln(I_0/I_1)$ where μ is the sample's absorption coefficient, x is the sample thickness and I_0 and I_1 are the incident and transmitted X-ray beam intensities.

be collected in the usual transmission geometry with a statistical noise component of the order of 10^{-4}. This fact is particularly important for studying liquid or disordered materials systems where the magnitude of the structural signal is in general quite small. The nature of the bending magnet source means that it is possible to collimate a small yet stable beam from the produced radiation fan which can be well matched with challenging sample environments such as pressure cells with their small X-ray windows.

Figure 1 illustrates the conventional transmission geometry for X-ray absorption measurements, highlighting the key optical components of polychromatic source, monochromator and collimating slits. I_0 and I_1 are beam intensity monitors that are most commonly of ion chamber type.

As mentioned in the introduction, a key development required in order to experimentally determine by X-ray absorption the structure of the hydrophobic hydration shell of a non-polar solute such as noble gas atom, was the development of a suitable pressure cell. By applying moderate pressure of the order of 100bar, the solubility of a noble gas in water can be raised sufficiently to facilitate a transmission geometry experiment. At this pressure a typical concentration would be of the order of 1 gas atom for a few hundred water molecules.

Figure 2 shows one of the early gas pressure cells that were developed in the course of this project. The cell was designed to be filled with approximately 0.25cm^3 of water and then directly pressurised by krypton gas. After approximately thirty minutes to one

Figure 2: An early transmission geometry liquid pressure cell. Note the small X-ray windows (1.0×0.1cm width×height) that allow X-ray transmission. The sample volume (≈0.25cm^3) was delimited by two, 0.4cm thick epoxy resin windows.

hour, the concentration of gas in the water reaches a sufficiently stable equilibrium for an X-ray absorption measurement, which typically would take 20 minutes.

Typical experiments are performed when the ESRF synchrotron is operating in high intensity 2/3 fill mode, with storage ring currents ranging between 150 and 200mA, and at an electron energy of 6GeV. The monochromator was equipped with either a pair of Si(111) or Si(311) crystals detuned by 50% for harmonic rejection. Typical beam collimation was set so as to define a primary beam of 0.2mm vertical size by 6mm horizontal size under which conditions a typical energy resolution of ≈1eV is obtained. This can be compared with the intrinsic width of the Kr K absorption edge at 14.3236keV of 2.7eV. A spectrum measured under these conditions is shown in Figure 3.

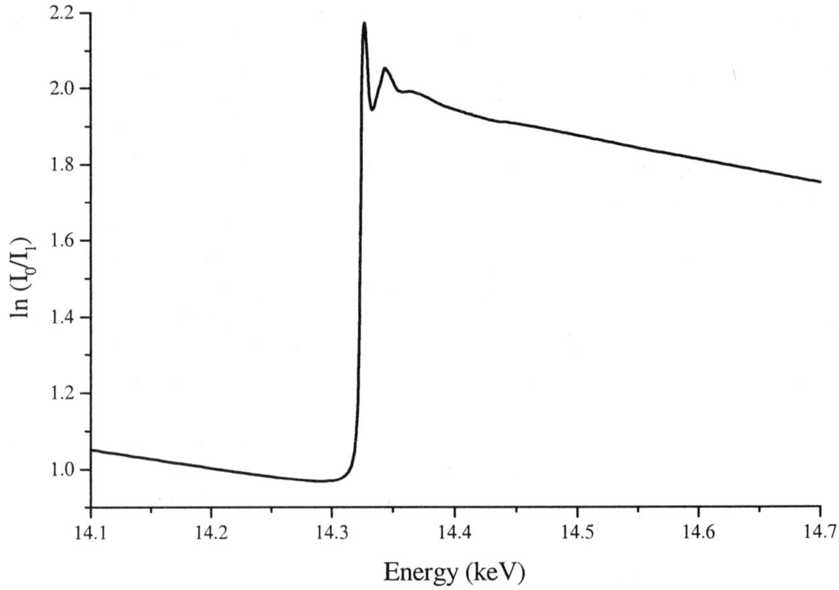

Figure 3: The measured Kr K edge X-ray absorption signal for krypton in water at 100bar and room temperature.

3. X-RAY ABSORPTION SPECTROSCOPY APPLIED TO LIQUID SYSTEMS

The link between a measured Extended X-ray Absorption Fine Structure (EXAFS) spectrum and the local atomic structure surrounding a photo-absorbing element was graphically demonstrated in the seminal paper of Sayers *et al* [8]. Their work illustrated the potential for Fourier inversion of a measured X-ray absorption signal that in turn raised the possibility of applying this technique to the structural characterisation of non-crystalline systems and a recent comprehensive review has been produced by Filipponi [9].

The potential of this method to directly address some of the key questions in the area of the structural characterisation of hydrophobic effects is immediately apparent. This spectroscopic technique is only sensitive to the short-range structural environment surrounding a selected atomic species. It is sensitive to inter-atomic separations, the number and type of atoms in the local environment, the geometrical distribution of these atoms and the degree of thermal and structural local disorder. At first view, this spectroscopy would therefore appear to be an almost perfect probe for looking at the hydration of a simple solute such as a noble gas atom. However, there are some serious complicating issues with regards to the data inversion scheme that first need to be addressed [10].

By and large, the most significant problems that need to be overcome when applying this spectroscopic technique to liquids and disordered materials are:

(1) the fact that the radial pair distribution function of atoms, $g_2(r)$, surrounding the photo-absorbing atomic sites is continuous and not simply a series of discrete shells, and

(2) the insensitivity to longer-range structure of the X-ray absorption spectrum and the consequent lack of any integral sum-rule that can link the measured structural signal to a macroscopic parameter such as a material's isothermal compressibility.

Some analysis scheme is therefore required to take account of these facts. If not, a simplistic structural inversion of the measured data achieved through a conventional peak fitting analysis will often lead to a significantly under-determined set of co-ordination numbers.

Though several analysis methods have been proposed in the literature[11, 12], the approach we have adopted here revolves around the refinement of a model X-ray absorption signal in terms of a difference pair distribution function of atoms[10], $\Delta g_2(r)$. This difference function, is derived by utilising a model for the atomic environment, $g_2^m(r)$, of a noble gas atom to satisfy long range structural constraints. The general scheme can be written as:

$$\Delta g_2(r) = g_2(r) - g_2^m(r) \tag{1}$$

This is then expanded as a Taylor series for data refinement purposes. In this way it is possible for structural information that would otherwise be completely absent in the experimental data, to be incorporated.

The above described data refinement procedure forms an integral part of the GNXAS data analysis package[13-15] that has been developed for the treatment of disordered materials X-ray absorption spectroscopy data. Below, various results will be highlighted.

Admittedly this method of data analysis could be considered somewhat model dependent, but we highlight that the additional information thus incorporated can in general be well quantified by independent computational or experimental means and remains fixed in the actual analysis procedure.

The key to this analysis procedure is thus to establish a reasonable model radial distribution function of atoms from which one can extract the long-range behaviour of the pair correlations, centred on the photo-absorbing atomic sites. This can be done in a number of ways, either experimentally by utilising an appropriate radial distribution function if available, such as from first order difference isotopic substitution neutron scattering methods[16], or from a constrained computer simulation[17]. A valid is why, if a suitable model radial distribution function is already available, a subsequent X-ray absorption spectroscopy refinement is required? The answer to this question is threefold.

(a) X-ray absorption spectroscopy is a technique that is very heavily weighted to the short-range structure around a photo-absorbing atom and certainly to a far greater extent than an analogous X-ray or neutron diffraction experiment. As such the spectroscopic result can be used to refine more precisely the short range structural correlations[18].
(b) The spectroscopic technique is ideally suited to follow systematic trends in the evolution of short-range structure as a function of temperature[19] and pressure[20]. Longer- range structure tends to be less sensitive to such perturbations. So, a given starting model parameterised under different conditions can largely provide sufficient constraints for a spectroscopic determination of local structure under conditions quite different from those in the initial model.
(c) The third advantage of the method that is often overlooked is that the same experimental technique and data analysis procedure can be used for both crystalline and disordered systems[21]. Direct comparisons can thus be made between crystalline or disordered state local structure, without excessive problems regarding the nature of different systematic errors that may affect and distort independent data analysis methods attempted in a comparable fashion.

In the analyses presented here, an initial model distribution function of Ar-O[16], for argon in water, has been used to create a reasonable model for the longer range atomic pair correlations that would be expected in the krypton water system: the principal subject of this research project.

Figure 4 shows the final result of a data refinement for krypton gas in water. This illustrates how two signals plus a long range model contribution have been combined to produce a determination of the hydration shell of the noble gas from the viewpoint of the structural correlations with water oxygen atoms. Figure 5 shows how this decomposition is matched to the measured experimental data.

The refined radial distribution function in Figure 4, thus illustrates the ability of this technique to address the local structure of the hydrophobic hydration shell of krypton gas. The results are consistent with a solvation number for krypton in water of ≈24 water molecules and an average Kr-O_{water} distance of ≈3.8Å. The width of the peak illustrates that though well defined, the hydrophobic hydration shell contains considerable scope for structural disorder with regards to the precise localisation of the water molecules in the shell. The average Kr-O_{water} distance, in combination with the number of solvent molecules in the shell is consistent with a largely "tangential" orientation for the solvent water molecules i.e. with both water oxygen and hydrogen atoms in the same radial shell at 3.8Å. However, this can by no means be considered the exclusive geometry, though no evidence was found of photoelectron multiple scattering that is characteristic of co-linear Kr-H-O or Kr-O-H atomic configurations.

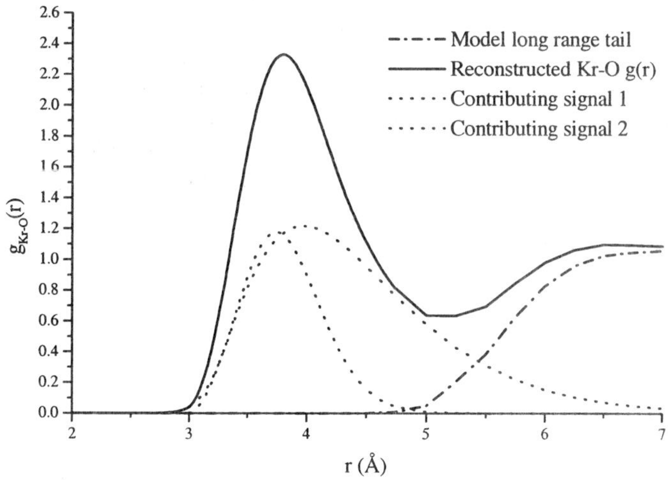

Figure 4. Reconstructed Kr-O_{water} aomic pair distribution function, showing the long range tail derived from a model Ar-O pair distribution function and the two model peaks refined to be consistent with the X-ray absorption spectroscopy data when combined with the contribution of the long range tail.

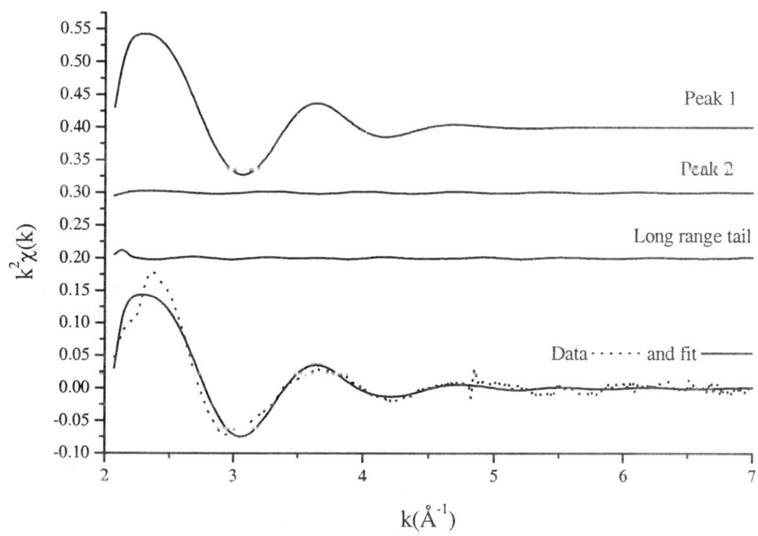

Figure 5: Calculated model signals fitted to the experimental data to reconstruct the radial distribution function shown in Figure 4.

4. SUBTLETIES OF THE ELECTRONIC EXCITATION BACKGROUND

The electronic excitation background that contributes to a raw X-ray absorption spectrum is a considerable challenge when refining a model structure. The krypton in water system offers us a rare chance to investigate the effects of a non-bonded chemical environment on the distribution of the electronic states of the noble gas atom that can be compared to those found in its free gaseous state. Gaining an appreciation of the magnitude of these effects is useful for gauging their likely importance on the physical and chemical properties of the system.

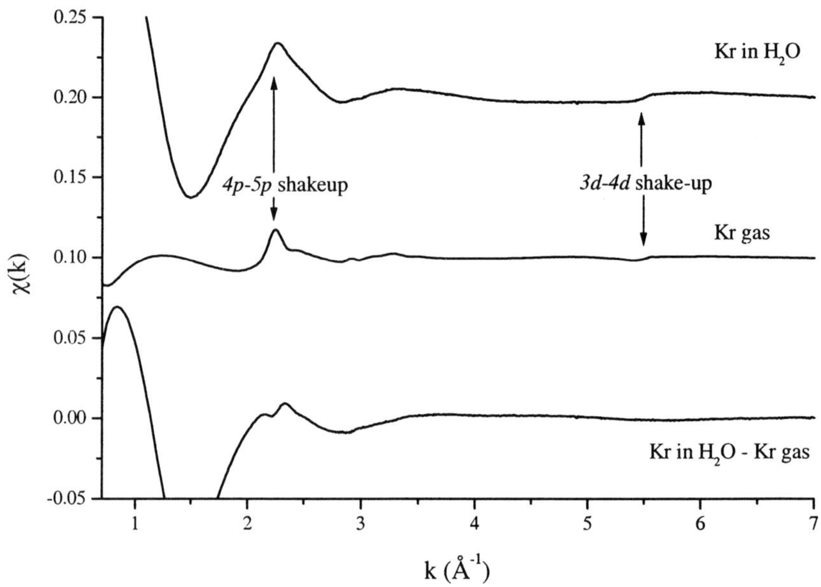

Figure 6: Illustration of the multi-electronic excitation background that accompanies the Kr K-edge excitation (*1s-continuum)*. From top to bottom are shown the as-measured experimental spectrum of krypton gas in water, the experimental spectrum for krypton gas and the difference spectrum. Note the perfect compensation of the *3d-4d* shake-up feature whereas the *4p-5p* feature is still evident in the background subtracted spectrum.

The multi-electronic excitations in the krypton gas and liquid system have been studied extensively[22, 23] and the various contributions to these features calculated. Figure 6 illustrates the difficulty of assuming complete transferability of the bound electronic states from a gaseous electronic excitation model to a condensed phase. This is evident in the incomplete cancellation of the *4p-5p* shake-up feature at 2.2$Å^{-1}$, on gas background subtraction. In contrast the *3d-4d* shake-up feature at 5.5$Å^{-1}$ is almost perfectly compensated. This difference can be rationalised by an examination of the radial extent of the appropriate krypton and oxygen electron wave functions.

In Figure 7, we show the ground state wave functions with the oxygen *2p* wave function offset by 3.8Å i.e. the atomic separation that would be found when the gas is dissolved in water. It is immediately clear that whereas the krypton *4p* wave function overlaps significantly with the oxygen *2p*, the krypton *3d* state does not, being more localised about the atomic nucleus. Hence when attempting to correct the experimental krypton in water data for the parasitic electronic excitations visible in the gas phase data, it is only possible to obtain a reasonable compensation for the electronic states that are not highly perturbed by the atomic environment present in the condensed phase. If a more precise determination of the likely effects is to be performed the wave functions ought to be calculated allowing for the existence of the Kr *1s* core hole with the concomitant excitation of the *3d* or *4p* level. However, for the purposes of this illustration, it is not necessary.

Fortunately, with regards to the structural analysis of the current system, krypton in water, the bulk of the structural signal to be analysed is in the EXAFS region of the absorption spectrum i.e. k>2.5$Å^{-1}$ (see Figure 5). The poorly corrected *4p* transition region, thus has little impact on the final result.

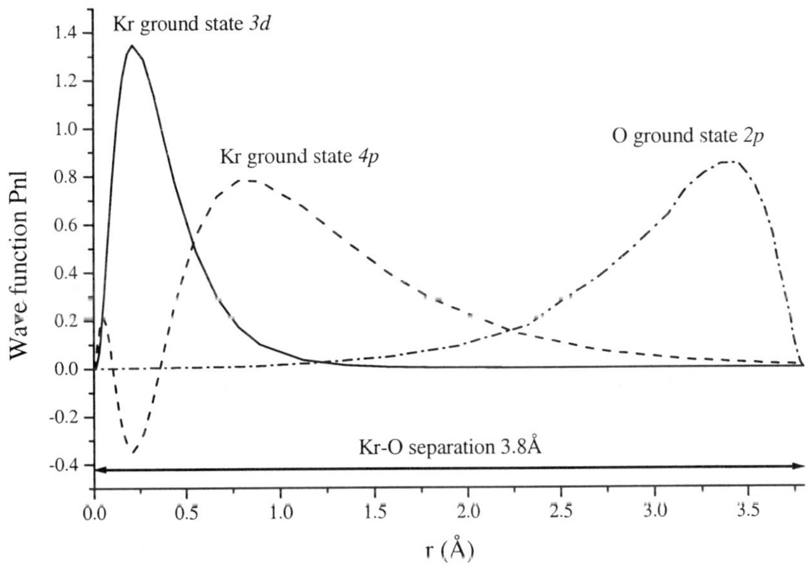

Figure 7: Illustration of the ground state wave functions calculated for Kr *3d*, Kr *4p* and O *2p*. The oxygen wave function has been displaced by 3.8Å, to represent the structural separation that is found when krypton is solvated by water molecules. The overlap between the O *2p* and Kr *4p* wave functions is obvious and in marked contrast to the situation found for the Kr *3d* orbital.

5. STRUCTURAL INSIGHTS GAINED INTO THE NATURE OF THE HYDROPHOBIC HYDRATION OF NOBLE GASES

Following the initial demonstration of the power of the X-ray absorption spectroscopy technique to address the principal structural questions[18], a considerable amount of new insight has been gained relating to the structural nature of the hydration of ideal hydrophobic moieties. The effect of temperature in the range from 277K to 368K was studied[19], and the results clearly demonstrated that a purely hydrophobic hydration shell is very sensitive to thermal disruption. This experimental finding contrasted considerably with simulation results that suggested that the hydrophobic hydration shell would be thermally robust in this range[24]. The disparity between experimental results and classically simulated models highlights the importance of performing direct structural measurements in this field of research. Hydrophobic phenomena, as mentioned in the introduction, most often arise in systems where there is a fine balance between enthalpic and entropic thermodynamic components. At the present time, classical parameterisations of atomic and molecular interaction potentials are not yet refined enough as to be able to reproduce accurately the phase behaviour and intricate structural aspects of these chemical phenomena. Another example that illustrates how the experimental results confounded conventional expectations of non-polar solute hydration is in the response of the hydrophobic hydration shell to applied pressure. The pressure dependence of hydrophobically driven processes such as protein folding and unfolding, has resulted in considerable simulation-based investigations of these structures[25,26]. Initial considerations suggested that some kind of local structural perturbation would result as pressure was applied in the kilobar range, and which has been supported by some simulation results[26]. This perhaps could then explain why applied pressure can increase parameters such as noble gas solubility. Surprisingly a systematic experimental study of the effect of applied pressure in the 150bar to 1kbar range on the krypton gas hydration shell[20] showed this initial assumption to be false. If the effect of pressure in this range is indeed to modify the structure of the liquid mixture, it does not do this in the region of a purely hydrophobic hydration shell. This structure, in marked contrast to its temperature sensitivity, possesses a highly resilient nature with respect to applied pressure. That modelling based methods can capture the broad structural picture is not in dispute. Rather the experimental findings simply serve to caution against over reliance on the precise of details and phase positioning of temperature and pressure dependent phenomena observed in simulation based investigations of hydrophobic systems.

A key result from this project came with the comparison of the local structure of the solid state clathrate hydrate of krypton and its liquid state hydration shell. The aim of the experiment was to directly test the validity of the long-standing and widely accepted "iceberg" model of hydrophobic hydration. Though initially put forward as a speculative means to explain the observed thermodynamic characteristics of mixtures of non-polar solutes with water, this idea of an "ice-like" enhancement of the water network solvating the extreme form of non-polar solutes has over the years grown to the status of an accepted truth. In particular, the idea that the structural enhancement follows a template of a clathrate hydrate has become particularly popular in the area of biochemistry[5]. The X-ray absorption technique provided an ideal means by which this hypothesis could be tested and the results showed quite clearly that the liquid state hydration structure is considerably more disordered than the clathrate ice structural form[21]. This finding suggests very strongly that the structural source of the negative entropic contribution that is observed on hydrating a non-polar solute must be sought elsewhere in the liquid mixture. This conclusion has recently received tentative support from an isotopic substitution neutron scattering study of an amphiphilic alcohol–water solution[27]. That study identified a previously unobserved *intermediate* range structural perturbation in the solvent water network.

6. CONCLUSIONS

In this paper we have shown how modern techniques in X-ray absorption spectroscopy analysis have enabled us to perform a detailed investigation of the hydration shell structure of a non-polar solute. The radial distribution function analysis scheme for X-ray absorption analysis is widely applicable to both liquid and non-crystalline systems. Though still under development this technique has provided a rational means by which disordered systems EXAFS analysis may now be pursued. The scientific results obtained for the krypton in water system have rationalised a great deal of the speculative ideas relating to the structural nature of hydrophobic hydration. The identification of the presence or absence of the local structural perturbations now motivates considerable further studies to enable us to replace the hypotheses that have been shown to be inadequate in the light of experimental results.

ACKNOWLEDGEMENTS

We would like to acknowledge the invaluable analytic and theoretical contributions to this work made by Adriano Filipponi. All the experimental work discussed in this paper was performed at Beamline BM29 at European Synchrotron Radiation Facility under the competitive proposal mechanism. We wish to acknowledge the skilled technical assistance provided by Mr. Sebastien Pasternak and Mr. Ralf Weigel, the beamline technicians who designed and built much of the pressure cell equipment.

REFERENCES

1. W.Blockzijl and J.B.F.N.Engberts, *Angew. Chem. Int. Ed. Engl.* **32**, 1545 (1993)
2. W.Kauzmann, *Adv. Prot. Chem.* **14**, 1 (1959)
3. B.Guillot, Y.Guissani and S.Bratos, *J. Chem. Phys.* **95**, 3643 (1991)
4. H.S.Frank and M.W.Evans, *J. Chem. Phys.* **13**, 507 (1945)
5. D.W.Urry, *Angew. Chem. Int. Ed. Engl.* **32**, 819 (1993)
6. R.P.Kennan and G.L.Pollack, *J. Chem. Phys.* **93**, 2724 (1990)
7. A.Filipponi, M.Borowski, D.T.Bowron, S.Ansell, A.Di Cicco, S.De Panfilis and J-P.Itié, *Rev. Sci. Instr.* **71**, 2422 (2000)
8. D.E.Sayers, E.A.Stern and F.W.Lytle, *Phys. Rev. Lett.* **27**, 1204 (1971)
9. A.Filipponi *J. Phys.: Condens. Matter* **13,** R23 (2001)
10. A.Filipponi *J. Phys.: Condens. Matter* **6**, 8415 (1994)
11. A.Filipponi and A. Di Cicco, *Phys. Rev. B*, **51** 12322 (1995)
12. G.Bunker, *Nucl. Instru. Methods,* **207**, 437 (1983)
13. A.Filipponi, A.Di Cicco and C.R.Natoli, *Phys. Rev. B* **52**, 15122 (1995)
14. A.Filipponi and A. Di Cicco *Phys. Rev. B* **52**, 15135 (1995)
15. A.Filipponi and A. Di Cicco *Task Quarterly* **4**, 575 (2000)
16. R.D.Broadbent and G.W.Neilson, *J. Chem. Phys.* **100**, 7543 (1994)
17. B.Guillot and Y.Guissani, *J. Chem. Phys.* **99**, 8075 (1993)
18. A.Filipponi, D.T.Bowron, C.Lobban and J.L.Finney, *Phys. Rev. Lett.* **79**, 1293 (1997)
19. D.T.Bowron, A.Filipponi, C.Lobban and J.L.Finney, *Chem. Phys. Lett.* **293**, 33 (1998)
20. D.T.Bowron, R.Weigel, A.Filipponi, M.A.Roberts and J.L.Finney, *Mol. Phys.* **99**, 761 (2001)

21. D.T.Bowron, A.Filipponi, M.A.Roberts and J.L.Finney, *Phys. Rev. Lett.* **81**, 4164 (1998)
22. S.J.Schaphorst, A.F.Kodre, J.Ruscheinski, B.Crasemann, T.Åberg, J.Tulkki, M.H.Chen, Y.Azuma and G.S.Brown, *Phys. Rev. A*, **47**, 1953 (1993)
23. A.Di Cicco, A.Filipponi, J.P.Itié and A.Polian, *Phys. Rev. B*, **54**, 9086 (1996)
24. N.T.Skipper, *Chem. Phys. Lett.* **207**, 424 (1993)
25. G.Hummer, S.Garde, A.E.Garcia, M.E.Paulaitis and L.R.Pratt, *Proc. Natl. Acad. Sci. USA* **95**, 1552 (1998)
26. P.L.Chau and R.L.Mancera, *Mol. Phys.* **96**, 109 (1999)
27. D.T.Bowron, A.K.Soper and J.L.Finney, *J. Chem. Phys.,* **114**, 6203 (2001)

HIGH RESOLUTION DIFFUSE X-RAY SCATTERING BY PROTEIN CRYSTALS – FROM *hkl* TO 000

Richard J. Matyi, Gabrielle G. Long and Heather M. Volz [*]

1. INTRODUCTION

Much of our knowledge of the structure of protein and related macromolecules has come from X-ray diffraction analyses of protein crystals. Typically a protein is crystallized; the crystal is irradiated, and the angular positions and intensities of the diffracted beams are recorded. The directions of the diffracted beams depend on the positions of the atoms within the unit cell, while the intensities of the diffracted X-ray beams are determined by the types of atoms that occupy specific sites within that unit cell. By measuring the intensities of a large number of X-ray reflections and comparing the measured data set to that calculated from an assumed structure, it has been possible to describe in detail the atomic structures of an increasing number of protein and other biologically important molecules.[1,2]

For this method to work, however, the protein molecule must be coaxed into a crystalline form.[3,4] Only when the molecules are organized in a regular array with sufficient long-range order can the amplitudes of the radiation scattered by each of the hundreds of atoms in the structure be added together to form a diffraction pattern. In the case of inorganic materials, it is relatively easy in most cases to form a crystal with sufficient long range order to permit detailed X-ray analyses. This is not the case with protein molecules, since their high molecular weight and complex chemical structure make it difficult to attain the high degree of order that is usually associated with crystalline solids.

The geometric difficulty associated with protein crystallization is only complicated by the complex physical chemistry of protein solutions. Factors such as protein concentration, solution pH, salt concentration, and temperature all interact in a complex fashion during the crystal growth process. Small solvent molecules (*e.g.* water) in concentrations that can exceed 50% are typically located in the interstitial spaces between the protein molecules. A further complication arises due to the heterogeneous nature of the protein molecules themselves. In most cases it is difficult to obtain a strictly monodisperse collection of protein molecules, a distribution of molecular species is the typical result of protein synthesis. This lack of molecular purity only increases the difficulty in growing high quality crystals with good structural quality.

* Richard J. Matyi and Gabrielle G. Long, National Institute of Standards and Technology Gaithersburg, MD 20899. Heather M. Volz, Materials Science Program University of Wisconsin, Madison, WI 53706 (current address: IBM Microelectronics, Essex Junction, VT 05452)

From Semiconductors to Protein, edited by S.J.L. Billinge and M.F. Thorpe
Kluwer Academic/Plenum Publishers, 2002

The lack of structural perfection in protein crystals has a very serious effect on subsequent analyses of molecular structure, since the displacements caused by defects reduce the intensities of the diffracted beams generated by the crystal. The presence of structural defects thus reduces the resolution, or the minimum spacing between crystallographic planes, of the X-ray analysis. The fewer reflections measured, the poorer the resolution, and the lower the accuracy in determining atomic positions within the protein molecule.

One of the signatures of the presence of structural defects is the re-direction of scattered intensity in an X-ray analysis – the presence of defects diverts intensity away from the nominal peak positions into off-peak angular directions. This off-peak scattering is strongly affected by the magnitude of the lattice distortions generated by a given defect and exhibit a well-known reciprocal relationship where the angular displacement of the scattering is inversely proportional to the spatial extent of the defect's strain field. By characterizing the off-peak scattering in both high resolution (in the immediate vicinity of a Bragg peak) and in low resolution (between fundamental peaks) one should be able to develop a picture of the types of structural defects that are present.

Here we will discuss experimental methods characterized by high angular resolution that are useful for probing the scattering that is generated by structurally defective protein crystals. Our general approach will be to use the scattering in the immediate vicinity of a reciprocal lattice point as a diagnostic of the structural perfection. Two distinct methods for doing this have been examined – high resolution triple axis X-ray diffraction (HRTXD) analysis of the diffracted intensity in the vicinity of a high order (*hkl*) Bragg reflection, and ultra small angle X-ray scattering (USAXS) to characterize the diffuse intensity generated by structurally defective crystals of hen egg white lysozyme (HEWL) near the origin of reciprocal space (*i.e.* the main beam direction). The ultimate goal of this work will be to provide structural feedback to the crystal growth process and help guide subsequent modifications to improve the structural quality of protein crystals.

2. DEFECTS IN PROTEIN CRYSTALS – LONG-RANGE AND SHORT-RANGE DISTORTIONS

When discussing structural defects in protein crystals it is useful to classify them in terms of the spatial extent of their distortion fields. Here we will be concerned primarily with defects that give rise to relatively *long-range distortions*; such defects are often referred to as *extended defects*. Typical extended defects that have been observed in protein crystals include growth striations,[5,6] dislocations,[7-10] twins,[11] and both inclusions and stacking faults.[12] One of the most well known extended defects that can occur in any crystal (and especially protein crystals) is *mosaic structure*.[13] Even when single crystal growth is favored thermodynamically, kinetic factors such as material transport to the growth interface and molecular transfer from the liquid into the solid phase can make it difficult for the crystallization process to occur uninterrupted over large distances. Instead, it is often more favorable for the crystal to form mosaic blocks, or regions with a relatively high degree of internal order that are slightly misoriented with respect to each other. The mosaic spread, or average misorientation between two blocks, is an important measure of the degree of structural perfection in a crystal, protein or otherwise.

It is also worth mentioning that there is a wide array of defects that are characterized by distortions with a shorter range of spatial extent. In contrast with the nature of extended defects discussed above, we can refer to this second class as *localized* defects. It is well known that in even highly perfect crystals (inorganic and protein) there can be both static and dynamic perturbations in the periodic average structure. The displacements may arise from many sources, such as the acoustic and optical phonon modes in the crystal, positional or rotational disorder of individual molecules, or from missing (vacancy) or extra (interstitial) molecules.

In protein crystals the diffuse scatter that exists between reciprocal lattice points can be relatively strong due to the weak intermolecular bonding within the crystal and the natural difficulty in producing a highly perfect three dimensional packing with large and complex molecules. Despite the difficulty in fully understanding its origin, a number of investigations have studied long-range diffuse scatter in protein crystals; an extensive review has been published recently by Benoit and Doucet.[14] It should be noted that, in the case of protein systems, the analysis of intra-peak diffuse scatter is complicated by the presence of other factors, such as scattering by both the solvent and the glass capillary, Compton scatter, and fluorescence from the sample, solvent, and capillary. Unlike the case with inorganic crystals, it is difficult to experimentally remove all of these contributions to the diffuse background in a protein crystal analysis.

Despite the experimental challenges, significantly more work has been performed on the observation and analysis of long-range diffuse scatter generated by localized defects than corresponding studies of the intensity in the immediate vicinity of a reciprocal lattice point generated by extended defects. In the discussion given below we will therefore concentrate on the experimental aspects related to high resolution diffuse scattering analyses of structurally defective protein crystals.

3. HIGH RESOLUTION X-RAY DIFFRACTION METHODS

Although our principal concern involves the characterization of structural defects in protein crystals, the desire to relate the types and densities of growth-induced defects to the details of crystal growth processes is a phenomenon that bridges all classes of materials. In systems where high degrees of structural perfection can routinely be achieved (most notably in the case of single crystal semiconductor materials), methods based on high resolution[†] X-ray diffraction are routinely used to articulate the relation between the growth details and the resultant defect structure. Here, our purpose will be to demonstrate that the experimental techniques that were originally developed for highly perfect inorganic crystals can be successfully applied to organic crystals with a much lower degree of structural perfection.

3.1 High resolution double axis X-ray diffractometry

One of the most established methods for assessing the structural perfection in a single crystal is double axis X-ray diffractometry. In a conventional double axis diffractometer (schematically illustrated in Figure 1) an X-ray beam is first reflected from a highly perfect reference crystal (or crystals[‡]) in order to reduce its angular spread in the plane of the

[†] It is unfortunate that the term "resolution" is used by both the protein crystallography community and the defect characterization community to denote two completely different concepts. In defect characterization resolution refers to the *angular precision* at which an experiment can be performed, while to protein crystallographers it refers to the *spatial precision* to which a structure can be determined. When necessary, we will use the term *atomic resolution* to emphasize the protein crystallography sense.

[‡] Due to the proliferation of multiple crystal monochromators, the term "double axis diffraction" is now preferable to the more established term "double crystal diffraction"; the same logic applies for the same reason to the term "triple axis diffraction" discussed below.

diffractometer (the "horizontal" divergence of the beam). Typically there is little if any conditioning of the beam's vertical divergence (*i.e.* in and out of the plane of the diffractometer), and the intensity from the sample is recorded with a wide-open detector. The sample crystal is rotated through the angular range for a Bragg reflection to generate a "rocking curve" of a particular Bragg reflection.

At this point it is useful to discuss an important distinction in diffraction theory. Most treatments of X-ray diffraction presuppose that one is working at or near the kinematical limit. Under kinematic conditions, it is assumed that the incident X-rays interact only weakly with the crystal and that each X-ray photon experiences only a single elastic scattering event. Other approximations in kinematic diffraction are that the incident and diffracted waves do not interact, the crystal is small and imperfect, the crystal-to-detector distance is large, and the diffracted intensity is low. Alternatively, if the diffracting crystal is nearly perfect, it may be necessary to invoke the dynamical theory, where the incident and diffracted X-ray wavefields are assumed to interact strongly with the crystal and experience multiple scattering events. For example, electron diffraction is primarily dynamical due to the strong interaction between electrons and most materials. Both kinematical and dynamical diffraction effects can be seen in high-resolution diffraction.

According to kinematic theory, the reflection profile from a large crystal should be an infinitely narrow δ-function. In contrast, the theoretical peak breadth (denoted as the full width at half maximum or FHWM) determined from the dynamic theory of X-ray diffraction is typically on the order of arcseconds under most conditions. The dynamic peak breadth typically will be observed only if the diffracting crystal is both (1) relatively structurally perfect and (2) large in extent when compared to the extinction distance ξ defined as[15]

$$\xi = \frac{V_C}{r_o C |F_h| \lambda} \tag{1}$$

where C is the polarization factor (1 or $\cos 2\theta$ depending on whether the incident X-ray electric field is perpendicular or parallel, respectively, to the scattering plane and θ is the diffraction angle), V_C is the unit cell volume , r_o is the classical electron radius, $|F_h|$ is the structure factor amplitude, and λ is the X-ray wavelength. Because of the relatively small magnitude of the structure factor in protein crystals (due to the low atomic number of atoms such a oxygen and carbon) as well as the typically large unit cell, the value of ξ in crystals such as HEWL is rather large. Otalora *et al.* have estimated an average value for the structure factor of approximately 596 electrons[16]. Using this value with a unit cell volume of 239.7 nm^3 for tetragonal HEWL, a polarization factor $C = 1$ and 1.54Å CuKα radiation, we calculate an extinction distance of 0.9 mm. This agrees with estimates of the extinction distance given by Fourme *et al.*[17] and Helliwell.[18]

The dynamical rocking curve width in the symmetric transmission geometry is then

$$\Delta\theta = \frac{2|C||\chi_h|}{\sin 2\theta} = \frac{2\lambda}{\pi\xi \sin 2\theta} \tag{2}$$

where $\chi_h = -F_h r_o \lambda^2 / \pi V_C$. Using this expression for the HEWL (440) reflection we calculate an extremely narrow dynamic rocking curve FWHM of about 0.2 arcseconds. Again, this is in agreement with similar calculations by Fourme *et al.*[17] and Helliwell.[18]

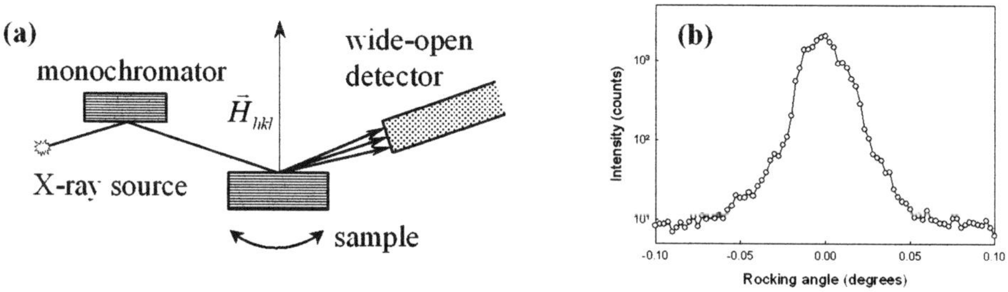

Figure 1. High resolution double axis X-ray diffraction. (a) typical experimental geometry; (b) (440) rocking curve from a typical (structurally defective) HEWL crystal.[21]

The rocking curves of most real crystals (and almost all protein crystals) rarely exhibit their theoretical FWHM but are instead broadened due to perturbations in crystal structure caused by the presence of defects. In a typical double axis analysis, the rocking curve from a nearly-perfect crystal shows the dynamical diffraction reflectivity characteristics of that crystal – that is, the characteristics of the Bragg peak. The mosaic spread of the crystal will be manifested by deviations in the shape and breadth of the profile from that predicted from dynamical theory. A small mosaic spread will merely broaden the peak, while a highly defective crystal may exhibit multiple peaks as individual mosaic blocks "light up" as they are rocked though the incident beam. Off-peak diffuse scattering due to other structural defects will be superimposed on top of the perfect (or nearly perfect) crystal reflection, making separation of the two difficult or impossible.

One of the earlier systematic analyses of mosaic spread in protein crystals was performed by Shaikevitch and Kam in 1981.[11] Using slit-collimated radiation from a sealed X-ray tube they reported a mosaic spread of about 60 arcseconds for a lysozyme crystal at $2\theta \approx 15°$. Recently the advent of synchrotron sources with intense radiation and low divergence has led to a number of other X-ray rocking curve analyses.[17-20] Rocking curve FWHM's as narrow as 8.3 arcseconds from the (776) reflection have been reported for lysozyme crystals grown in microgravity[20]; this value is far superior to the approximately 47 arcsecond FWHM observed in a similar crystal grown on earth. Figure 1b illustrates a more typical double axis rocking curve recorded from an HEWL crystal (denoted HEWL-A in this case[21]). The measured FWHM of the rocking curve is 77 arcseconds, a value that is similar to other rocking curves recorded HEWL crystals.[17,18]

Very recently, the issue of mosaic spread has been examined in detail by Borgstahl and coworkers.[22] Using a novel "superfine ϕ-slicing" data collection procedure, the mosaic spreads of 260 reflections from a protein crystal were measured; the mosaic spread was found to be anisotropic in the principal crystallographic directions. The FWHM values were also found to be very small in this work (less than 0.014° and usually much smaller), indicating the fact that the crystal studied was of high structural perfection.

3.2 High resolution triple axis X-ray diffractometry

As mentioned above, the diffracted intensity in a double axis experiment is recorded with a wide-open detector. However, further angular conditioning can be achieved if the diffracted beam is re-diffracted by an analyzer crystal before it is detected. The effective angular acceptance range is thus determined by the intrinsic reflection range of the analyzer crystal – typically several arcseconds, as opposed to several degrees as is the case in double axis measurements with a wide-open detector. In triple axis diffraction, the inclusion of the

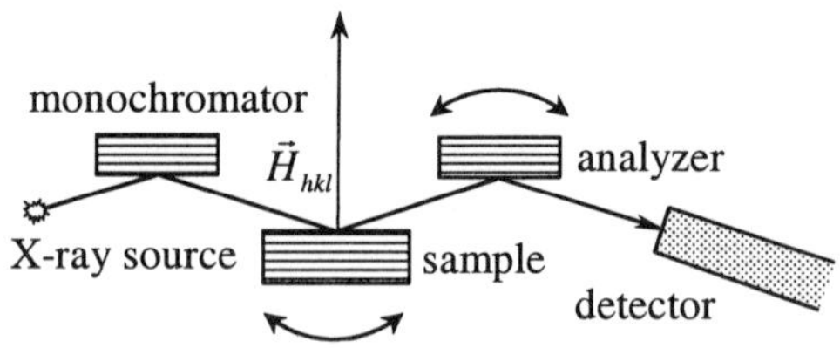

Figure 2. High resolution triple axis X-ray diffraction geometry

analyzer crystal permits the intensity generated by defects to be discriminated from the perfect crystal intensity, thus leading to a significant increase in the amount of information that can be extracted from an analysis. The combination of a highly perfect analyzer crystal with a monochromator crystal of similar structural perfection is necessary for this to be possible; while this arrangement permits reciprocal space to be mapped out with high resolution, it has the liability of reducing the total X-ray flux that arrives at the detector. As the structural perfection of the sample crystal decreases, the angular spread of its reflectivity increases, thus reducing the net signal from the sample.

High resolution triple axis diffraction can be understood through reference to Figure 3. Angular deviations of vectors shown in the Figure are controlled in a triple axis experiment to a level that is commensurate with the diffraction characteristics of perfect crystal optics. In other words, very small changes in diffracted intensity in close proximity to a Bragg reflection, or reciprocal lattice point, are examined with this technique. As mentioned above, the angular breadth of a Bragg reflection from an ideally perfect crystal is on the order of arcseconds. Hence for an experiment to be considered high resolution, (1) the angular distribution in the incident wavevector ($\vec{S}_o$) direction, (2) the angular distribution in the diffracted wavevector ($\vec{S}$) direction, and/or (3) the orientation of the reciprocal lattice vector, $\vec{H}_{hkl}$ must be controlled with optics to arcsecond precision.

Defects introduced into a crystal will alter the characteristics of a reciprocal lattice vector (and hence the intensity distribution about a reciprocal space point) in two principal ways as shown in Figure 3. First, compositional variations and/or strain will locally change the lattice parameters of the sample. This will be manifested in a change in the length of the

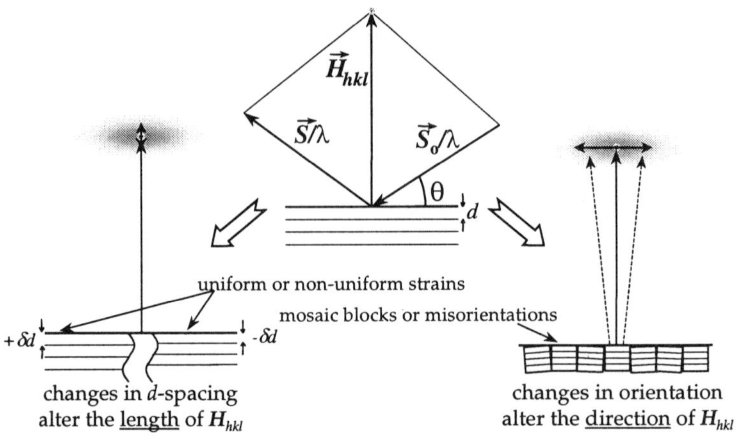

Figure 3. The basic kinematic diffraction geometry (center) and the effects of strains (left) and misorientations (right) on a reciprocal lattice vector $\vec{H}_{hkl}$

reciprocal lattice vector (equal to $1/d_{hkl}$), which will result in a redistribution of intensity away from the exact Bragg condition in the θ/2θ direction (parallel to the direction of the reciprocal lattice vector in a symmetric geometry as shown in Figure 3). Secondly, mosaic spread in the sample will alter the angular distribution of $\vec{H}_{hkl}$ and will create extra intensity away from the Bragg angle in the direction that is perpendicular to the reciprocal lattice vector.

The off-peak scattering that is associated with both factors is usually graphed as a plot of equal-intensity contours (often called a "reciprocal space map") with one axis representing a rotation in ω and the other representing movement in θ/2θ. The contours represent equal diffracted intensity regions in reciprocal space, and the axes are converted into units of reciprocal distance. Small deviations from the exact Bragg condition parallel and perpendicular to the reciprocal lattice vector are often represented by $\boldsymbol{q}_z$ and $\boldsymbol{q}_x$ defined as

$$\boldsymbol{q}_z = \beta \cos\theta_B / \lambda ; \quad \boldsymbol{q}_x = (2\alpha - \beta)\sin\theta_B / \lambda \tag{3}$$

where α and β are the angular deviations (in radians) of the sample and analyzer crystals, respectively, from the exact Bragg angle θ_B.

Figure 4 shows a typical reciprocal space maps from a HEWL crystal.[21] In the Figure one can observe (1) a faint distribution of intensity in the transverse direction (i.e., horizontally in the Figure), (2) a inclined, relatively intense distribution extending from the upper left and through the reciprocal lattice point towards the lower right, and (3) a weaker intensity distribution also extending from upper left to lower right in the figure, but at a steeper angle. This third intensity feature is probably the signature of parasitic scattering from our X-ray optical system superimposed on a possible analyzer streak. The other, more intense inclined feature is believed to be an inclined surface streak, which arises from the truncation of the crystal at the surface.[21]

The horizontal spread in intensity is believed to be the mosaic distribution in the HEWL crystal. Interpreted in this way, the mosaic spread of the crystal would be approximately 150 to 180 arcseconds. It is important to note that this value corresponds to the maximum extent of the transverse off-peak scatter recorded from this crystal. In other words, it represents the full width at approximately 0.2% of the maximum intensity. Due to the logarithmic intensity scale of the reciprocal space maps, the full width at half maximum would correspond to a

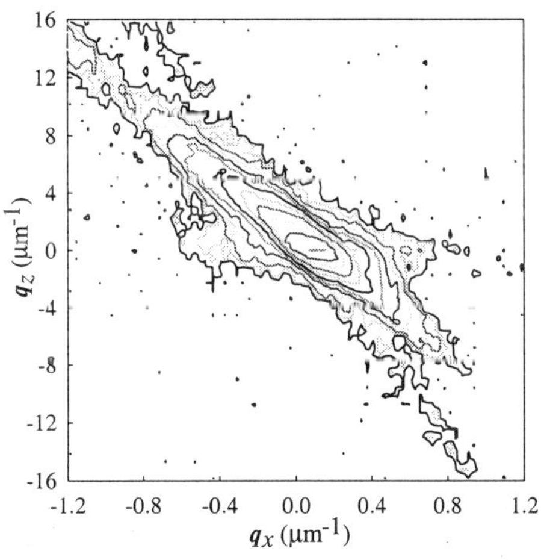

Figure 4. High resolution triple axis reciprocal space map from crystal HEWL-A.[21]

level that is between the second and third contour levels down from the maximum intensity and thus represents a value for the mosaic spread comparable to others in the literature.

X-ray topography, where a zero-magnification image is formed by the diffracted intensity from a sample crystal, deserves mention at this point, since it has also been used to explore the nature of grown-in defects in protein crystals. X-ray topographic studies of protein crystals have been reported recently by Fourme *et al.*,[17] Izumi *et al.*,[23] and Stojanoff *et al.*[24,25] Virtually all studies have employed synchrotron radiation; both perfect crystal monochromated topographs[17,24,25] and white beam topographs[23-25] have been reported. Stojanoff and coworkers also showed that a decrease in the defect density as observed through topographic imaging did correlate with an increase in resolution in subsequent crystallographic analyses[25]. More recent X-ray topographic studies by Thorne[26-28] have examined the effects of cryocooling, dehydration, and changes in the solution pH and salt concentration on the defect structure in tetragonal lysozyme. While these analyses clearly revealed alterations in the otherwise highly perfect crystal structure, no quantitative assessments were made. This points out one limitation of topographic imaging: it is primarily a qualitative characterization tool. Topography is very useful at revealing the presence of defects, but with a few notable exceptions it is very difficult to extract detailed information regarding the number or extent of specific defects.[29-31] In addition, topographic methods become less successful in materials with very high defect densities, where the images of individual defects may overlap and thus may not be interpretable.

3.3 Small angle X-ray scattering

The preceding discussion has been presented under the assumption that a protein can be coaxed into a form that is sufficiently ordered to deserve the designation "crystal". Sadly, in many cases, a crystal that appears by optical evaluation to be promising for subsequent analyses turns out to be so poorly diffracting that no useful structural data can be extracted. Further study is then not possible, leading to a depressing state in which the protein crystal grower may have limited (or no) options available to guide the development of a crystallization regimen.

Small angle X-ray scattering (SAXS) may provide an alternate approach. As shown in Figure 5, in SAXS an X-ray beam that is highly collimated (both in angle and spatially) is directed at a sample, and the scattering in the direct vicinity of the transmitted beam is examined. Note that the transmitted beam direction defines the 000 point in reciprocal space; hence SAXS measures the intensity distribution around this 000 point, in contrast with high resolution reciprocal space mapping discussed earlier where the intensity near a higher order *hkl* reciprocal lattice point is examined. The SAXS intensity from an inhomogeneous sample arises from spatial variations in electron density:

$$I(\mathbf{s}) = (\rho - \rho_0)|\Sigma(\mathbf{s})|^2 \tag{4}$$

where the scattering vector $\mathbf{s} = 2\sin\theta/\lambda$, and the average density of the scattering medium is ρ while the scattering objects have a density $\rho - \rho_0$. $\Sigma(\mathbf{s})$ is the Fourier transform of the form factor $\sigma(x)$ of the scattering objects (defined by unity within an object and zero outside it). At small angles, the transform of the form factor can be written as[32]:

$$\Sigma(\mathbf{s}) = \int_V \exp(2\pi i\mathbf{s}\cdot\mathbf{x})\,dv_x \approx \int \sigma(x)\,dx - 2\pi^2 s^2 \int x^2\sigma(x)\,dx \tag{5}$$

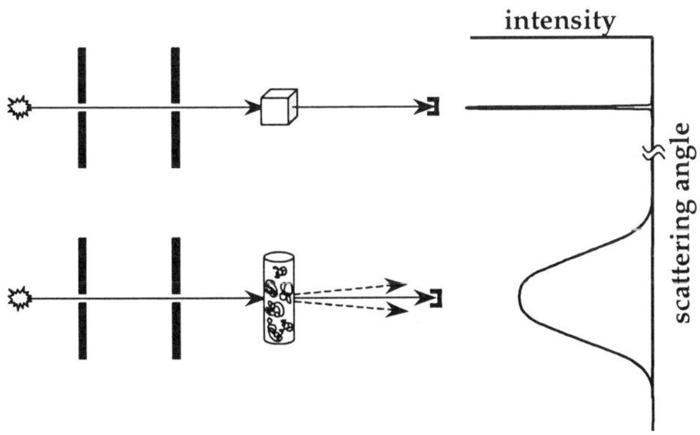

Figure 5. Schematic technical illustration (left) and anticipated observed result (right) of a small-angle X-ray scattering experiment. (a) large perfect crystal; (b) protein molecules in solution.

Consideration of Eq. (5) reveals that a "large" perfect crystal should generate a compact intensity distribution in reciprocal space that should not be observable. In other words, a homogeneous protein crystal of macroscopic size with high levels of structural perfection would be expected to generate zero-order scattering that would be immeasurably small in its extent.[32] In contrast, highly imperfect crystals would be expected to generate a SAXS signal.

Further utility of SAXS lies in the fact that variations with respect to the average electron density can be detected at large characteristic length scales that may range up to the micrometer (μm) range. Moreover, the material under analysis does not have to be crystalline; in fact, useful information can be extracted from amorphous or other non-crystallizing systems. For example, SAXS has been widely used for the study of protein molecules and their complexes in solution. The size of a molecule and the shape that it attains in solution will alter the angular dependence of the small-angle scattering. Information on the molecule's shape can be derived from the radius of gyration, a value that is obtained from the limiting slope of a plot of the logarithm of the intensity versus the square of the scattering angle q^2 as q approaches zero.[32,33]

SAXS is a powerful technique for the analysis of colloidal and non-crystalline systems, so it is legitimate to ask: how successful might SAXS be as a tool for characterizing highly defective protein "crystals" when that crystalline order is sufficiently poor that all reflections other than the 000 are extinguished? For a 0.1 mm crystal or larger, this scattering would be far too small to be observed and would be lost in the transmitted beam. Even if the protein crystal lattice is distorted due to fluctuations in the positions of atoms (the so-called "paracrystal" model[32]) there will also be no observable effect on the SAXS intensity – the breadth of a particular *hkl* reciprocal lattice point will increase with increasing distance from the origin, but the 000 point itself will be unaffected.

Now consider the case of a very highly defective protein "crystal". Let us assume that the structure consists of small (under 100 nm) mosaic blocks that are grossly misoriented (by, for instance, several arcminutes) with respect to each other; we will also assume that the interstitial spaces between individual blocks are filled with solvent (e.g. water). It is doubtful that an entity of this sort would be very useful for a structural determination, but based its contrast with the on the above discussion on relatively perfect crystals, *we would expect it to generate observable intensity in an SAXS experiment.* Specifically, the mosaic blocks in a highly defective crystal described above should behave as a dense ensemble of independent scatters that, we believe, will generate detectable small-angle scattering.

Alternatively, it is possible that the defects (specifically, misfit dislocations) at the interface between mosaic blocks can generate SAXS scattering; such scattering has been predicted[34] and experimentally observed[35] in metallic single crystals. It should be noted that the amplitude scattered by a dislocation is directly proportional to the magnitude of the dislocation's Burgers vector.[34,35] Since the Burgers vector of a dislocation in a protein crystal is one to two orders of magnitude greater than that that found in a metallic crystal, the scattering from dislocations in highly defective protein crystals should be relatively intense, particularly if a synchrotron-based state-of-the-art SAXS system is used.[36]

Although it appears that there have been no prior examinations of protein crystals (structurally perfect or otherwise) using SAXS, there is good reason to have optimism for this approach. In a classic study performed in 1961, Luzzati and coworkers[37] observed significant SAXS intensity from dilute solutions of lysozyme in water. Even at concentrations of only 1%, the difference in electron density between the protein (425 electrons nm^{-3}) and water (335 electrons nm^{-3}) was sufficient to allow quantitative analyses of both the molecular weight and the dimensions of the lysozyme molecule to be made with a sealed-tube SAXS system.

4. EXPERIMENTAL ASPECTS OF HIGH RESOLUTION X-RAY ANALYSES

4.1 Protein crystals preparation

The HEWL crystals examined by high resolution X-ray methods discussed here were grown by microdialysis. Lysozyme starting material was obtained from Sigma Chemical[§], and the concentration used was 40 g/L in water. Dialysis was performed at pH 4.7 using 0.020 M sodium acetate and 0.850 M sodium chloride. Crystals of appreciable size typically grow within two days. The HEWL crystals were tetragonal (space group $P4_32_12$, a = 79.3Å, c = 38.1Å) with {101}-type facets. Each optically transparent crystal was mounted with a small amount of mother liquor in a glass capillary (0.5 mm to 1 mm diameter), whose ends were then sealed with wax before mounting with modeling clay on an eucentric goniometer. The HEWL crystal's orientation matrix was determined on a Syntex $P2_1$ diffractometer (upgraded with Bruker-Siemens *P*3 software). Once the orientation matrix had been solved, the specimen was then transferred to the high resolution triple axis diffractometer. Reflections were chosen for reciprocal space mapping based on their intensities on the Syntex diffractometer. Since low intensities were a problem, reflections with relatively low 2θ values were chosen instead of the low *d*-spacing reflections that are normally of most interest to protein crystallographers.

4.2 Experimental approaches to laboratory-based high resolution X-ray analyses

There are several advantages to being able to perform high resolution diffraction analyses of protein crystals in a conventional laboratory environment as opposed to a synchrotron radiation source. Ideally, if an X-ray technique such as double axis diffraction could be established in the protein community as a method to rapidly test protein crystal quality using the same X-ray equipment that crystallographers already have in their home

---

§ Certain commercial equipment, instruments, or materials are identified in this paper to foster understanding. Such identification does not imply recommendation or endorsement by the National Institute of Standards and Technology, nor does it imply that the materials or equipment identified are necessarily the best available for the purpose.

laboratories, systematic studies into producing higher quality crystals could be undertaken more regularly. Ultimately, high resolution analyses could lead to a better understanding of the conditions that limit structural quality in protein crystal growth. Characterization of defects would lead to better methods of reducing their densities, thereby increasing crystalline quality and improving the structural determinations that are central to the study of protein function.

There are problems with laboratory-based analyses, however. When compared with inorganic or semiconductor brethren, protein crystals exhibit much higher defect densities, and they scatter X-rays much more weakly. Although the experimental work presented below will show that the first concern can be accommodated, the intensity issue turns out to be a significant problem. The reason for this becomes apparent when one considers the dependence of the diffracted intensity in the kinematic limit:

$$I(hkl) \propto \frac{|F_h|^2}{V_C^2} e^{-2M} \qquad (6)$$

where e^{-2M} is the Debye-Waller temperature factor. In very general terms, inorganic crystals are relatively easy to study with X-rays – they have small unit cells consisting of relatively high-Z elements that scatter X-rays strongly, and they typically exhibit strong intermolecular bonding. In contrast, organic materials are far more difficult to analyze with X-rays due to the fact that large organic molecules typically crystallize into large unit cells (often hundreds of atoms in a basis). Moreover, elements such as C, H, O, N have fewer electrons than metallic or semimetallic elements, while the weak intermolecular bonding and solvent environment that are commonly present in organic (and especially protein) crystals lead to large Debye-Waller factors. As a result, high angular resolution measurements on protein crystals, using extensive beam conditioning and performed in a laboratory environment, will need some modification.

Figure 6 illustrates the system used in the present work to perform high resolution X-ray analyses of protein crystals in a laboratory setting.[21,38] Cu Kα X-rays were obtained using a Rigaku RU-200 rotating anode generator configured in a point focus geometry (0.5 mm × 10

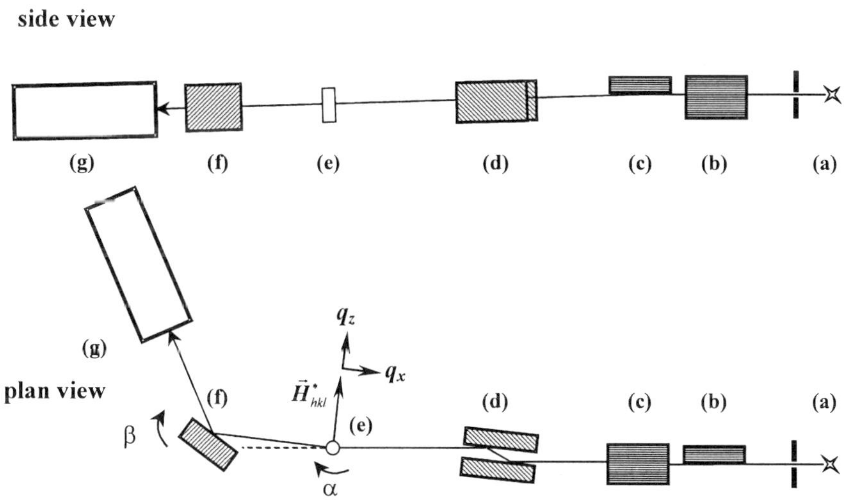

Figure 6. Schematic illustration of the high resolution X-ray diffraction system used for protein crystal characterization (letters refer to components described in the text).

mm focus). The X-rays from the rotating anode source (a) were conditioned by a pair of graded parabolic X-ray mirrors (b,c).[39] The crossed mirrors, which were situated in a novel X-ray optical system,[40] produced a quasi-parallel X-ray beam (nominally three arcminutes horizontal and vertical divergence) with an approximate beam dimension of 1 mm × 1 mm. The use of this crossed-mirror beam conditioner resulted in a significant increase in the incident beam X-ray flux as well as providing spatial collimation of the incident beam.

Following the crossed parabolic mirrors, the incident X-ray beam was further refined by diffraction from a grooved silicon crystal (d).[41] The two (220) reflections served to condition the horizontal divergence of the X-ray beam from three arcminutes to several arcseconds. The conditioned incident beam was then diffracted in transmission geometry from a protein crystal (e). The sample was carried on a Huber 511/424 four-circle diffractometer modified to permit a minimum step size of 0.00125° (4.5 arcseconds). The diffracted beam was analyzed by a single (004) reflection from a high quality Ge (001) single crystal (f) before being recorded with a scintillation detector (g).

Reciprocal space maps of the HEWL reflections were obtained by performing $\theta/2\theta$ scans with an increment in the sample angular position ω between scans. A count time of 5 seconds at each point was used at a generator power of 50 kV and 200 mA; each full data collection scan lasted about one day. The reciprocal space maps were generated as equal-intensity contours with the intensity placed on a logarithmic scale. The contours were plotted with four intensity levels per decade of intensity (i.e. an increment of $10^{0.25}$ s^{-1} between contours) with a minimum contour level of $10^0 = 1$ s^{-1}. Further analysis involved extracting the intensities at $\theta/2\theta$ values corresponding to the peak of the reciprocal space map. When these constant $\theta/2\theta$ data were plotted as a function of ω they represented intensity scans transverse to a reciprocal lattice vector direction. Measurement of the full width at half maximum (FWHM) and the full width at 1% of maximum (FWC^{-1}M) were used to describe the angular breadth of the Bragg peak and the off-peak scattering.

4.3 Small-angle X-ray scattering analyses

Small angle X-ray scattering data were collected using the NIST ultra-small angle X-ray scattering (USAXS) beamline (33-ID) at the Advanced Photon Source at Argonne National Laboratory. The Bonse-Hart design (shown schematically in Figure 7) operates with a six reflection, channel-cut monochromator and a two crystal, six-reflection analyzer to achieve a minimum $q = (4\pi/\lambda)\sin\theta = 0.00015$Å^{-1} or 1.5 μm^{-1}. The system was used in conjunction

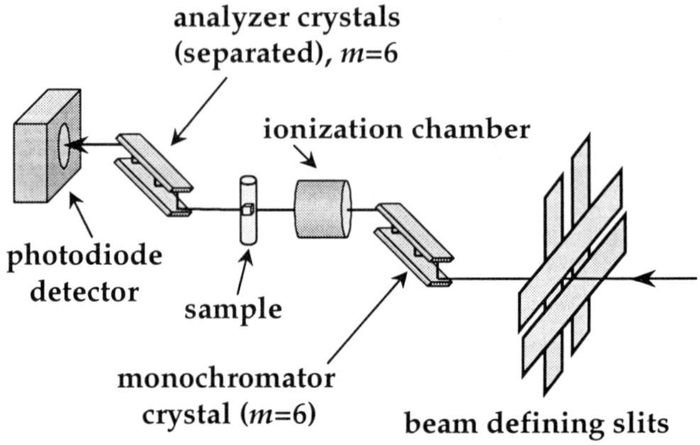

Figure 7. Schematic illustration of the high resolution small-angle X-ray scattering system used for protein crystal characterization.

with a double crystal monochromator (for photon selection) and two mirrors (for harmonic rejection). The photon energy chosen for this analysis was 10 kV with a flux in excess of 10^{12} s^{-1} in the beam cross section. Horizontal and vertical slits before the monochromator restricted the beam dimensions to be a 0.5 mm × 0.5 mm, or smaller than the protein crystal under study. A photodiode detector with an effective dynamic range of approximately 10 orders of magnitude was used. All data were corrected for parasitic scatter generated by the air/liquid/glass system that constituted the sample capillary.

5. HIGH RESOLUTION TRIPLE AXIS RECIPROCAL SPACE MAPPING OF LYSOZYME CRYSTALS

Figure 8 and 9 show a representative high resolution diffraction data recorded from a lysozyme crystal sample that we denoted HEWL-D.[38] The three different reflections examined from this crystal (the (440), (160) and $(1\bar{2}0)$ reflections) were chosen based on their strong intensities that they exhibited while being oriented. The FWHM of the rocking curves from these reflections were narrower (45, 48, and 50 arcseconds, respectively, for the (440), (160) and $(1\bar{2}0)$ reflections), suggesting a higher crystalline quality than that seen in earlier work.[21]

Examination of the (440) reciprocal space map shows that two intensity features can be clearly seen. As discussed previously, the horizontal streak arises from the mosaic spread in the crystal. In addition, an intense, angled streak of X-ray intensity attributed to the extremely high incident intensity that was generated by the crossed mirror beam conditioner is again seen. Interestingly, there is an extra off-peak intensity feature (indicated by an arrow) located at a deviation from the exact Bragg condition (q_x) of approximately +1 μm^{-1}. This type of feature is frequently observed in reciprocal space maps from defective semiconductor materials and represents a rather large mosaic block whose size starts to approach that of another grain in the crystalline microstructure. Notice that this type of detail could not easily be discerned from the double axis rocking curve in Figure 8a, since extra features in such curves are not distinguishable from spurious peaks due to compositional variations or inadequately filtered Kβ radiation, for example. The value of reciprocal space mapping is therefore its ability to separate out the components contributing to the diffracted intensity, which are unobservable with other diffraction techniques.

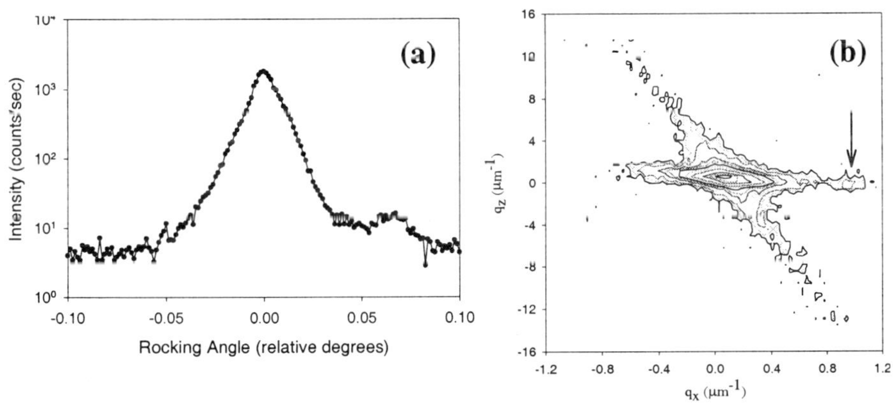

Figure 8. (440) double axis rocking curve (a) and triple axis reciprocal space map (b) from protein crystal HEWL-D. [38]

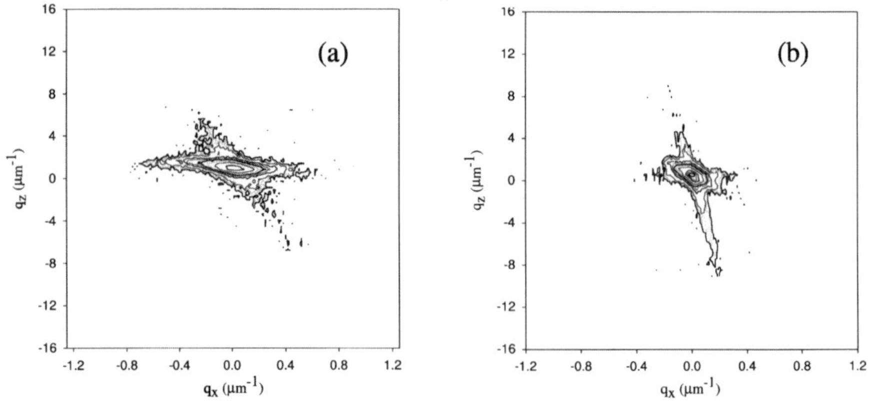

Figure 9. Triple axis reciprocal space maps from the (a) (160) and (b) $(1\bar{2}0)$ reflections from HEWL-D.[38]

Figure 9a shows the triple axis reciprocal space map generated by the (160) reflection. A comparison with Figure 8 indicates that the (160) reciprocal space map is quite similar to that recorded from the (440) reflection, except that the extra coherently diffracting domain is no longer detectable. The benefit of obtaining data from multiple reflections before drawing conclusions is thus apparent.

Figure 9b (recorded from the $(1\bar{2}0)$ reflection) initially appears to be quite different in appearance from the (440) and the (160). However, the Bragg angle for the (440) and (160) are similar ($\theta_{(440)} = 3.16°$ and $\theta_{(160)} = 3.40°$), while $\theta_{(1\bar{2}0)} = 1.25°$. Hence the trigonometric transformation in Equation (3) tends to compress the apparent intensity distribution in the $\boldsymbol{q}_x$ direction. Mosaic spread is again evident as a horizontal streak, as is the parasitic scatter that is part of the instrument function. Between these two, another feature streak appears. This may represent an inclined surface streak similar to that seen in Figure 4. Surface streaks are often seen in high resolution triple axis reciprocal space maps of semiconductor crystals; when a high quality crystalline lattice is truncated by a flat, nearly perfect surface. If the diffraction planes are not perpendicular to the highest quality surface (an asymmetric reflection), the surface streak will appear at an angle in reciprocal space, since this streak will always be perpendicular to the terminating surface. The {101} facets observed optically on the tetragonal HEWL crystals could be of sufficient quality to generate surface streaks.

6. HIGH RESOLUTION TRIPLE AXIS DIFFRACTION ANALYSES OF RADIATION DAMAGE

The applicability of high resolution reciprocal space mapping to the characterization of defects in protein structures is illustrated by a study on the physical changes that accompany radiation damage in HEWL crystals.[42,43] Initially a freshly-grown crystal (designated HEWL-I) irradiated on our high resolution diffractometer continuously, with reciprocal space maps taken periodically. After nearly 5 days of continuous radiation at maximum power, however, virtually no radiation damage (indicated by a change in the intensity distribution in the (440) reciprocal space map) was detected. It was also found that consecutive reciprocal space maps did not show a significant change in the amount of radiation damage, despite each map requiring up to 24 hours of continuous irradiation by the X-ray beam.

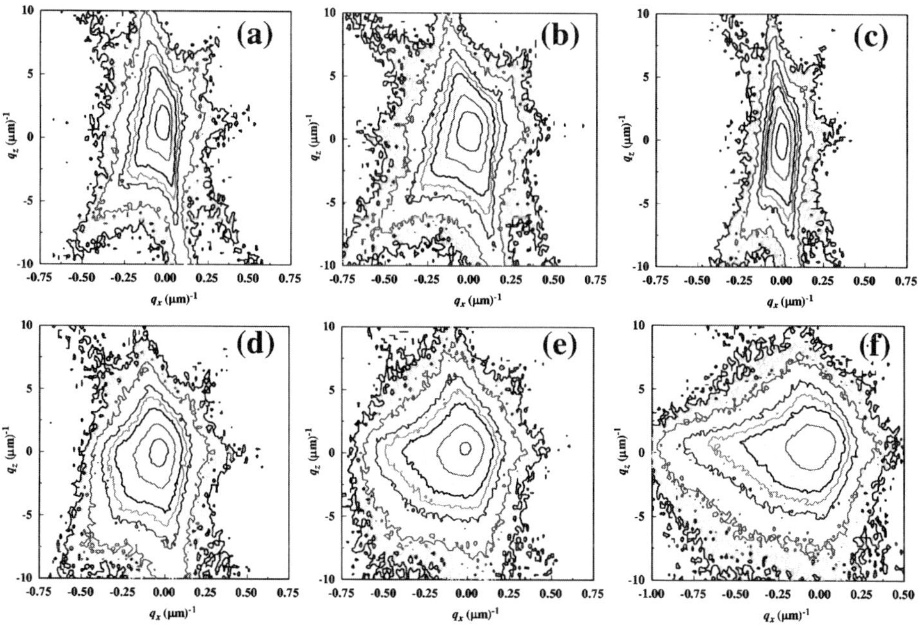

Figure 10. High resolution (440) triple axis reciprocal space maps from crystal HEWL-G (a) prior to intentional radiation damage; (b) after 25 h irradiation; ; (c) after 45 h irradiation;. (d) after 75 h irradiation; (e) after 99 h irradiation; ; (f) after 126 h irradiation. [42]

The lack of radiation damage was probably due to the extensive incident beam conditioning needed to achieve high angular resolution on our triple axis diffractometer system. In order to increase radiation damage, a new crystal (designated HEWL-G) was transferred back to our orientation diffractometer where the Cu Kα flux (measured by attenuation with brass foils) was 8.4×10^6 mm^{-2} s^{-1} at 20kV/10mA. Irradiations were then performed at 45kV/30mA in increments ranging from 5 to 15 hours, with the sample subsequently transferred back to the high resolution system for reciprocal space mapping. We found that the mounting and demounting of the goniometer on the two diffractometers had no observable effect on the intensity distribution around the (440) reflection.

Figure 10 shows a series of reciprocal space maps that were recorded from HEWL-G as a function of X-ray irradiation time.[42] The Figure reveals that a number of systematic changes occur as the cumulative irradiation time increases. Early in the exposure history (up to about 25 hours of irradiation) the angular extent of the diffuse scatter increases, but at intermediate times (most notably at 45 hours cumulative irradiation) the angular breadth of the diffuse scatter apparently goes through a minimum. Continued exposure causes the angular extent of the scatter to again increase.

In order to develop a more quantitative picture of the changes in structure caused by X-ray irradiation, we re-examined the original high resolution triple axis data. The maximum peak intensity in each of the reciprocal space maps was located, and the data with the peak's value of θ/2θ (i.e. corresponding to an ω-scan of the sample) were extracted. The integrated intensity under this "virtual rocking curve" was then determined. Next, we fit the virtual ω-scan rocking curve with a five-parameter Gaussian of the form

$$y = y_o + a\exp\left(-0.5\left[\left|x - x_o\right|/b\right]\right)^c \qquad (6)$$

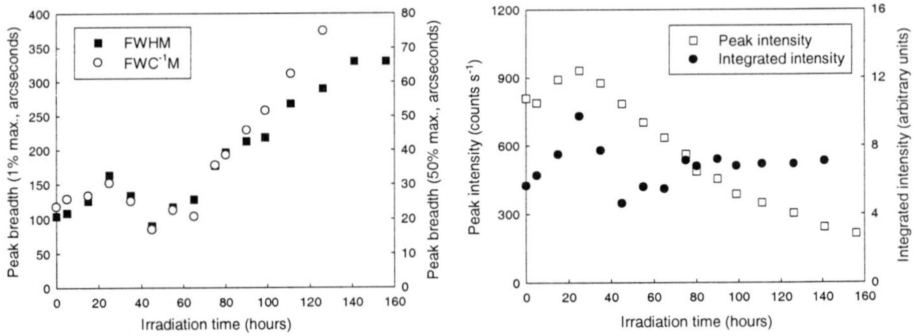

Figure 11. Full width at 50% and 1% of maximum intensity (left) and peak and integrated intensities (right) for HEWL-G as a function of irradiation time.[42]

using conventional regression analysis software where y is the intensity at angular position x and a, b, c, x_o and y_o are adjustable parameters. In order to determine the full width at half maximum (FWHM) of a particular rocking curve, the data were handled on a linear intensity scale, while the full width at 1% of the maximum ($FWC^{-1}M$) was determined from the logarithm of the intensity values. It should be noted that the goal of this effort was not to precisely fit the entire rocking curve profile with a single curve, but only to reproducibly determine the angular positions corresponding to the peak breadths.

Figure 11 shows the trends in the observed intensity (maximum peak intensity and integrated intensity) as well as the transverse peak breadths of each profile at one-half of the maximum intensity (FWHM) and at 1% of the maximum ($FWC^{-1}M$) as a function of X-ray irradiation time. During the initial stages of irradiation (up to 25 hours) we observed the anticipated increase in the angular extent of the off-peak diffuse scatter, a trend that would be consistent with an increase in structural imperfection with increasing irradiation time. In contrast, increasing the irradiation time from 25 to 45 hours of irradiation caused the angular breadth of the intensity to decrease to a local minimum value. Continued irradiation of the crystal up to 150 hours was accompanied by an increase in the angular extent of the diffuse scatter. The behavior of the peak intensity is surprising in that it increased with irradiation from zero to 25 hours; for greater times the peak intensity decreased. The integrated intensity also showed a local maximum at about 25 hours of accumulated irradiation. For longer times where the peak intensity was decreasing while the peak breadth was increasing, the integrated intensity remained approximately constant.

These results raise two obvious questions – could they arise due to time-dependent changes that are independent of X-ray irradiation, and are they reproducible? To answer the first question, a new lysozyme crystal (denoted HEWL-P) was characterized both in its freshly-grown state and then again about one month later. Figure 12 shows reciprocal space maps recorded using the $(5\bar{2}1)$ reflection of this crystal. The only irradiation seen by this crystal was that needed for orientation (less than 5 hours), while during the month between scans, the crystal was left undisturbed at room temperature. The $(5\bar{2}1)$ reflection shows little change after 28 days, with both the peak intensity and the FWHM remaining approximately the same. There is some additional background intensity that is probably attributable to slightly increased disorder in the crystal and/or solvent over time. Very

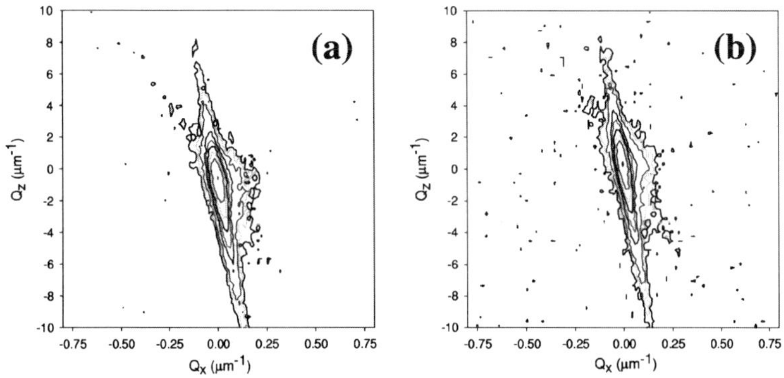

Figure 12. High resolution $(5\bar{2}1)$ triple axis reciprocal space maps from crystal HEWL-P: (a) as-grown crystal; (b) after 28 days at room temperature without intentional irradiation.

similar results were obtained from the $(\bar{4}40)$ reflection; hence we conclude that the changes illustrated in Figure 10 from HEWL-G are not simply an aging effect.

Figures 13 shows reciprocal space maps from the $(4\bar{4}0)$ (top row) and the $(2\bar{5}1)$ (bottom row) reflections from a different crystal (designated HEWL-Q) as a function of irradiation time. Comparison of this Figure with the data from HEWL-D in Figure 10 shows that different *hkl* reflections from the same crystal (not to mention different reflections from different crystals) have obviously different characteristics. This behavior is not unexpected; similar differences have been seen in other materials such as highly perfect semiconductor crystals.

The evolution of the peak intensity from samples G and Q with irradiation time along with the peak breadth of sample HEWL-Q is plotted in Figure 14. As with HEWL–G, transverse scans were fitted with a modified Gaussian to reproducibly identify the FWHM

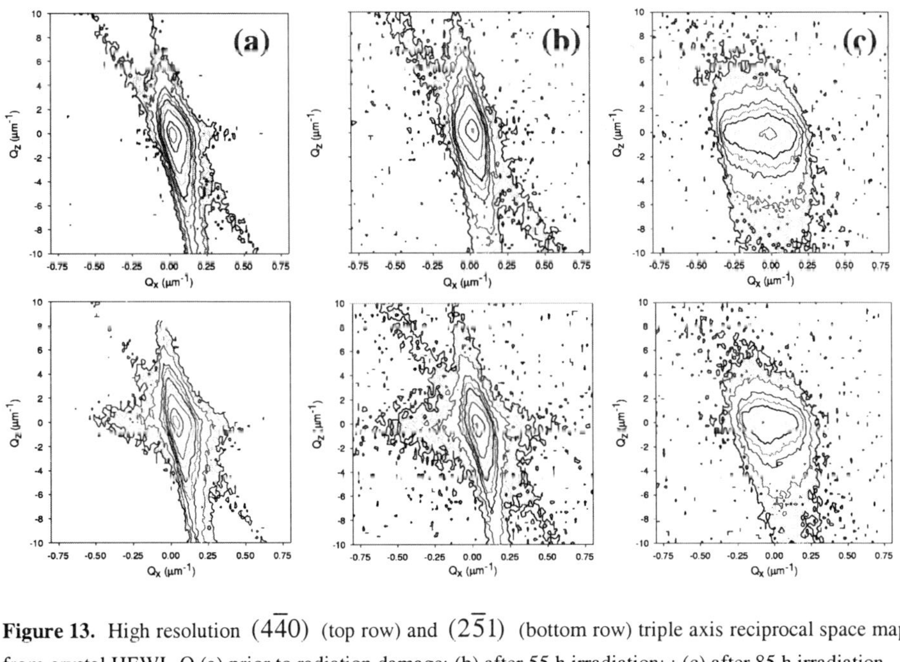

Figure 13. High resolution $(4\bar{4}0)$ (top row) and $(2\bar{5}1)$ (bottom row) triple axis reciprocal space maps from crystal HEWL-Q (a) prior to radiation damage; (b) after 55 h irradiation; ; (c) after 85 h irradiation.

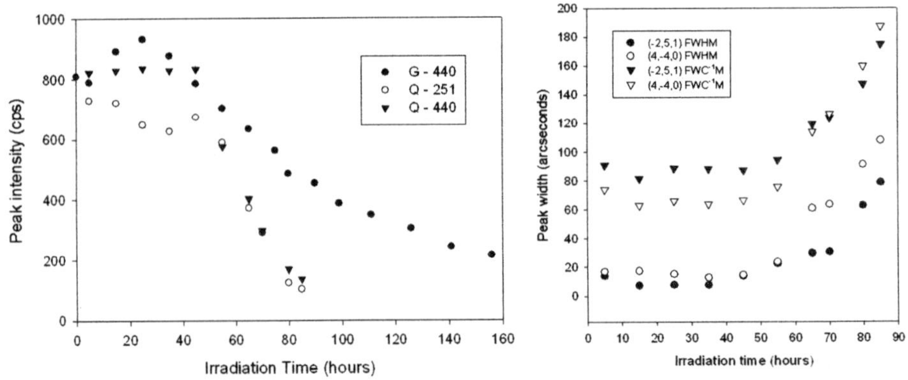

Figure 14. Comparison of the effects of irradiation time on HEWL-G and HEWL-Q in terms of the maximum Bragg peak intensities (left) and the transverse peak breadths (FWHM and $FWC^{-1}M$) (right).

and $FWC^{-1}M$ from each reflection. Once again the longitudinal scan's FWHM (data with constant $\theta/2\theta$ value passing through the Bragg peak) remained approximately constant throughout the irradiation study with FWHM values remaining between 20 and 25 arcseconds. The transverse peak breadths at long irradiation times (greater than 45 hours) shown in Figure 14 markedly increase, while the Bragg peak intensities decreases, which is expected in a sample with increasing disorder. At shorter times, a slight narrowing of the FWHM was again observed.

7. SMALL-ANGLE X-RAY ANALYSES OF PROTEIN CRYSTALS

The triple axis reciprocal space maps that are shown above demonstrate a significant practical shortcoming in the use of reciprocal space mapping– the diffracting crystal must exhibit a sufficient level structural perfection if the diffuse scatter is to be observed in high resolution. Figures 10 and 13 graphically illustrate that as the structural perfection of a protein crystal is progressively degraded (in this case, due to radiation damage), the off-peak diffuse scatter is spread over an increasing volume of reciprocal space. At extremely high levels of disorder, the angular extent of the Bragg scattering would be expected to become quite large with an accompanying decrease in intensity. As discussed in Section 3.5, this

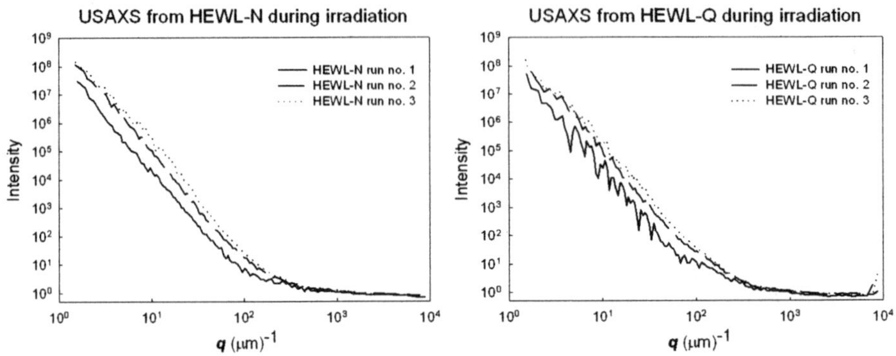

Figure 15. USAXS data from HEWL-N (initially unirradiated crystal, left) and HEWL-Q (initially heavily irradiated, right).

state of affairs represents a regime where high resolution triple axis diffraction would have limited utility – but it also represents the conditions under which small angle X-ray scattering might be observable. Figure 15 illustrates SAXS data that were obtained from two protein crystals: HEWL-N, which had not been intentionally irradiated in prior experiments, and HEWL-Q, which had experienced extensive radiation damage by the conclusion of the experiments illustrated in Figures 13 and 14. To the best of our knowledge, these data represent the first SAXS data reported from protein crystals in the solid state.

The net scattering curves shown in Figure 15 represents the difference between the scattering generated by the protein crystal within the glass capillary and the capillary by itself (measured after translating the capillary along its axis so that the crystal was no longer in the 0.5 mm × 0.5 mm X-ray beam). After the scattering curve for one crystal was recorded, that crystal (mounted in its capillary on a standard goniometer) was removed, and the other crystal was inserted for its data collection; this back-and-forth sample exchange was used to generate the six curves from the two crystals. In both crystals we observe a trend of an increase in scattered intensity with increasing X-ray dose. While the change in the amount of scattering with synchrotron X-ray radiation dose is relatively modest, it does follow exactly the behavior expected from SAXS theory as discussed in Section 3.5.

Figure 16 compares scattering curves obtained from the two crystals (initially unirradiated HEWL-N and intentionally radiation damaged HEWL-Q) after each of three synchrotron radiation cycles. Again, it must be remembered that the crystal and its capillary was inserted into the X-ray beam immediately prior to each data collection run and was removed immediately afterwards. The data in the Figure show that at each synchrotron irradiation cycle, the initially damaged crystal exhibits more intense scattering than does its initially unirradiated colleague. In addition, the increasing ordinate in each plot in the Figure clearly shows the increase in the scattered intensity as exposure time in the synchrotron beam increases. Finally, the scattering curve from the HEWL crystal with presumably the least radiation damage (HEWL-N, first cycle) shows what may be an oscillatory behavior; the first cycle scattering of initially damaged HEWL-Q exhibits less intense oscillations. As synchrotron radiation proceeds, however, the scattering curves seem to "smooth out" even as their net intensity increases.

These first results from a SAXS analysis of protein crystals in the solid state suggest that useful scattering data can be obtained from these materials. Moreover, the data are apparently consistent with the thesis that an increasing level of structural disorder (induced in this case by radiation damage) results in an increase in the magnitude of the SAXS intensity. The apparently oscillatory angular dependence in the least-damaged crystals may be due to effects that are related to interparticle interference;[32] such effects would be expected to be most prominent in the least-damaged protein crystals and should dissipate as the degree of

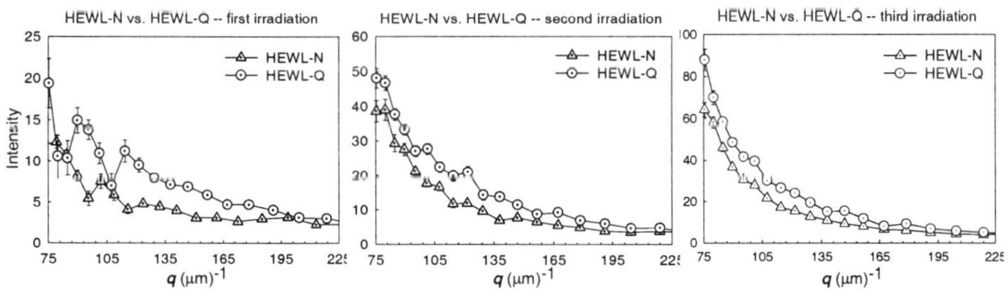

Figure 16. USAXS data from HEWL-N (initially unirradiated) and HEWL-Q (initially heavily irradiated) at different stages of synchrotron X-ray irradiation. Left: first cycle; middle: second cycle; right: third cycle. Note the change in the scale of the ordinate from left to right.

structural perfection increases. It is critical to note that these are only first results, and much more work is needed in order to confirm the utility of SAXS for the characterization of highly defective protein crystals.

8. QUANTITATIVE ANALYSIS OF HIGH RESOLUTION X-RAY DATA

It should be evident that the interpretations of the high resolution data presented above have been primarily qualitative in nature. The main reason for this comes from the fact that (at least initially) there was no guarantee that the desired experimental approaches would produce useful results. In the case of reciprocal space mapping, our initial concern was that the extensive beam conditioning that is needed to achieve high angular resolution, when applied to protein crystals with an inherently low diffracting power, would make it impossible to observe off-peak diffuse scattering in a measurable signal could be seen at all.

One of the advantages of X-ray scattering as an analytical tool is that the process can in principle be well described mathematically, particularly in the kinematic limit. Thus one should be able to model the defect scattering by calculating the angular distribution of the off-peak intensity from the crystallographic distortions associated with a particular crystal defect and the number of those defects that are present. Through a comparison of the calculated scatting from an ensemble of defects with the observed scattering, a quantitative assessment of the defects present in a crystal is possible.

While fully quantitative modeling of off-peak diffuse scattering can be difficult, some simplifications are possible. For instance, in the case of relatively simple crystal structures such as nearly perfect semiconductors, the transition between one functional dependence of the off-peak scattering on the deviation q_x from the exact Bragg condition to another can be used to estimate the size of the defect that generates the scattering.[44,45] Experimentally, this can be achieved by taking a high resolution triple axis diffraction data set and examining the decay of the intensity as a function of q_x.

Figure 17 illustrates this process, applied to the off-peak diffuse scatter from the (440) reflection of HEWL-G. Although the large number of data points may make individual scans difficult to pick out, it can be seen that the overall trend in the data is that for short irradiation times away from the Bragg peak, the diffracted intensity as a function of q_x falls off with roughly the same functional dependence (i.e. the same slope in a log I – log q plot). At longer irradiation times, however, the trend is for the data to exhibit two different slopes: one at a shallower slope than that for early radiation times, and one steeper slope that

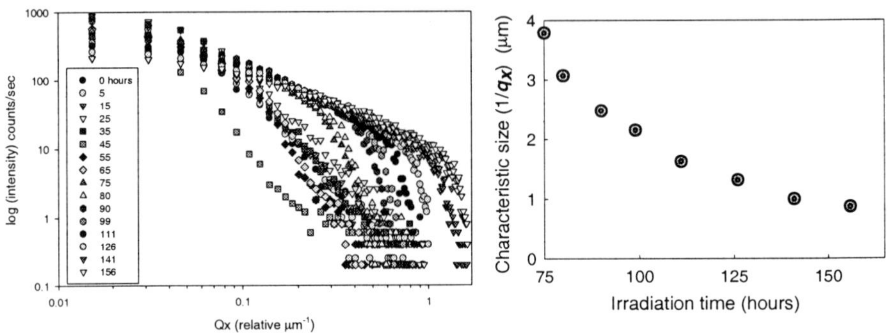

Figure 17. Double logarithmic plot of transverse scans of the off-peak diffuse intensity from HEWL-G (left) and the characteristic size extracted from the breakpoints in the transverse scans (right).

dominates at higher q_x values. As the irradiation time increases, the second, steeper slope starts appearing at larger and larger q_x values.

Figure 17 also shows a plot of the transition point for the data collected at long radiation times with dual slopes in terms of the reciprocal of the scattering vector corresponding to this change in slope. Because $1/q_x$ has units of length, we refer to this value as a measure of a "characteristic size" that decreases with increasing irradiation time. It is interesting that these potential defects are on the order of microns, which is quite large. Nevertheless, since a unit cell of lysozyme is roughly 237 nm^3 and X-ray diffuse scattering measurements are highly sensitive to the strain fields around a defect and not limited to the size of the defect itself, this number may be an indicator of the size of an undamaged region during X-ray irradiation.

Of course, a more extensive modeling task might involve the calculation of the diffuse scattering expected from an assumed set of structural defects and then comparing this calculated scattering profile with experimental data. Such an approach has proven to be very useful in characterizing defects in semiconductor materials.[46] For this to be successfully applied to protein crystals, however, an estimate of the types and densities of the array of structural defects expected to be present in a protein crystal would have to be quantitatively known; in addition, the types of strain fields and their spatial extents would be needed.

As a first approximation, it may be possible to calculate the diffuse scatter that is expected in the vicinity of a Bragg reflection by modeling the protein crystal as an ensemble of (approximately) uniform distributions of electron density (*i.e.* the roughly egg-shaped protein molecule) that is immersed in a crystal with an adjustable degree of rotational disorder. This approach may prove to be useful for modeling SAXS data from protein crystals; here, the disorder induced by radiation damage could be simulated by altering the respective orientation and separation of mosaic blocks (which contain ellipsoidal electron clouds) that are immersed in an aqueous environment.

9. CONCLUSIONS

In this work we have tried to show that high resolution X-ray scattering methods – those that examine the intensity distribution about a high-order *hkl* reflection as well as those which concentrate about the origin of reciprocal space – can yield valuable insights into the nature of physical defects in protein crystals. In the semiconductor community, it has taken several years for these high resolution methods to elevate their status from that of laboratory curiosity to an established and important analytical technique. Hopefully, with continued development (particularly in the area of modeling of the diffuse intensity) a similar progression will occur to the benefit of the protein crystal growth community.

REFERENCES

1. D.E. McRee, *Practical Protein Crystallography*. San Diego: Academic Press (1993).
2. J. Drenth, *Principles of Protein X-Ray Crystallography, Second Ed.* New York: Springer-Verlag (1999).
3. M.L. Pusey, *J. Cryst. Growth*, **122**, 1 (1992).
4. F. Rosenberger, P.G. Vekilov, M. Muschol, and B.R. Thomas, *J. Cryst. Growth*, **168**, 1 (1996).
5. P.G. Vekilov, L.A. Monaco, B.R. Thomas, V. Stojanoff, and F. Rosenberger, *Acta Cryst.*, **D52**, 785 (1996).
6. P.G. Vekilov and F. Rosenberger, *J. Cryst. Growth*, **158**, 540 (1996).
7. P.G. Vekilov, M. Atake, and T. Katsura, *Acta Cryst.*, **D51**, 207 (1995).
8. A. McPherson, A.I. Malkin, and Y.G. Kuznetsov, *Structure*, **3**, 739 (1995).
9. Y.G. Kuznetsov, A. Malkin, A. Greenwood, and A. McPherson, *J. Struct. Biol.*, **114**, 184 (1995).
10. Y.G. Kuznetsov, A. Malkin, W. Glantz, and A. McPherson, *J. Cryst. Growth*, **168**, 63 (1996).

11. A. Shaikevitch and Z. Kam, *Acta Cryst. A*, **37**, 871 (1981).
12. A. Malkin, Y.G. Kuznetsov, and A. McPherson, *J. Struct. Biol.*, **117**, 124 (1996).
13. A.A. Chernov, *J. Cryst. Growth*, **174**, 354 (1997).
14. J.P. Benoit and J. Doucet, *Quart. Rev. Biophysics*, **28**, 131 (1995).
15. D.K. Bowen and B.K. Tanner, *High Resolution X-Ray Diffractometry and Topography*. Bristol, P.A.: Taylor & Francis (1998).
16. F. Otalora, J.M. Garcia-Ruiz, J.A. Gavira, and B. Capelle, *J. Cryst. Growth*, **196**, 546 (1999).
17. R. Fourme, A. Ducruix, M. Ries-Kautt, and R. Capelle, *J. Synch. Rad.*, **2**, 136 (1995).
18. J.R. Helliwell, *J. Cryst. Growth*, **90**, 259 (1988).
19. M. Colapietro, G. Cappuccio, C. Marciante, A. Pifferi, R. Spagna, and J.R. Helliwell, *J. Appl. Cryst.*, **25**, 192 (1992).
20. E.H. Snell, S. Weisgerber, J.R. Helliwell, E. Weckert, K. Hölzer, and K. Schroer, *Acta Cryst. D*, **51**, 1099 (1995).
21. H.M. Volz and R.J. Matyi, *Philos. Trans. R. Soc. London, Ser. A*, **357**, 2789 (1999).
22. H.D. Bellamy, E.H. Snell, J. Lovelace, M. Pokross, and G.E.O. Borgstahl, *Acta Cryst.*, **D56**, 986 (2000).
23. K. Izumi, S. Sawamura, and M. Ataka, *J. Cryst. Growth*, **168**, 106 (1996).
24. V. Stojanoff and D.P. Siddons, *Acta Cryst.*, **A52**, 498 (1996).
25. V. Stojanoff, D.P. Siddons, L.A. Monaco, P. Vekilov, and F. Rosenberger, *Acta Cryst.*, **D53**, 588 (1997).
26. C.L. Caylor, I. Dobrianov, S.G. Lemay, C. Kimmer, S. Kriminski, K.D. Finkelstein, Z. Zipfel, B.R. Thomas, A.A. Chernov, and R.E. Thorne, *Proteins-Structure Function and Genetics*, **36**, 270 (1999).
27. I. Dobrianov, K.D. Finkelstein, S.G. Lemay, and R.E. Thorne, *Acta Cryst.*, **D54**, 922 (1998).
28. I. Dobrianov, C. Caylor, S.G. Lemay, K.D. Finkelstein, and R.E. Thorne, *J. Cryst. Growth*, **196**, 511 (1999).
29. K. Kranjc, *J. Appl. Cryst.*, **2**, 262 (1969).
30. V.V. Aristov and E.C. Shulakov, *J. Appl. Cryst.*, **8**, 445 (1975).
31. C.C. Wu and R.W. Armstrong, *Phys. Stat. Sol. A*, **29**, 259 (1975).
32. A. Guinier, *X-ray Diffraction in Crystals, Imperfect Crystals and Amorphous Bodies*. San Francisco: W.H. Freeman (1963).
33. M. Forstner, *J. Appl. Cryst.*, **33**, 519 (2000).
34. R. Thomson, L.E. Levine, and G.G. Long, *Acta Cryst.*, **A55**, 433 (1999).
35. G.G. Long, L.E. Levine, and R. Thomson, *J. Appl. Cryst.*, **33**, 456 (2000).
36. G.G. Long, A.J. Allen, J. Ilavshy, P.R. Jemain, and P. Zschack. *The Ultra-Small-Angle X-ray Scattering Instrument on UNICAT at the APS*. in *Synchrotron Radiation Instrumentation: 11th U.S. National Conference*. 2000.
37. V. Luzzati, J. Witz, and A. Nicolaieff, *J. Mol. Biol.*, **3**, 367 (1961).
38. H.M. Volz and R.J. Matyi, *Acta Cryst. D*, **56**, 881 (2000).
39. Osmic, Inc., Troy, Michigan, USA..
40. R.J. Matyi, P.D. Moran, W.W.D. Hagquist, and H.M. Volz, *Rev. Sci. Instrum.*, **71**, 2292 (2000).
41. Crystal Specialties, Ltd., Newcastle, UK..
42. H.M. Volz and R.J. Matyi, *J. Cryst. Growth*, **232**, 502 (2001).
43. H.M. Volz, Ph.D dissertation, University of Wisconsin (2001).
44. M.A. Krivoglaz, *Theory of X-Ray and Thermal Neutron Scattering by Real Crystals*. New York: Plenum (1969).
45. A.N. Morozov and V.T. Bublik, *J. Cryst. Growth*, **97**, 475 (1989).
46. V. Holy, U. Pietch and T. Baumbach, *High-Resolution X-Ray Scattering from Thin Films and Multilayers*. Berlin: Springer-Verlag (1999).

PARTICIPANTS

Gassem Al-Zoubi
Department of Physics and Astronomy
Michigan State University
East Lansing, MI 48824-1116, USA
al-zoubi@pa.msu.edu

Philip Anfinrud
Lab of Chemical Physics, NIDDK
National Institutes of Health
5 Memorial Drive, Building 5, Rm. 134
Bethesda, MD 20892-0520, USA
anfinrud@nih.gov

Rosa Barabash
Oak Ridge National Laboratory
Metals & Ceramics Division
1 Bethel Valley Road
Oak Ridge, TN 37831-6118, USA
barabashr@ornl.gov

Simon Billinge
Department of Physics and Astronomy
Michigan State University
East Lansing, MI 48824-1116, USA
billinge@pa.msu.edu

Daniel Bowron
ISIS Facility
CLRC Rutherford Appleton Laboratory
Chilton, Didcot
Oxon, OX11 0QX, UK
D.T.Bowron@rl.ac.uk

Emil Bozin
Department of Physics and Astronomy
Michigan State University
East Lansing, MI 48824-1116, USA
bozin@pa.msu.edu

Branton Campbell
Argonne National Lab
9700 South Cass Avenue
MSD/223
Argonne, IL 60439, USA
branton@anl.gov

W.I.F. David
ISIS Facility
Rutherford Appleton Laboratory
Chilton
Oxon, OX11 0QX, UK
Bill.David@rl.ac.uk

Sofia Díaz Moreno
European Synchrotron Radiation Facility
B.P. 220
38043 Grenoble Cedex 9, France
moreno@esrf.fr

Takeshi Egami
LRSM, University of Pennsylvania
3231 Walnut Street
Philadelphia, PA 19104, USA
egami@seas.upenn.edu

John Finney
Department of Physics and Astronomy
University College London
Gower Street
London WC1E 6BT, UK
j.finney@ucl.ac.uk

Henry C. Foley
158 Fenske Laboratory
Pennsylvania State University
University Park, PA 16802-4400, USA
hankfoley@engr.psu.edu

Matthias J. Graf
Theoretical Division (T-11)
Los Alamos National Lab
Mail Stop B262
Los Alamos, MN 87545, USA
graf@lanl.gov

Alex Hannon
ISIS Facility
Chilton, Didcot
Oxon, OX11 0QX, UK
a.c.hannon@rl.ac.uk

Simon Hibble
Department of Chemistry
University of Reading
Whiteknights
Reading RG6 6AD, UK
s.j.hibble@rdg.ac.uk

Václav Holý
Institute of Condensed Matter Physics
Faculty of Science, Masaryk University
Kotlarska 2
611 37 Brno, Czech Republic
holy@physics.muni.cz

Ilkyoung Jeong
Condensed Matter & Thermal Physics
Los Alamos National Lab
MST-10 K764
Los Alamos, MN 87545, USA
jeong@lanl.gov

Mercouri Kanatzidis
Department of Chemistry
Michigan State University
East Lansing, MI 48824, USA
kanatzid@cem.msu.edu

Ming Lei
Department of Physics and Astronomy
Michigan State University
East Lansing, MI 48824-1116, USA
lei@pa.msu.edu

Valentin Levashov
Department of Physics and Astronomy
Michigan State University
East Lansing, MI 48824-1116, USA
levashov@pa.msu.edu

Jeff Lynn
NIST Center for Neutron Research
Building 235
National Institute of Standards and Technology
Gaithersburg, MD 20899-8562, USA
jeff.lynn@nist.gov

Richard J. Matyi
National Institute of Standards and Technology
100 Bureau Drive, Stop 8422
Gaithersburg, MD 20899-8422, USA
richard.matyi@nist.gov

Jan Meinke
Department of Physics and Astronomy
Michigan State University
East Lansing, MI 48824-1116, USA
meinke@pa.msu.edu

Geoff Mitchell
Polymer Science Centre
JJ Thomson Lab
University of Reading
Whiteknights
Reading RG6 6AF, UK
G.R.Mitchell@reading.ac.uk

Simon Moss
Department of Physics
University of Houston
4800 Calhoun
Houston, TX 77204-5506, USA
Smoss@uh.edu

Inmaculada Peral
Chemical Engineering
University of Delaware
150 Academy Street, Colburn Lab
Newark, DE 19716, USA
peral@che.udel.edu

Valeri Petkov
Department of Physics and Astronomy
Michigan State University
East Lansing, MI 48824-1116, USA
petkov@pa.msu.edu

David Price
Argonne National Lab
Bldg. 223
9700 S. Cass Avenue
Argonne, IL 60439, USA
dlp@anl.gov

Xiangyun Qiu
Department of Physics and Astronomy
Michigan State University
East Lansing, MI 48824-1116, USA
xiangyun@pa.msu.edu

Dale Sayers
Department of Physics
North Carolina State University
Box 8202
Raleigh, NC 27695-8202, USA
dale_sayers@ncsu.edu

Friedrich Schotte
National Institutes of Health
Building 5, Room 136
Bethesda, MD 20892-0520, USA
schotte@nih.gov

Sunil Sinha
Department of Physics
University of California, San Diego
9500 Gilman Drive
La Jolla, CA 92093-0354, USA
ssinha@physics.ucsd.edu

Michael F. Thorpe
Department of Physics and Astronomy
Michigan State University
East Lansing, MI 48824-1116, USA
thorpe@pa.msu.edu

Matt Tucker
Department of Earth Sciences
Cambridge University
Downing Street
Cambridge, CB2 3EQ, UK
mtuc98@esc.cam.ac.uk

Richard Welberry
Research School of Chemistry
Australian National University
Canberra, ACT 0200, Australia
welberry@rsc.anu.edu.au

Hasan Yavas
Department of Physics and Astronomy
Michigan State University
East Lansing, MI 48824-1116, USA
yavas@pa.msu.edu

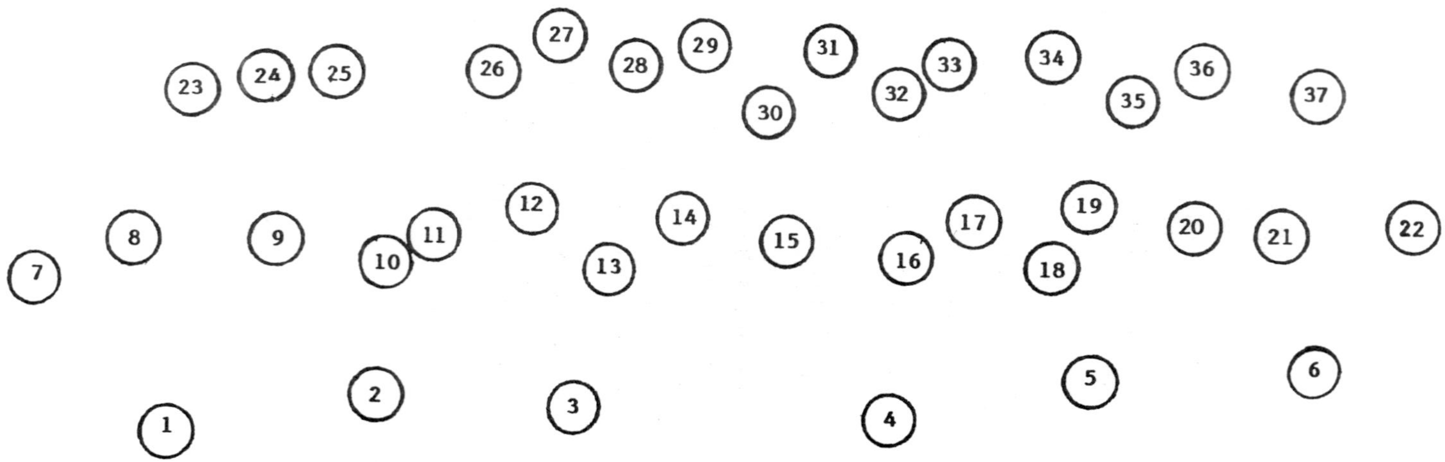

[1] I. Peral, [2] S. Billinge, [3] S. Moss, [4] R. Barabash, [5] R. Welberry, [6] W. David, [7] O. Barabash, [8] M. Tucker, [9] J. Finney, [10] R. Matyi, [11] D. Sayers, [12] J. Meinke, [13] S. Hibble, [14] M. Thorpe, [15] M. Kanatzidis, [16] G. Mitchell, [17] H. Foley, [18] A. Hannon, [19] F. Schotte, [20] D. Bowron, [21] J. Lynn, [22] V. Holy, [23] P. Anfinrud, [24] L. Neuman, [25] B. Campbell, [26] G. Al-Zoubi, [27] E. Bozin, [28] M. Graf, [29] V. Petkov, [30] S. Diaz Moreno, [31] H. Yavas, [32] I. Jeong, [33] X. Qiu, [34] V. Levashov, [35] T. Egami, [36] S. Sinha, [37] M. Lei

Not in photograph: D. Price

INDEX